D0266590

PHYSICAL TECHNIQUES IN BIOLOGICAL RESEARCH

Volume I, Part A

Optical Techniques

Second Edition

Contributors to This Volume

C. ALLEN BUSH

V. E. COSSLETT

T. A. HALL

L. C. MARTIN

GEORGE J. THOMAS, JR.

W. T. WELFORD

PHYSICAL TECHNIQUES IN BIOLOGICAL RESEARCH

SECOND EDITION

Edited by

Gerald Oster

MOUNT SINAI SCHOOL OF MEDICINE OF THE CITY UNIVERSITY OF NEW YORK
NEW YORK, NEW YORK

Volume I, Part A

Optical Techniques

NORTHWEST MISSOURI STATE
UNIVERSITY LIBRARY
MARYVILLE, MISSOURI 64468

 1971

ACADEMIC PRESS   New York and London

COPYRIGHT © 1971, BY ACADEMIC PRESS, INC.
ALL RIGHTS RESERVED
NO PART OF THIS BOOK MAY BE REPRODUCED IN ANY FORM,
BY PHOTOSTAT, MICROFILM, RETRIEVAL SYSTEM, OR ANY
OTHER MEANS, WITHOUT WRITTEN PERMISSION FROM
THE PUBLISHERS.

ACADEMIC PRESS, INC.
111 Fifth Avenue, New York, New York 10003

United Kingdom Edition published by
ACADEMIC PRESS, INC. (LONDON) LTD.
Berkeley Square House, London W1X 6BA

LIBRARY OF CONGRESS CATALOG CARD NUMBER: 66-24377

PRINTED IN THE UNITED STATES OF AMERICA

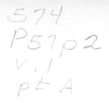
574
P51p2
v. 1
pt A

Contents

Chapter 1

The Light Microscope

L. C. MARTIN AND W. T. WELFORD

Chapter 2

Electron Microscopy

V. E. COSSLETT

136.30 art
JUL 24 '78

Chapter 3

The Microprobe Assay of Chemical Elements

T. A. HALL

Chapter 4

Infrared and Raman Spectroscopy

GEORGE J. THOMAS, JR.

Chapter 5

Optical Rotatory Dispersion and Circular Dichroism

C. ALLEN BUSH

List of Contributors

C. Allen Bush, Department of Chemistry, Illinois Institute of Technology, Chicago, Illinois

V. E. Cosslett, Department of Physics, Cavendish Laboratory, University of Cambridge, Cambridge, England

T. A. Hall, Department of Physics, Cavendish Laboratory, University of Cambridge, Cambridge, England

L. C. Martin, Department of Physics, Imperial College of Sciences and Technology, London, England

George J. Thomas, Jr., Department of Chemistry, Southeastern Massachusetts University, North Dartmouth, Massachusetts

W. T. Welford, Department of Physics, Imperial College of Sciences and Technology, London, England

PREFACE TO THE SECOND EDITION OF VOLUME I

Since the publication of the First Edition the growth of biophysical techniques has taken place at such a rapid rate that a second edition is warranted. Indeed, instrumentation such as the scanning electron microscopy and electron-probe X-ray microanalysis were unknown at the time of the First Edition. The invention of the gas laser in 1960 has revitalized Raman spectroscopy and commercial instruments are available in this field. The ensuing years have also seen the extension of the theory of optical rotation dispersion to macromolecules as has the theory of the image in the conventional light microscopy.

Clearly biological scientists are demanding more and more of the physicist to provide further refinements both in instrumentation and in theories to enable them to probe deeper into the complexities of biology. This too places a burden on biologists who must keep abreast of many current physical ideas and developments if they are to work most effectively. It is hoped that the present volume satisfies to some extent the needs of the biological scientist.

New York, New York   GERALD OSTER
November, 1970

PREFACE TO THE FIRST EDITION OF VOLUME I

These volumes have resulted from collaborative effort, in which a large number of physical techniques that have been found useful in biology are discussed briefly from both the theoretical and the practical viewpoint. No claim to encyclopedic range is made, but it is believed that from these volumes the reader can get a fairly comprehensive idea of the present place of the techniques of physics and of physical chemistry is biological research.

During the last two decades there has been unprecedented broadening of the scope of attack on fundamental problems of physiology. This is largely a consequence of the increasing use by biologists of modern physical techniques, some of which are of the most advanced types and characterized by precision and delicacy rarely employed in their application in physics. These refinements have not always resulted from the collaboration of a biologist and a physicist, as one might suppose. Fully as often the biologist has turned physicist and has himself adapted, or indeed sometimes developed, physical techniques suited to his needs. Thus, Martin and Synge developed adsorption experimental procedures into the enormously useful paper chromatography; Holter and Linderstrøm-Lang made from the diver of Descartes an apparatus delicate enough to obtain analytical data from single cells; while in the microtome electron microscopists have designed an engineering marvel that cuts slices about one order of magnitude above the range of monomolecular films.

Workers with any of the powerful new aids to biology from physics cannot avoid feeling strongly encouraged to continue to elaborate new methods and to broaden the range of biological problems to which the techinques are applicable. The present work has been conceived in the hope of accelerating such development and wider use. Every specialist in one of these techniques is constantly called upon to help his fellow biologist to decide whether, or how, a particular technique can serve a biological use more or less unlike that for which it was originally designed. It is also not an infrequent experience of a physicist or a physical chemist to have a biologist come to him with a difficulty which he hopes can be overcome if he can find the right physical approach. The success of such conferences depends to a large extent upon both participants having a fair knowledge of the details of the techniques and the biological uses to which they have already been put; and one of the aims of the present work is to serve such a need. For every such biologist who makes a vigorous effort to use new physical methods, there must be many who have a vague feeling that their researches might proceed better if reoriented in a physical direction, but who hestitate to attempt this because of timidity. For such workers it is felt that the many examples of simple methods will be helpful. Finally, in a broad sense, it is hoped that this treatise will serve as a real orientation for biologists and for chemists and physicists who may be potential biologists. In these volumes each author, an expert in his field, has written in such a way that a biologist can see whether he may start to employ the technique, or whether the application to his particular biological problem demands collaboration with a physicist or a physical chemist. The latter, on the other hand, should be able to access in realistic terms the possibility of fruitful and exciting application of his special training to the baffling problems of biology.

The arrangement of material has been determined primarily by biological considerations. Volumes I and II deal with theory and methods applied to relatively pure preparations of biological substances that are obtained from cells or other tissue elements. The optical approaches, so favored in biology, are in Volume I, while in Volume II are a wide variety of nonoptical techniques. Volume III deals with application of physical techniques to cells and tissues.

The editors wish to express their appreciation to the authors for their contributions.

New York, New York   GERALD OSTER
June 30, 1955   ARTHUR W. POLLISTER

CONTENTS OF VOLUME IB

CHAPTER 1

The Light Microscope

L. C. Martin and W. T. Welford

I. Principles of Lenses and Lens Systems

A. RAYS AND IMAGES

The light microscope is a good example of the class of optical instruments of which the main principles of operation can be explained in terms of the simple operations of lenses on light rays, although deeper concepts of interference and diffraction are needed to explain the existence of limitations to ultimate performance.

The concept of a light ray is intuitively simple in terms of the pinhole camera (Fig. 1). The pinhole selects a "single ray" from each point on the object, here suggested as a black and white drawing, and projects this on an image plane; if the distances of the object planes are respectively l and l', then the magnification is by similar triangles equal to l'/l and this has the correct negative sign for the image inversion if l and l' are measured as coordinates from the pinhole. If we were simply to make l' very much larger than l, it would appear that the essential problem of microscopy would be solved, but of course this does not work, for two reasons; the first is that if the pinhole is made small enough to select a narrow enough pencil of light to make a sharp image of fine detail in the object, then the image is not bright enough; the second reason is that, again if the pinhole is made very small, it is found that after a certain point the pencil of light selected as a "ray" no longer gets smaller but instead begins to spread out after the pinhole, a process known as diffraction.

The above discussion covers all the essential points in the evolution of microscope design. Elaborately designed lens systems are needed to

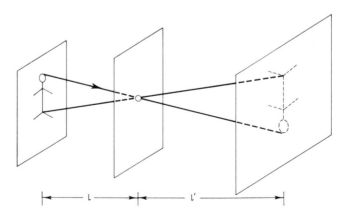

FIG. 1. The pinhole camera principle; the magnification is the ratio of the distances, l'/l.

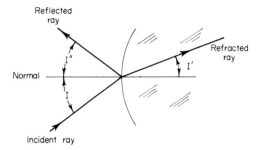

FIG. 2. Reflection and refraction of a ray at an air-glass interface.

produce well-defined images from pencils of light of large diameter and similarly elaborate condensers are needed to illuminate the object adequately. Nonetheless the simple concept of image formation by rays traveling in straight lines through pinhole apertures serves to illustrate many essential concepts. A lens system operates on bundles or pencils of rays from a single object point and brings them together to form more or less a single image point, but the ray concept is the same and the magnification formula is a very simple adaptation of the above.

B. REFLECTION AND REFRACTION OF RAYS

If a ray meets a smooth glass surface, it is partly reflected and partly transmitted or refracted; the ray may be considered as a path of energy flow of the light so that the flow is divided into two streams. Figure 2 illustrates the notation and nomenclature which are used to describe these effects. The incoming or incident ray meets the surface at the point of incidence and the normal or perpendicular to the surface can be drawn there. The incident ray, the refracted ray, and the reflected ray make angles with the normal I, I', and I'', respectively, and these are called the angles of incidence, refraction, and reflection.

The laws of reflection and refraction state that the refracted and reflected rays both lie in the plane containing the normal and the incident ray, that the angle of reflection is equal to the angle of incidence, and that the sine of the angle of refraction bears a constant ratio to the sine of the angle of incidence as I is varied; however, this ratio depends on the kind of glass and on the wavelength of the light. If the first medium is air (or more strictly vacuum), as suggested in Fig. 2, the ratio is called the refractive index of the glass, denoted by the symbol n.

Consider a ray passing through several media bounded by parallel planes, as in Fig. 3, which shows a microscope slide, coverslip, etc; with angles of incidence or refraction as indicated, it is found that

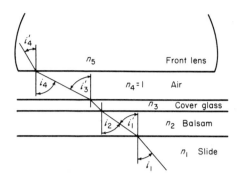

FIG. 3. Refraction through several surfaces.

$$n_1 \sin I_1 = n_2 \sin I_2 = n_3 \sin I_3 = \cdots \qquad (1)$$

Typical approximate refractive indices for light in the middle of the visible spectrum are given in Table I.

TABLE I
REFRACTIVE INDICES

Material	Refractive index
Air	1.0003
Water	1.33
Crown glass	1.52
Flint glass	1.65
Canada balsam	1.52

Ordinarily the refractive index of air is taken as unity and those of other materials are specified as relative to air. "Crown" and "flint" refer to large groups of different optical glasses and the values given are merely representative samples.

If a ray travels from glass to air as in Fig. 4, the law of refraction or Snell's law as it is usually called, gives

$$n \sin I = \sin I' \qquad (2)$$

It can be seen that as I increases a point is reached at which the emergent ray has $I' = 90°$, i.e., it grazes the surface; this happens when $\sin I = 1/n$. For larger angles of incidence, it is found that there is no refracted or emergent ray but all the light is reflected internally in the glass, the so-called total internal reflection. The angle at which this effect begins, given by $\sin I = 1/n$, is called the critical angle and it has applications in the determination of refractive index. Of course,

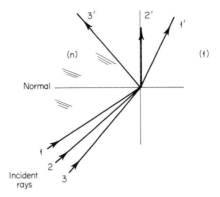

FIG. 4. Total internal reflection; ray 1 is refracted out of the glass in the usual way, ray 2 emerges at grazing angle, and ray 3 is internally reflected. For these three rays, $n \sin I$ is respectively less than, equal to, and greater than unity.

some of the light flux is reflected at angles less than the critical angle. If an empty test tube is immersed in water, the silvery appearance of the inner tube wall indicates that total internal reflection is occurring; the effect is used in right-angle prisms and similar devices in microscopes for turning the beam through an angle, as in Fig. 5.

The refractive index depends on wavelength, an effect known as dispersion; for ordinary optical materials and for wavelengths in the visible region, the refractive index increases with decreasing wavelengths. It is usual to specify refractive index for the wavelengths of certain Fraunhofer lines, which are conveniently obtained from discharge tubes, and for some laser wavelengths. The order of magnitude of dispersion over the visible spectrum for optical glasses is 0.01 to 0.03 in refractive index. The existence of dispersion complicates the design of optical systems with lenses, since all the properties of a system vary with wavelength and it is necessary to use combinations of lenses to *achromatize* the system.

C. LENSES AND IMAGE FORMATION

Figure 6 shows in section a simple lens; it has two surfaces which are symmetrical about the optical axis; usually these surfaces are spheri-

FIG. 5. Right-angle prism, using total internal reflection.

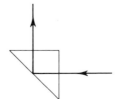

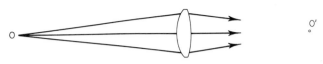

FIG. 6. A simple converging lens.

cal in shape, so that the centers of the two spheres lie on the common axis. It can be seen that rays from a point O on the axis will be refracted towards the axis, according to Snell's law, and thus they converge roughly towards a point O' on the axis. If one of the surfaces is given a suitable nonspherical shape, but still with revolution symmetry about the axis, it is found to be possible to make all the refracted rays pass exactly through the point O' and this is said to be the image of O. In general with spherical surfaces this exact re-coincidence of rays does not occur and there are said to be aberrations; however, if the aperture of the lens is sufficiently restricted by a suitable diaphragm, it is found that the resulting narrow pencil forms an exact image within adequate limits, as in Fig. 7. Thus an ideal image point O' exists and the rays more remote from the axis deviate more or less from this.

This ideal image formation by pencils close to the axis is called paraxial or Gaussian imagery. The complex structure of actual lens systems, such as high power objectives, arises because components with opposing aberrations have to be combined so that the aberrations are eventually canceled, to within suitable tolerances.

In some situations, the emergent rays do not actually intersect after emerging from the lens but they appear to have come from a common point O' in front of the lens, as in Fig. 8; the point O' is then said to be a *virtual image,* in contrast to the case in Figs. 6 and 7, where we have a *real* image. We can also, by an obvious extension of these ideas, have both real and virtual objects. If the object point O does not lie on the axis, there will be an off-axis image point, again subject to reservations about aberrations. The relations between the positions and sizes of objects and images in the paraxial approximation can be described by a few equations which represent simple generalizations of

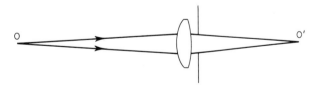

FIG. 7. Paraxial image formation; the diaphragm selects rays close enough to the axis to give a close approximation to ideal image formation.

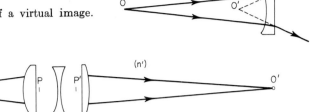

FIG. 8. Formation of a virtual image.

FIG. 9. The principal points and the conjugate distances.

the pinhole camera optics described in Section I.A. These relations apply to any optical system, not merely to a single lens.

Let the lens system be as in Fig. 9; it can be shown that two points P and P' can be found on the axis such that if l is the distance PO to an object O, then the corresponding image O' is at a distance l' from P' given by

$$\frac{n'}{l'} - \frac{n}{l} = K \tag{3}$$

In this equation n and n' are the refractive indices of the media containing the object and the image, respectively, and K is a constant of the system called the *power*. This is said to be a conjugate distance equation, l and l' being the conjugate distances. In the important special case where the object point is at an infinite distance, so that the incoming rays are parallel to the axis as in Fig. 10, the image is easily found from Eq. (3) to be at a distance n'/K from P'. This special point F' is called the image side principal focus and the distance $P'F' = n'/k$ is denoted by f', the image side focal length. There are similar formulas for the object side principal focus F and focal length f. The points P and P' are the *principal points*. We have $nf' = -n'f$.

It can be seen from Eq. (3) that there is an image point for any object point on the axis; in particular, if the object is at P, the image is found to be at P' so that P and P' are conjugates; it is also found

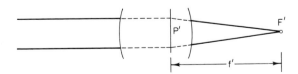

FIG. 10. The image-side principal focus.

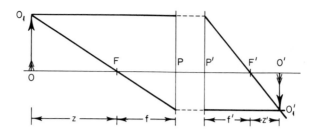

Fɪɢ. 11. Geometrical construction for conjugates.

that an extended object on a plane through P perpendicular to the axis is imaged at unit magnification at P'.

These properties lead to a simple graphical construction for images when the principal points and the foci are known. We draw what may be regarded as a skeleton representation of the optical system, as in Fig. 11; let OO_1 be an object of which the conjugate is to be found; the ray from O_1 parallel to the axis meets the first principal plane, i.e., the transverse plane through P in P_1 and since there is unit magnification between the principal planes this ray emerges through P_1' at the same distance from the axis as P_1; also since the ray came in parallel to the axis it must pass through F', so its path is determined as shown. In a similar way, we can determine the path of a ray from P_1 which goes through F and the intersection of these rays must be O_1', the image of O_1.

This construction can be represented by a simple equation connecting the distances z and z' of the object and the image from the respective foci:

$$zz' = ff' \tag{4}$$

This is called Newton's conjugate distance equation.

Let the object and image heights OO_1 and $O'O_1'$ be η and η' and let the magnification be $m = \eta'/\eta$. Then it is found that

$$m = -f/z = -z'/f' \tag{5}$$

An alternative formula for the magnification in terms of the distances l and l' from the principal points is

$$m = nl'/n'l \tag{6}$$

The relations stated above contain most of the essentials of Gaussian optics. They have to be interpreted with due regard to signs; it is implied that a quantity such as $l' = P'O'$ has a sign corresponding to the order of the symbols and also any such segment which goes from left to right is positive, whereas segments going right to left are negative. Again, a segment going vertically upwards is positive and vice versa, so that the signs are just as in ordinary coordinate geometry. It then appears that a point to the left of the optical system could be either a real object point or a virtual image point, depending on whether we assume that rays through it are about to reach the lens or have already passed through it; similar considerations apply to a point to the right of the system. We are thus led to the concept of the *object space* and the *image space,* both extending over all physical space and overlapping completely. Rays have to pass through the lens to get from one space to the other and they may have to be produced in either direction to find the image.

The above description of the operation of lens systems can be seen as a simple generalization of the elementary theory of thin lenses and of the pinhole camera. The rays arrive at the object-side principal point as if at the left-hand surface of a thin lens or the pinhole aperture and they emerge from the image-side principal point as if from the right-hand surface of the thin lens or pinhole. The space between P and P' is, as it were, a missing volume which does not enter into the general paraxial theory. The principal points can occur in any order along the axis and so can the foci, subject only to compliance with the relation mentioned above,

$$\frac{n'}{f'} = -\frac{n}{f} \tag{7}$$

Systems containing concave or convex mirrors can also have their properties described by these equations; however, reflecting systems are not much used in microscopy and we can omit the details here.

D. ABERRATIONS

The failure of an optical system to bring all rays from a given object point to a unique focus was described in Section I.C as aberration. For specifying the degree of correction of optical systems such as microscope objectives and eyepieces and for describing the results of tests on these systems, it is necessary to have a classification of aberrations into different types. We shall now describe the main aberration types, concentrating on the outstanding properties and omitting mathematical details.

Aberrations can be classified first as monochromatic, i.e., those effects which occur when light of only one wavelength is used, and chromatic, which applies to effects which change with wavelength. There is a differ-

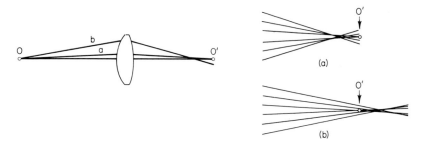

Fɪꜰ. 12. Spherical aberration: (a) undercorrected, (b) overcorrected.

ent but equally important classification into point-imaging aberrations, in which the rays from a single object point do not all pass through one image point, and aberrations of image shape; the latter term applies to a situation in which reasonably recognizable point images are formed but, for example, a plane object appears as a curved image (*field curvature*), or the image is not geometrically similar to the object (*distortion*). Of these, distortion is of almost no interest in microscopy and field curvature is of importance mainly in photomicrography. Of the monochromatic aberrations, that called spherical aberration is the simplest to explain. Let O and O' in Fig. 12, be axial conjugates for paraxial rays such as that labeled a; a ray such as b at larger aperture will not in general meet the axis at O' but at some point nearer to or further from the lens, as indicated to the right of the figure. If the rays through the outer zones of the lens focus short, as in Fig. 12a, the spherical aberration is said to be undercorrected, since this is the usual situation with a singlet lens of positive power (converging lens); the opposite case, Fig. 12b, is called overcorrection. In a complex system such as a high-power microscope objective, there would actually be a more complicated situation in which rays from some zones focus short and others long, as a result of the balancing of opposite effects from different components of the objective; however, there is still the general property that the aberration pattern is symmetrical about the axis and this is what characterizes spherical aberration, an aberration patch on the image plane which has circular symmetry.

To a first approximation, spherical aberration appears the same all over the field of view, if it is present at all, and it is the only monochromatic aberration which can appear in axial images, because of the essential symmetry of the system. Other aberrations appear for off-axis object points and the resultant spread of the image increases with distance from the axis. Figure 13 shows the image plane with the axial image point O'; a narrow paraxial pencil from an off-axis object point

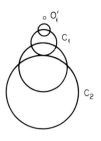

FIG. 13. Coma: rays through each ring in the lens aperture concentric with the axis meet the image plane in the dis-placed rings shown. The overall effect is a strongly asymmetric image.

might form an image at O_1' but it could then occur that the hollow cone of rays at, say, half the full aperture of the lens meets the image plane in the ring C_1 instead of all coinciding at O_1' and the cone at full aperture forms the ring C_2. Thus the point is imaged as a comet shape with a bright head at O_1' and a blurred tail formed by the light from the outer zones of the lens. This aberration is called coma. The scale of the coma patch increases linearly with the distance of the object point from the axis, so that it can be apparent in the image very close to the center of the field; thus, in order to get a sharp image of extended objects, it is essential to have both spherical aberration and coma well corrected. Coma is in many circumstances a particularly objectionable aberration because the point image is so unsymmetrical; it has only symmetry about one line, the line through the center of the field. As with spherical aberration, there can be combinations of different kinds of coma which partly compensate each other; for our purposes, any point image with this minimal symmetry may be said to have coma.

The remaining point imaging monochromatic aberration, astigmatism, has two lines of symmetry, so that it is intermediate between spherical aberration (circular symmetry) and coma; it increases with the square of the distance of the object point from the axis, so that it is not objec-tionable at small field angles; it is sometimes ignored completely in the design of conventional microscope objectives, i.e., those which do not have a flat field. The general characteristic of astigmatism is illus-trated in Fig. 14; instead of a point focus, there is a short line at one plane of focus and at a neighboring plane there is a similar line at right angles to the first; at the midplane between these two, the rays form a bright disc, the *disc of least confusion*.

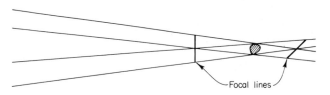

Focal lines

FIG. 14. Astigmatism: rays from all points in the pupil pass through the two focal lines, of which one points towards the lens axis (sagittal focal line) and the other at right angles (tangential focal line).

It should at this stage be pointed out that we have so far been discussing optical image formation and aberrations purely on the basis of geometrical optics, i.e., assuming that the light energy travels precisely along the ray paths as calculated by Snell's law. In fact, diffraction effects take place in the lens aperture which are like those mentioned for the pinhole camera and these have the effect of changing considerably the appearances of point images. We shall return to this in a later section and show examples of actual point images.

There are two main types of chromatic aberration. In the first type, it is found that images formed by light of different wavelengths appear at different positions along the axis, as in Fig. 15, where the images formed from a single axial object point are shown; this is called longitudinal chromatic aberration or, more briefly, longitudinal color. The result is that the image of, say, a pinhole illuminated by white light appears as a broad disc with radially varying color. Some approach to correction of longitudinal color can be made by using glasses of different dispersions for the components of a system but in practice it is not possible to obtain perfectly balancing dispersions and there is always some residual longitudinal color; it is frequently the aberration which ultimately limits the performance of highly corrected systems such as microscope and telescope objectives.

The second effect, transverse chromatic aberration or transverse color, appears as a change in magnification with wavelength, so that an off-axis point image appears as a short radial spectrum; the length of the spectrum is proportional to the distance from the axis of the object point.

The process of optical design involves partly trial and error procedures based on extensive tracing of rays according to Snell's law; it would not be possible to describe this here, but on the other hand we need

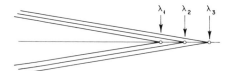

λ_1 λ_2 λ_3

FIG. 15. Longitudinal chromatic aberration. Foci for three wavelengths are shown.

the above brief description of aberrations in order to understand what corrections are possible in microscope systems.

E. APLANATISM

In the discussion of aberrations in Section I.D, it appeared that spherical aberration and coma are the most important monochromatic aberrations to be corrected in a microscope objective. A system corrected for these two aberrations is said to be aplanatic; it is possible to express the condition for aplanatism very simply in terms of the rays from an axial object point. Figure 16 shows two rays at full aperture from the axial object point of a microscope objective; the *convergence angles* of one of these rays with the axis are shown as U and U'. Now let the object point O move away from the axis a small distance η_s at right angles to the plane of the diagram; the image point O' moves correspondingly η_s' in the same plane and it is found that η_s and η_s' are related by the equation

$$n\eta_s \sin U = n'\eta_s' \sin U' \tag{8}$$

This is called the optical sine law and it gives the magnification for this pair of rays in a section which is perpendicular to the plane containing the rays; this is called a *sagittal* section, from which the subscript s is derived. Now coma can be regarded as a variation of magnification with the convergence angle U or, what amounts to the same thing, with the zone of the lens through which a particular hollow cone of rays goes. Thus the condition for absence of coma is that the sagittal magnification should be the same for all values of U; also there will be no spherical aberration if all the rays at different convergence angles U meet the axis at the same image point O'. For paraxial rays, we must have, by letting the angles U and U' in Eq. (8) become very small,

$$n\eta u = n'\eta'u' \tag{9}$$

where u and u' are paraxial angles; this is a paraxial magnification law, and if the sagittal magnification is constant for all values of U,

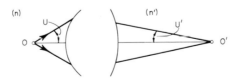

FIG. 16. Aplanatism: in an aplanatic system corrected for spherical aberration, the axial pencils must comply with the Abbe sine condition.

it must have the value implied by this equation. Thus if there is no spherical aberration, the condition for absence of coma is

$$\frac{\sin U}{u} = \frac{\sin U'}{u'} \tag{10}$$

This is the celebrated *sine condition* discovered by Abbe (1879). Any microscope objective or other system which has to form images of extended objects with a large aperture must satisfy this condition to within reasonable tolerances.

It will appear in a later section that a large value of the convergence angle U of the objective is essential for obtaining high resolution and then fulfillment of the sine condition becomes very important.

F. Geometrical Optics of the Complete Microscope

We can now see how the general layout of the microscope developes. We can start with a simple magnifier as in Fig. 17; if the object O is placed a small distance to the right of the object side focus F, it can be shown by the construction in Fig. 11 that an enlarged virtual image is formed at O' to the left of the object, and this can be viewed by the eye placed as shown. The object can actually be placed at the focus, when the image will appear at infinity and can be viewed with relaxed accommodation, but it seems that the more usual, natural action is to set the image at the near point, i.e., the least distance of distinct vision, which is generally taken as 250 mm away. We can define the magnification for the present purpose as the angular subtense of the image divided by the angular subtense which the object would have if viewed at the near point *without* the magnifier; than if f is the focal length of the magnifier in millimeters, the magnification is found to

$$\text{Visual magnification} = 250/f \tag{11}$$

It might now be asked, why not simply use a very short focal length magnifier and get all the magnification required, bearing in mind the need for a large convergence angle mentioned at the end of Section I.E? In fact there is no objection in principle to doing this but there are practical difficulties concerned with illumination, the need for me-

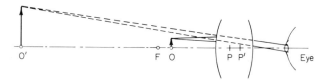

Fig. 17. Action of magnifying glass or simple microscope.

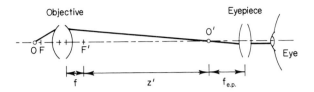

Fig. 18. Gaussian optics of the compound microscope.

chanical stability, the problems of making lenses with very short focal lengths, and the positioning of the eye, which lead to the microscope system as now used, the compound microscope. The magnifier of Fig. 17 becomes the *objective*, as in Fig. 18; the object is now placed slightly to the left of the focus, so as to form a magnified (and inverted) real image at O' and this so-called primary image is then re-imaged by a second lens system, the *eyepiece*, as a further magnified virtual image at O''; here it is conveniently placed to be viewed by the eye.

The magnification produced by the objective is, by Eq. (5), equal to $-z'/f'_{obj}$ where z' is the distance $F'O'$; this distance is the *optical tube length* and in modern objectives its nominal value is usually about 160 mm. The eyepiece magnification is, by Eq. (11), equal to $250/f'_{e.p.}$, so that the overall magnification of the system is

$$-250z'/f'_{obj}f'_{e.p.} \quad \text{(dimensions in mm)} \quad (12)$$

It may now be remarked that if the objective and eyepiece shown in Fig. 18 are regarded as a single optical system, this does actually have the very short focal length needed to get large magnification and the eye can be placed at a convenient position, so that we have really only a simple magnifier. However, it is in practice convenient to make objectives and eyepieces as separate components which are to a certain extent interchangeable, so that the concept of the compound microscope is valid.

The remainder of the system of the compound microscope includes, of course, a condenser, which is an optical system of relatively simple optical design whose purpose is to concentrate illumination on the object.

For deeper discussions of geometrical optics in general and of the optics of the microscope in particular, reference can be made to, e.g., Martin (1955), Welford (1962), or Martin and Welford (1966).

II. Physical Theory

A. PHYSICAL OPTICS AND THE LIMIT OF RESOLUTION

The treatment of image formation outlined in Section I was based on rays and the law of refraction, i.e., geometrical optics. These simple

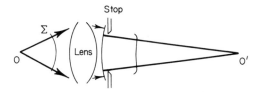

FIG. 19. Diffraction at a lens aperture.

ideas suggest that when a microscope is set up as in Fig. 18 there is
no reason why the system should not form perfectly sharp images of
as fine detail as may be specified, provided the aberrations are well
enough corrected. In fact, it is found that this is not the case; if a
lens forms an image of a bright point source of light and if there are
no aberrations, it is found that the image is not a point but a blurred
patch surrounded by diffraction rings; the dimensions of the patch are
related to the wavelength of the light and to the aperture of the optical
system. Thus there is a lower limit to the size of detail which may
be seen.

Figure 19 shows a lens system, which may, to fix our ideas, be thought
to be a microscope objective, forming at O' an image of the object
O. In order to understand how the diffraction image is formed we recall
that visible light can be regarded as electromagnetic waves, the wavelength being in the range 0.4 to 0.76 μm. It is not important that the
wave motion be thought of as electromagnetic; it is enough to realize
that it is a wave motion. The object O then emits diverging spherical
waves; these we indicate by a line Σ, which is a surface of constant
phase, or a wavefront; the objective converts these into converging wavefronts such as Σ' and, if there are no aberrations, these are spherical
and centered on the geometrical image O'. Now the angular extent of
the wavefronts accepted by the objective is limited by an aperture stop
somewhere in the system; it may be the rim of one of the component
lenses or it may be a stop put in deliberately, as in photographic objectives; wherever it may be in the system, it can be seen from O as
a virtual image in some of the components and this virtual image, which
is in the object space, in effect limits the incoming pencil; it is called
the entrance pupil. Similarly the emerging pencil appears to be limited
by the exit pupil, indicated in Fig. 19. If we imagine, then, a wave
of unlimited extent emerging from the objective, it is limited by the
exit pupil and it starts to spread out again after passing through the
pupil. This spreading out of a limited wave is called diffraction; our
problem is to determine the effect near the geometrical image point
O'. For this purpose, we assume that the wave propagates in the following

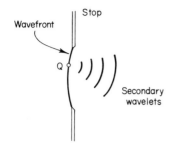

Fig. 20. The Huygens–Fresnel secondary wavelets.

way; at a given instant, each element of the wavefront is regarded as a source of divergent spherical waves which propagate as if from a point source, as in Fig. 20; these systems of diverging waves all have constant phase relationships with each other, since they originate in the parent wavefront at the pupil, so that they interfere where they overlap, i.e., they are *coherent*; the image is formed as the combined interference effect of all the secondary wavelets. This concept was originated by Huygens (1690) and it was developed by Fresnel (1816) to the point at which it could be used to calculate the diffraction effects by integrating the contributions from all the secondary wavelets to the disturbance at a given point. It is worth stressing that the Huygens–Fresnel theory is a very convenient model which correctly predicts a wide range of effects, but it is no more than that; the secondary wavelets can be said to represent reality only in the sense that the components obtained in the Fourier analysis of complex waveforms have reality. However, formulas for diffraction effects obtained from a deeper analysis of the wave motion can be broken down into components in a way which represents approximately the Huygens–Fresnel theory.

The diffraction calculation is carried out in terms of the *complex amplitude* of the light, which is a quantity which provides a measure of the size and phase of the wave disturbance at any point. If we wish to represent a spherical wave, we can say that its *amplitude* is

$$(a/r) \cos 2\pi(\nu t - r/\lambda) \tag{13}$$

where a is the value of the effect (e.g., the electric field strength in an electromagnetic wave) at unit distance from the source, ν is the frequency, t the time, λ the wavelength, and r the distance from the source point. The *intensity* of the light, which is what is actually observed, is the average in time of the square of this quantity, so that we have the inverse square dependence on r; however, in considering interference effects between coherent disturbances, we have to deal in amplitudes, not intensities, because the effect at a given point is obtained by summing the amplitudes of all disturbances arriving at that point with due regard to phases. The frequencies of these disturbances are all equal, since they all originate in the same source point, and it therefore turns out that in the final calculation of intensity as the square of the summed amplitudes the time term νt disappears, so that this can be omitted. It also turns out to be more convenient to do the calculation in terms of a *complex amplitude* of which the actual amplitude we want is the real part; thus we use

$$(a/r) \exp\left(-2\pi i r/\lambda\right) \tag{14}$$

instead of

$$(a/r) \cos\left(2\pi r/\lambda\right) \tag{15}$$

to represent a spherically diverging disturbance. Then if the final complex amplitude is A, which is some function of position near the image, it is found that the intensity is given by AA^* where A^* is the complex conjugate of A; this is called the squared modulus of A, written $|A|^2$, because it is the square of the magnitude of A considered as a two-dimensional vector.

Returning now to our actual calculation, we assume that the complex amplitude over the portion of the wavefront transmitted by the pupil is known; we take coordinates x,y in the exit pupil (Fig. 21) and ξ,η in the image plane and we have to integrate the effects of secondary wavelets from all elements $dx\,dy$ at a point O_1' with coordinates (ξ,η); the result is the complex amplitude A at O_1'. It is found that

$$A\left(\frac{\xi}{\lambda R}, \frac{\eta}{\lambda R}\right) = \iint\limits_{\text{pupil}} a(x,y) \exp\left[-2\pi i\left(\frac{\xi}{\lambda R}x + \frac{\eta}{\lambda R}y\right)\right] dx\,dy \tag{16}$$

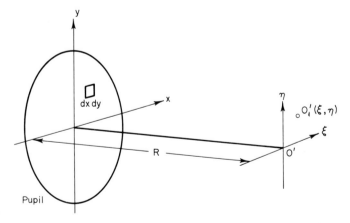

FIG. 21. Coordinate systems for calculating the diffraction image.

where a is the amplitude of the wave in the pupil and R is the distance from the pupil to O'. Before stating the form of the result for the most important special case, we shall note that this expression can be rewritten in a slightly changed form which gives it a more general significance. The function $a(x,y)$ is the complex amplitude of the wave arriving at the exit pupil and it is, of course, zero outside the pupil; we can therefore extend the limits of integration in Eq. (16) to $\pm \infty$ on the understanding that $a(x,y)$ is explicitly given as zero outside the pupil area. We then have

$$A\left(\frac{\xi}{\lambda R}, \frac{\eta}{\lambda R}\right) = \int_{-\infty}^{\infty}\int_{-\infty}^{\infty} a(x,y) \exp\left[-2\pi i\left(\frac{\xi}{\lambda R}x + \frac{\eta}{\lambda R}y\right)\right] dx\, dy \quad (17)$$

In this form, the equation states that the complex amplitude A in the image is the two-dimensional Fourier transform of the complex amplitude a in the pupil. The Fourier transform is used frequently in the modern theory of the microscope and it is necessary to have a general understanding of its significance. Two functions $F(u,v)$ and $f(x,y)$ are connected by the Fourier transform relation if

$$F(u,v) = \int_{-\infty}^{\infty}\int_{-\infty}^{\infty} f(x,y) \exp\left[-2\pi i(ux + vy)\right] dx\, dy \quad (18)$$

and this implies the reciprocal relationship,

$$f(x,y) = \int_{-\infty}^{\infty}\int_{-\infty}^{\infty} F(u,v) \exp\left[2\pi i(xu + yv)\right] du\, dv \quad (19)$$

Suppose now that $f(x,y)$ represents some function of position in the x,y plane; Eq. (19) states that $f(x,y)$ may be represented as a synthesis of components of spatial frequencies u and v and the amplitude of the component with frequencies u and v is proportional to $F(u,v)$; Eq. (18) shows how $F(u,v)$ is computed from $f(x,y)$ and it shows also that $f(x,y)$ similarily represents the amplitude of a Fourier component of $F(u,v)$. Thus the process is very similar to Fourier analysis applied to periodic functions, but the Fourier integral applies to nonperiodic functions; it can also be formally interpreted in such a way as to apply to periodic functions also, with due regard to certain problems of convergence of the infinite integral, so that all cases are covered.

B. The Image of a Point Object

Suppose that, as is customary, the exit pupil is circular and let there be no aberration, so that the complex amplitude in the pupil $a(x,y)$ is constant inside a circle of radius b, say, and zero outside. Under these conditions we can compute the diffraction image of a point from

Eq. (17); the complex amplitude turns out to be

$$\frac{2J_1[(2\pi b/\lambda R)(\xi^2 + \eta^2)^{\frac{1}{2}}]}{(2\pi b/\lambda R)(\xi^2 + \eta^2)^{\frac{1}{2}}} \tag{20}$$

where J_1 is a Bessel function; the properties and numerical values of
Bessel functions are given in mathematical reference books; for our
purposes, it is sufficient to know that J_1 behaves rather like a sine function of decreasing amplitude. The expression (20) has been multiplied
by a numerical constant so that its value is unity at the geometrical
image point, i.e., when $\xi = \eta = 0$; this normalization is usual in dealing
with image formation. We have to take the squared modulus to obtain
the light intensity in the image; before doing this, we note that the light
distribution obviously has symmetry about the optical axis, so that it

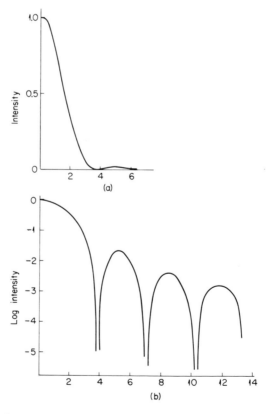

Fig. 22. The point spread function or Airy pattern; (a) linear and (b) logarithmic
ordinates.

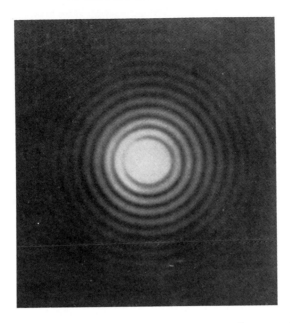

FIG. 23. The point spread function; the photograph has been considerably over-exposed to show the outer rings [From Harris (1964), p. 909].

is sufficient to write it as a function of one variable only, say η; we then have for the intensity

$$I(\eta) = \left[\frac{2J_1[(2\pi/\lambda_0)n\eta \sin U]}{(2\pi/\lambda_0)n\eta \sin U} \right]^2 \tag{21}$$

In this expression we have substituted $\sin U$, the sine of the convergence angle, for b/R and we have put λ_0/n for λ, where λ_0 is the wavelength of the light in vacuum and n is the refractive index of the medium in which the image is formed. We thus have the definitive form of the aberrationless point image or, as it is sometimes called, the point spread function. Figure 22 shows the intensity across a diameter of the distribution as a function of the quantity $z = (2\pi/\lambda_0)n\eta \sin U$ and Fig. 23 is a photograph of a point image. This image is sometimes also called the Airy pattern or Airy disc, after the astronomer Sir G. B. Airy who first developed the theory. It is, of course, the form which the image of a star would take if the telescope had no aberrations and if there were no disturbances of the image by atmospheric turbulence. We can see from the fact that the expression on the right is a function of z that the scale or size of the pattern is proportional to the wavelength and inversely proportional to the sine of the convergence angle; a convenient measure of the scale is the radius of the first dark ring, 3.83 in z units; in actual units of distance it is

$$\text{Radius of first dark ring} = 3.83\lambda_0/2\pi n \sin U$$
$$= 0.61\lambda_0/n \sin U \tag{22}$$

The next dark rings occur at z values, 7.02, 10.17, . . . , etc.; the maximum intensities in the first, second, . . . , bright rings are 0.017, 0.0058, . . . , relative to unit intensity at the center of the image. The proportion of the total light flux which falls inside the first dark ring is 84%.

Finally it is of some interest for our purposes to note the appearance of the defocused point image. The Airy pattern can be seen very easily with a microscope if a silvered slide containing pinholes (star-test slide) is examined with strong illumination and with sufficient magnification; if the image is racked out of focus, there is an appearance as of a sequence of rings spreading out from the center of the pattern. Figure 24 shows some defocused images and Fig. 25 shows the results of a

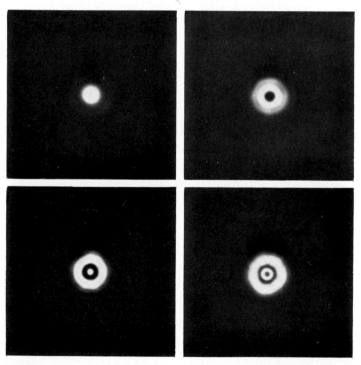

FIG. 24. Out-of-focus Airy patterns; the images are, respectively, $\frac{1}{2}$, 1, $1\frac{1}{2}$, and 2 wavelengths out of focus at the rim of the pupil [Plate I in Martin and Welford (1966)].

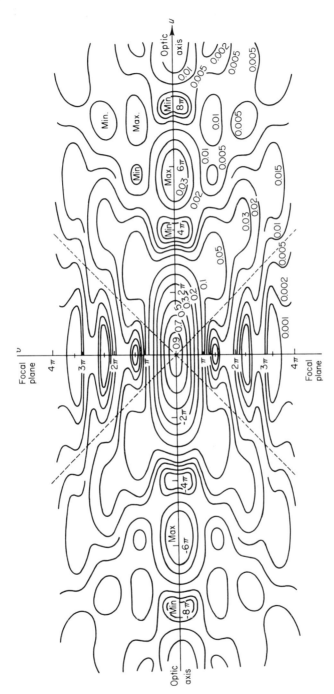

FIG. 25. Contours of constant light intensity near the focus of a pencil. The v axis is in the focal plane and the u axis is along the principal ray; the broken lines are the boundary of the geometrical cone of rays [Figure 8.41 in Born and Wolf (1965)].

calculation, using an appropriate form of Eq. (17), of the light distribution at planes near the focus. The appearances are completely symmetrical on either side of the correct focal plane and there are actually zero values of light intensity at certain positions along the axis. These remarks apply, of course, to a point image in monochromatic light; if there are several different wavelengths or if there is a continuous spectrum, each wavelength forms its own Airy pattern scaled according to its own wavelength and we see a superposition of these; usually there is a further complication from chromatic aberration, which causes the patterns to be formed at different positions along the axis, but even so the main features of the Airy pattern can easily be recognized with some practice.

C. Resolving Power

Suppose that we form an image of two bright points, say pinholes in a star-test slide, which are close together; then the resulting image is formed by adding together the two intensity distributions according to Eq. (21) and Figs. 22 and 23. There is an implied assumption here that the two pinholes are incoherently illuminated, i.e., the two beams of light from them will not interfere, and this is in fact not often true in microscopy; however, it makes a simple treatment and we return to the complications later on. The result for two equally bright points

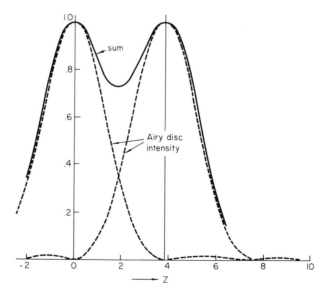

Fig. 26. Image of two incoherently illuminated points at the resolution limit [Figure 8.41 in Martin (1966)].

is shown in Fig. 26, which shows the intensity along a line joining the centers of the patterns for separation 3.83 z-units. It will be seen that, as the centers approach, the dip in intensity disappears and at this point it is probably not easy to tell by casual inspection that there are two bright points, not one. One cannot, of course, say precisely at what point the resolution is lost but a convention has grown up of taking the least resolvable separation as the distance corresponding to the radius of the first dark ring. From Eq. (22) we thus have

$$\text{Resolution limit } \eta_{\lim} = 0.61\lambda_0/n \sin U \qquad (23)$$

This distance, which is measured in the same units as the vacuum wavelength λ_0, is between the centers of the two Airy patterns, i.e., between the image points according to geometrical optics. We should like to know what it corresponds to in the object space and we can get this immediately from the magnification formula, Eq. (8); this shows us that, provided we put in the object space convergence angle and refractive index, we have just the same formula:

$$\text{Resolution limit of mieroscope objective } \eta_{\lim} = 0.61\lambda_0/n \sin U \qquad (23)$$

We see that the resolution limit decreases with decreasing wavelength and with increasing values of the quantity $n \sin U$; this quantity was called by Abbe the numerical aperture, usually abbreviated to NA. The dependence on wavelength is the reason for the use of ultraviolet and electron microscopy: electrons have an effective wavelength about a thousand times less than visible light. The dependence on NA leads to the very large angle cones collected by high power objectives, sometimes exceeding 60° semiangle, and to the use of oil immersion to increase the NA; these points will be explained in the discussion of objectives.

Other definitions of resolution arise when different forms of test object are considered and when the illumination of the object is not incoherent; these definitions all lead to expressions which have the same dependence on wavelength and NA as in Eq. (23) but with a slightly different numerical factor. Thus Eq. (23) gives a useful guide to what is needed for high resolution in all cases.

Table II gives the nominal resolution limit in micrometers (microns) for some typical microscope objectives, calculated according to Eq. (23).

D. DEPTH OF FOCUS

We can use similar notions to those of Section II.C to arrive at a figure for focal depth. If we refer to Fig. 25, we see that the light intensity at the center of the image decreases with defocus, drops to zero, and rises again with smaller oscillations; the actual expression for this

TABLE II

RESOLUTION LIMITS AT WAVELENGTH 0.55 μm

Numerical aperture	Usual magnification	Resolution limit (μm)
0.25	10	1.35
0.65	40	0.52
0.95	40	0.35
1.3	100	0.26

central intensity is

$$\left[\frac{\sin \left[(2\pi/\lambda_0)n\zeta \sin^2 \frac{1}{2}U \right]}{(2\pi/\lambda_0)n\zeta \sin^2 \frac{1}{2}U} \right]^2 \tag{24}$$

where ζ is the actual defocus distance and as before U is the convergence angle. Again this expression holds for both object and image spaces with appropriate values of U and n. If we adopt as a criterion of the limit of focal depth that this quantity shall not fall below **0.8**, its value being unity at correct focus, we find that the permitted focal shift is given by

$$\zeta_{\text{lim}} = \frac{\pm \lambda_0}{8n \sin^2 \frac{1}{2}U} \tag{25}$$

This quantity can become very small for large numerical apertures. Table III gives values of the focal range for some typical values of

TABLE III

FOCAL DEPTH AT WAVELENGTH 0.55 μm

Numerical aperture	Focal depth (μm)
0.1	± 27
0.3	± 3
Dry	
0.6	± 0.7
0.8	± 0.35
0.9	± 0.25
Immersed in $n = 1.52$	
0.6	± 1.1
0.8	± 0.60
0.9	± 0.46
1.0	± 0.36
1.3	± 0.19

NA. It is worth noting that the focal range for an immersion objective can be nearly twice as great as for the dry objective of the same NA. This is of course, because the actual convergence angle U is less for given NA if the object space is immersed.

These focal ranges can be converted to image space distances by a longitudinal magnification formula which is similar to the transverse magnification formula of Eq. (8); the formula is

$$n\zeta \sin^2 \tfrac{1}{2}U = n'\zeta' \sin^2 \tfrac{1}{2}U' \tag{26}$$

It can be seen that, for small convergence angles and ignoring effects of different indices, the longitudinal magnification is the square of the transverse magnification.

Image formation in the microscope is discussed in detail by Martin (1966) and the general physical theory of image formation is treated by Born and Wolf (1965).

III. The Eye in Relation to Microscopy

A. OPTICAL SYSTEMS AND DETECTORS

We have so far discussed in Section I and II two basic ideas related to microscopy, namely magnification, a geometrical optics concept, and resolution, a concept from physical optics. We saw that the microscope can easily be made to give any required magnification by suitable choice of focal lengths, but that in order to ensure that resolved detail is present in the image it is necessary to have adequate numerical aperture in the objective. Thus resolution depends on the objective NA and magnification on the geometrical optics of the rest of the system.

We thus arrive at the concept of an image carrying detail of size down to the resolution limit and it is then necessary to ensure that whatever registers or detects the image shall be able to record all the required fine detail; in other words, we have to consider the resolution of the detector. We speak in general terms of the "detector" since this may be a photographic or a photoelectric system, but the most important detector for light microscopy is the eye and it is therefore appropriate to describe the action of the eye in some detail as far as it concerns microscopy. The essential ideas can then easily be extended as far as we need them to other detectors.

B. THE OPTICAL SYSTEM OF THE EYE

Figure 27 shows the essentials of the optical system of the eye with relaxed accommodation, i.e., focused on objects at infinity. It is very

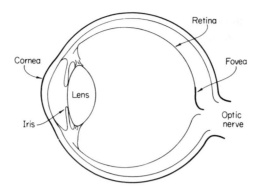

FIG. 27. Schematic cross section of the eye.

like a camera but with the image formed on a curved surface and in a medium with refractive index not equal to unity. The retina registers and presents to the brain a series of nerve signals which are interpreted as an image. In accommodation, the shape of the lens is changed by muscular action and the result is to move the principal foci in such a way that objects nearer the eye are brought into focus. In fact, accommodation is a more complicated process in detail because the principal points also move slightly, but this description is adequate for our purposes. From Eq. (5) we can write

$$\frac{\eta'}{\eta} = \frac{-f_e}{z} \tag{27}$$

where f_e is the first focal length of the eye, z is the distance from F to the object, η is the size of the object, and η' is the size of its retinal image; for real objects, z will be negative and also f_e is negative with the sign conventions implied by our equations. It is found that f_e is about -17 mm for the relaxed eye, i.e., accommodation for infinity, and it changes to about -14 mm in accommodating to the near point. This latter is generally taken as about 250 mm from the eye and it is assumed that in the normal use of the microscope the instrument is focused so that the image appears at the near point.

From Eq. (27) it can be seen that the size of the retinal image depends inversely on z, the distance from the object to the front focal point F of the eye; thus, strictly the angular subtenses of objects and images which are referred to in Section I.F should be taken from F, but in practice the difference between using this point and another such as the center of the eyeball is negligible.

The aperture stop of the eye is the iris (Fig. 27), a stop which is set by involuntary muscular action to give entrance pupil diameters ranging between about 2.5 and 8 mm, depending on the brightness of the scene; in microscopy, the optical system of the microscope provides an external pupil which is rarely more than about 0.5 mm in diameter, so that the iris does not actually limit the cone of rays accepted by the eye.

The field of view of the eye is very large, exceeding 90° from the axis in some azimuths; again this is not made use of in microscopy.

When used at full aperture the eye has quite appreciable aberrations but when it is in effect stopped down by an external pupil from a microscope the aberrations are probably negligibly small and it usually functions as a diffraction limited system up to the point at which the image is to be recorded by the retina.

C. RESOLUTION OF THE EYE AND CHOICE OF MAGNIFICATION

In microscopy at high powers, the resolution of the eye is determined by the structure of the receptor network of the retina, the rods and cones. At the *fovea centralis*, the receptors are all cones and the separation between adjacent cones subtends about half a minute of arc when projected back into object space. Clearly this angle must be related to some absolute minimum perceptible angular separation, regardless of the purely optical performance of the eye, and in fact it is found that the actual least perceptible angular subtense under optimum conditions is about 1′ of arc. It increases under adverse conditions such as low illumination, poor contrast, or fatigue. In practice it is found that a pair of bright points or lines ought to be separated by more than 1′, to ensure that they are easily perceived as separate; 1 mrad, about $3\frac{1}{2}′$, is a good round value to take.

This important datum enables us to choose the appropriate magnification in any situation. We know from Table II what NA must be used to resolve a given size of detail and we can say that we must also use enough magnification to make this detail subtend about 0.001 rad in the final image presented to the eye. Less magnification could mean that the eye could not see all the detail being picked up by the objective; more magnification would be "empty" in the sense that it would merely magnify diffraction structure in the image and would not show more actual detail in the object. Clearly if η is the size in millimeters of the just resolved detail in the object and if m is the required magnification, we must have

$$m\eta/250 \sim 0.001$$

or

$$m \sim 0.25/\eta \quad (\eta \text{ in mm}) \tag{28}$$

From Table II, the smallest resolvable detail for visible light is about 0.00025 mm, so that the largest justifiable magnification is about 1000.

(This size corresponds to the ribbed structure in the diatom *Amphipleura pellucida,* a test object frequently used by microscopists to assess the performance of objectives of the highest NA.)

The above-quoted value of 1′ for the resolution limit of the normal eye applies only under optimal conditions, i.e., an adequate general level of illumination with good contrast in the detail. If the detail has poor contrast or if it is perceptible chiefly through color differences, as in some stained preparations, the visual resolution limit may be much greater and consequently much larger magnification would be needed. The illumination level can, of course, cover a very wide range and still be optimal in the above sense. The retinal illumination is measured in trolands, 1 troland being the illumination on the retina when a surface with a luminance of 1 cd/m² is viewed with a pupil of 1 mm² area. The range of retinal illumination over which full acuity is retained is from about 100 trolands to perhaps 10,000, above which level there is some discomfort due to glare; from 100 trolands to 10, the decrease in acuity is relatively small, but below 10 there is a rapid fall. As a rough guide, a mat white surface illuminated by a 100 W tungsten lamp at 1 m distance would produce a retinal illumination of order 20 trolands, if viewed through a 1 mm² pupil.

The above discussion shows how the choice of magnification depends on the size of detail to be resolved; alternatively, if we have a given objective, it tells us the maximum useful magnification, i.e., that magnification which will show all detail resolved by the objective. Other considerations sometimes intervene: It may not be necessary to strive for high resolution but the largest field of view for a given magnification may be required; in this case, it is found that the lowest power objective with adequate NA should be combined with a high power eyepiece. Again, as the magnification increases, the diameter of the exit pupil of the microscope decreases, so that the pencils entering the eye are very narrow, perhaps ¼ mm or less in diameter; such narrow beams show up defects in the eye such as local opacities in the vitreous, surface damage on the retina, stray floating cell membranes, etc., as shadows in the field of view. These effects are minimized by using as low a magnification as possible or alternatively by using projection microscopy. More details and further references on the eye are given by Martin and Welford (1966).

D. Response of the Eye

The visual sensation, like hearing, is approximately logarithmic in response, i.e., if the retinal illumination is increased from B to aB, then

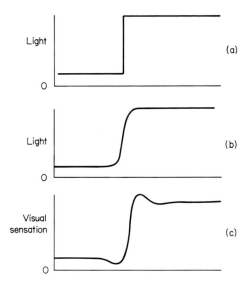

FIG. 28. Sharpening of an image by the Mach effect: (a) the geometrical image, (b) the physical image, (c) the visual image.

the increase in visual sensation corresponds to an increment proportional to log a. This seems to hold for retinal illumination levels over many orders of magnitude so that comfortable vision is possible over a wide range of light levels, provided the eye is given adequate time to adapt. The lowest convenient level for microscopy is that at which resolution begins to fall off, namely 100 trolands (see Section III.C).

The above statement about logarithmic response applies to the visual sensation from relatively large, uniformly illuminated areas. If the response to varying illumination in fine detail is examined, more complex effects are found. Consider, for example, the sharply defined edge between two regions of different brightnesses; Fig. 28a shows graphically how the illumination varies across the edge in the object, which may be imagined as a transparency. When this is viewed through a microscope, diffraction effects and possibly aberrations also cause the retinal illumination distribution to be as in Fig. 28b; the scale of the blurring depends on the magnification of the microscope but if enough magnification is used to take full advantage of the resolving power of the objective, according to the criteria of Section III.C, the blur will be easily perceptible. The actual image as perceived by visual process is, however, more as in Fig. 28c; the visual channel, i.e., the path from the retinal receptors to the brain, sharpens the edge and puts in spurious maxima and minima to make it more easily perceived. This is known as the

Mach effect, first noted in ordinary, nonmicroscopical vision by Ernst Mach about 100 years ago. It is of considerable importance in microscopy because it can introduce artifacts or distortions in the image additional to those due to diffraction and aberrations. The Mach bands or fringes could be taken for membranes or similar structures by an unsophisticated observer, while a more experienced microscopist might ascribe them to diffraction; in fact, experiments have shown that Mach bands can be seen when the conditions are such that no actual diffraction fringes are formed according to photoelectric measurements.

The Mach effect is thus a source of artifacts in fine detail in microscopy. It varies somewhat between individuals but the general features of the effect have now been determined for a number of cases of interest in microscopy; a survey is given by Welford (1968).

E. Visual Refractive Errors

The usual refractive errors of the eye are normally corrected by wearing glasses, but the eye clearance of most eyepieces is not enough to allow a full field of view with glasses. Some manufacturers list specially designed eyepieces with large eye clearance for spectacle wearers but these are more expensive and they have never come into general usage. In the cases of hyperopia and myopia, it is easy enough to work without glasses by simply changing focus appropriately. In the case of astigmatism, this cannot be done but it is possible to have a cylindrical lens of the right power mounted so that it can be slipped over the eyepiece. The correction required should be determined with a small pupil as it may differ from the value found for a normal pupil. Astigmatism of half a diopter or less would normally be ignored as it does not give trouble in microscopical viewing.

With adequate correction as above, if needed, prolonged monocular viewing without discomfort is possible, but some practice and attention to working conditions is desirable. A level of ambient illumination similar to that in the microscope is desirable, as in the use of other optical instruments. The eye in use should be relaxed and accommodated to a comfortable distance. The unused eye should not be closed but the observer should ignore the image from it; this can be done with a little practice but beginners find a screen around the eyepiece useful.

F. Binocular Vision

Vision is better in many ways when both eyes are used rather than one, particularly in the case of the entoptic effects mentioned above due to isolated defects in the eye. Entoptic effects in one visual field seem to be suppressed by the other. The gain is not only in stereoscopic depth perception: Even when no stereoscopy is involved, as in the case of ordinary binocular microscopes in which both eyes see the same image, there is a gain in comfort, clarity, and solidity of the image. Thus there is much to be said for a binocular instrument, although the necessary adjustments for convergence and accommodation must be carefully made to ensure comfort in prolonged observing periods. It can be maintained that a binocular instrument reveals nothing which cannot be seen by careful work with a monocular instrument, but there seems to be a psychological gain.

Actual stereoscopy, in which the two eyes view the object from different angles, can only be obtained at high powers with some loss of numerical aperture and therefore of resolution. It is not generally considered that stereoscopy at high powers has any particular advantage and it is usually possible to obtain the relevant information about the three dimensional aspects of an object simply by studying different focal planes. On the other hand, low power stereo instruments have considerable use with, e.g., micromanipulators and in other cases where rapid stereo perception is desirable.

IV. The Optical and Mechanical Systems of the Microscope

A. Optical and Mechanical Components

Our discussion in Sections I and II has shown that the optical performance of a microscope involves resolution, magnification, and light gathering power. To a certain extent these can be regarded as depending mainly on the objective, the eyepiece, and the condenser system, respectively, and so these three form well-defined separate systems, which can be combined in different ways, e.g., we can choose a range of eyepieces to use with a single objective according to circumstances. Thus, they are separate components which are designed and made independently. In addition to these optical components, we shall consider the mechanical aspects of the microscope, the stand with its stage for the object, substage for the condenser, etc.

B. Achromatic Objectives

The numerical apertures and magnifications of objectives are nowadays fairly standardized except for certain special purpose components. Table IV lists some of the usual values with their focal lengths.

The angular field of view of objectives does not normally exceed ±3° from the axis; this upper limit is due to the fact that the standard eyepiece tube diameter is 22 mm and the nominal tube length from

TABLE IV

MICROSCOPE OBJECTIVES

NA	Magnification	Focal length (mm)	Working distance (mm)	Object field diameter (mm)
0.12	5	32	30	4
0.25	10	16	5.5	2
0.54	20	8	1.4	1
0.65	40	4	0.60	0.5
0.95	40	4	0.25	0.5
1.30 (immersion)	100	2	0.25	0.2

(N. B. The values listed are typical but not standard.)

objective to eyepiece is about 160 mm. Thus the object space fields are arrived at as in Table IV. It should be remarked, however, that the diameter of the field of view is not a precisely defined quantity since it depends on the rate at which the aberrations deteriorate with increasing field angle and on the size of the field stop in the particular eyepiece being used.

We can therefore make the general statement that microscope objectives are systems with very small angular field and with large numerical apertures; even the lowest power standard objective, a $\times 10$ with NA about 0.15, has a fairly large aperture by the standards of photographic objectives, since NA 0.15 is equivalent to relative aperture $F/3.3$. In such systems, the most important aberrations to be corrected are spherical aberration, which affects the axial image; coma, which causes unsymmetrical point images in the field near the axis; and longitudinal chromatic aberration, which causes color halos around all point images. We discuss other aberrations as they arise in particular cases but the need to control these three aberrations leads to the broad types of objective design which we describe.

The lowest powers up to NA about 0.15 are made as simple cemented doublets or triplets; the combination of glass of low dispersion for the positive power element and high dispersion for the negative permits chromatic correction, and, if the refractive indices are properly chosen, spherical aberration and coma can also be corrected.

However, the word "correction" in the previous paragraph needs some qualification. It is found, for example, that while the rays from the center and the edge of the aperture can be made to focus together, so that the state of affairs in Fig. 12 is roughly remedied, there is still

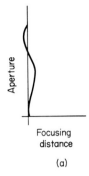

Focusing
distance

(a)

FIG. 29. (a) High-order spherical aberration. (b) Secondary spectrum.

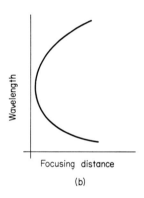

Focusing distance

(b)

a residual of aberration as shown by the fact that the rays from intermediate parts of the aperture do not quite come to the same focus; this effect, higher-order spherical aberration, is illustrated in Fig. 29a, which shows the focusing distance along the axis as a function of aperture in the lens. Similarly the chromatic correction may ensure that pencils of two different wavelengths come to a focus at the same axial point but intermediate wavelengths will still focus at slightly different distances, as in Fig. 29b. This is called secondary color or secondary spectrum. Such higher-order aberrations are small enough not to cause trouble with a simple doublet or triplet, if the numerical aperture does not exceed about 0.15, but they increase rapidly with numerical aperture and this is the reason for the greater complexity of design of higher power objectives.

For numerical apertures up to about 0.3, the Lister type objective, consisting of two separated doublets, is used (Fig. 30).

With still larger NA, the device of an aplanatically refracting surface is used to reduce the initially large divergence of the rays. Figure 31

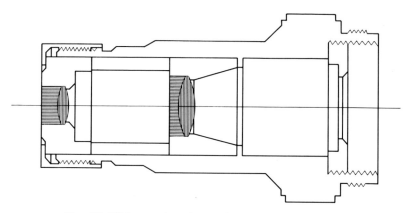

FIG. 30. Vickers ×10 achromatic objective, NA 0.25.

shows a spherical refracting surface with a real object and virtual image; for the particular conjugates related as indicated in the figure to the radius of curvature and the refractive index, it is found that the refraction is perfectly aplanatic, i.e., precisely no spherical aberration and no coma are introduced; there is, of course, some chromatic aberration from this single surface. This is used as in Fig. 32 in the Amici type objective, which is suitable for numerical apertures up to 0.65. The front component provides a good proportion of the refractive power and reduces the divergence of the rays without introducing coma; the doublets which follow provide the remaining power and correct the chromatic aberration of the front component (and also the spherical aberration introduced at the plane surface).

The next stage is the introduction of immersion, as in Fig. 33; there

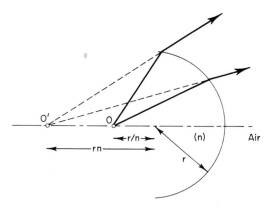

FIG. 31. Aplanatic refraction; the object is in the medium of refractive index n (taken as 1.6 in drawing the figure).

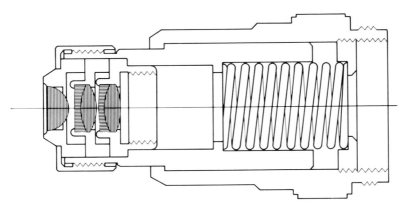

FIG. 32. Vickers ×40 achromatic objective, NA 0.65.

is again a front component with an aplanatic surface, but now by means of immersion oil the object is made to lie effectively inside the sphere, so that the refraction at the plane front is eliminated. A numerical aperture of up to 1.3 is obtainable in this way. The front component is hyperhemispherical to fulfill the condition indicated in Fig. 31 and this is followed by a meniscus component which the rays enter at normal incidence and leave under the aplanatic condition; then follow doublets to provide chromatic correction.

The secondary color of all these types of objectives is as in Fig. 29b; no significant improvement can be gained by choice of glasses within the range available today in glass catalogues. Oil immersion objectives show chromatic differences of magnification; the scale of the blue images is usually 1–2% greater than that of the red images. For critical work, this is overcome by using a compensating eyepiece (see Section IV.F).

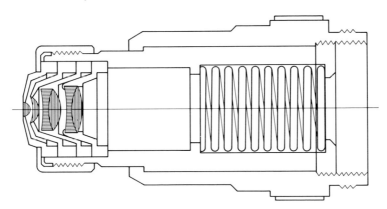

FIG. 33. Vickers ×100 achromatic objective, NA 1.3.

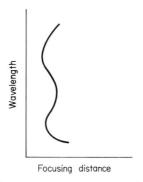

FIG. 34. Secondary spectrum in an apochromatic objective.

Focusing distance

C. APOCHROMATIC OBJECTIVES

The residual chromatic aberration in the objectives described in the previous section is sometimes found objectionable for extremely critical work on objects with delicately tinted structures. A reduction of the secondary spectrum below that level requires an optical material with a dispersion not found in optical glasses and the only material which is found to be suitable is fluorite, CaF_2. Objectives which incorporate fluorite in place of some of the crown glass components are known as *apochromats;* they have greatly reduced secondary spectrum, as in Fig. 34, and when a good apochromat is used with an appropriate compensating eyepiece, the image has very fine color correction. Figure 35 shows a typical apochromat.

An incidental advantage of the use of fluorite in the design is that slightly better spherical correction is found to be possible than with ordinary glasses and thus slightly higher NA, about 1.35, can be reached with a 2-mm objective. On the other hand, most fluorite scatters light more than optical glasses, both from true scatter in the body of the

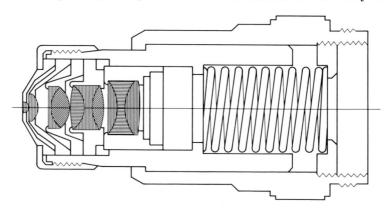

FIG. 35. Vickers ×80 apochromat objective, NA 1.32. The diagonally hatched parts are fluorite.

material and from the surface, since it is impossible to get such a high polish on fluorite as on glass. Thus, if an apochromatic objective is to be used for work where scatter is important, e.g., central dark ground illumination or vertical incident illumination, it should be tested under the conditions of use and compared with a good achromat to see if the fluorite scatters appreciably.

A third class of objectives, intermediate in color correction (and price) between apochromats and achromats, is the class of semi-apochromat or fluorite objectives. These contain less fluorite than a fully apochromatized objective but they still show greatly improved color correction over achromats. Apochromats and semi-apochromats are usually high-power objectives (2-mm focal length) but a few lower power objectives are made with fluorite for special purposes.

D. Flat-Field Objectives

All the objectives so far described have strong field curvature; the field in object space corresponding to a plane image surface is concave

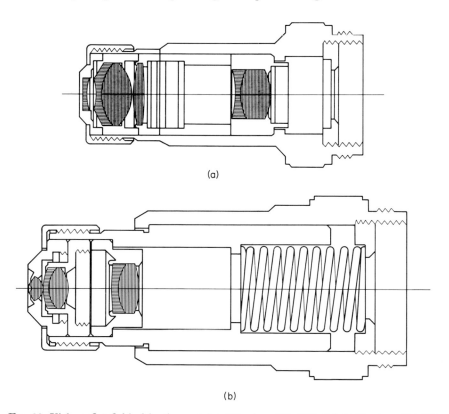

(a)

(b)

Fig. 36. Vickers flat field objectives: (a) ×10 achro, NA 0.25, (b) ×40 achro, NA 0.7.

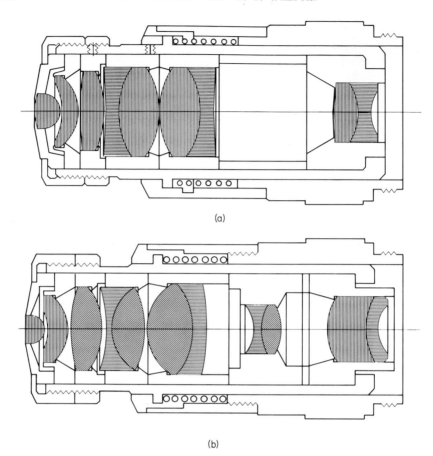

(a)

(b)

FIG. 37. Vickers flat-field objectives: (a) ×100 achro, NA 1.25, (b) ×100 apochromatic, NA 1.3.

towards the objective and it has a radius of curvature of the order of magnitude of the focal length of the objective. This field curvature is an inevitable consequence of the design of these systems, which all consist of a sequence of elements of positive power. It follows from the theory of aberrations that it is impossible to correct field curvature unless the system contains also some elements of negative power spaced apart from the positive elements. Figures 36 and 37 show some flat-field objective designs, both achromatic and apochromatic. It can be seen that they are more complex than ordinary objectives and so they are more difficult to make and more expensive. However, flat-field objectives are of great value in photomicrography, since a large field can be simultaneously in focus at one setting.

E. Other Objectives and Some Points of Design

We have covered in the preceding sections the types of objective of most interest to the biologist. Here we mention briefly some other kinds which may sometimes be useful.

Reflecting objectives, in which the image is formed as in Fig. 38, have the advantages of no chromatic aberration, transparency to the spectrum from about 200 nm to the far infrared, constant focal setting for all wavelengths, and very long working distance compared to the corresponding refracting objectives. However they are expensive to make and difficult to keep in adjustment; they have, therefore, never been very popular except for special purposes such as ultraviolet microscopy.

An objective of higher NA than 0.25 must be designed with the thickness of the coverglass and the immersion liquid (if any) included in the calculation. *Metallurgical objectives* are designed for use without a coverglass; also they are particularly carefully antireflection coated, although nowadays almost all manufacturers antireflection coat microscope objectives as a routine.

Water-immersion objectives of NA about 1.0 are occasionally made; these are designed for use with water and no coverglass.

Very high NA dry objectives (0.95) are very sensitive to the coverglass thickness and they usually incorporate a correction collar; this is a device for varying the spacing between two groups of components in order to compensate the spherical aberration due to different coverglass thicknesses. The requirements for accurate centering of the components are so critical that it is very unusual to find an objective with a correction collar which gives good test results (see Section V.B) at all settings.

Long working distance objectives consist of a conventional objective together with a mirror system to re-image the object as in Fig. 39. They are used with thermostated enclosures and in other situations where the ambient conditions are not suitable for an ordinary objective.

Interference and phase-contrast objectives are described in the Chapter on Phase and Interference Microscopy in Vol. II.A, 1st edition.

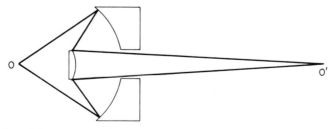

Fig. 38. Principle of a reflecting objective. One surface must be aspheric to obtain good spherical aberration correction and both must be aspheric for aplanatism.

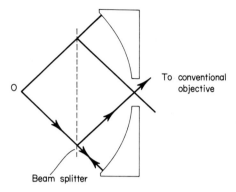

FIG. 39. Principle of Dyson's long working distance objective.

On a subject related to that of coverglass correction, it is appropriate to note here that any objective is designed to work at a specific magnification and aberrations are introduced if the wrong magnification is set by lengthening the microscope draw-tube. Objectives of NA 0.65 and more are quite sensitive to magnification change and with some NA 0.95 dry objectives it is possible to choose the optimum draw-tube setting to within a few millimeters. It is also possible to compensate coverglass thickness by change of magnification: The draw-tube is extended to allow for a thinner coverglass.

F. EYEPIECES

It will be recalled from Section I.F that the objective forms a magnified real image of the object and the eyepiece then presents a further magnified virtual image to the eye. Figure 40 shows this process in detail for a *Ramsden eyepiece*. The two components, field lens and eye-

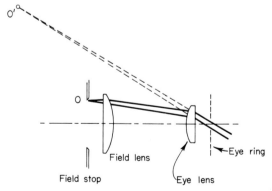

FIG. 40. Ramsden eyepiece. The intermediate real image O is magnified to become the final virtual image O′.

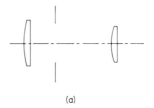

(a)

FIG. 41. (a) Huygens eyepiece, (b) compensating eyepiece.

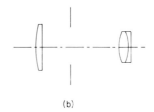

(b)

lens, appear in slightly different forms in nearly all microscope eyepieces. The field lens is near the primary image; it does little to magnify the image but it takes the pencils diverging from the objective and bends them in towards the axis. The eyelens does most of the magnifying. The pencils cross the axis at a point behind the eyelens which is known as the eye ring; it is the exit pupil of the complete system and it is the point at which the observer's eye has to be placed in order to see the whole field of view. The eye clearance usually varies between 10 and 20 mm, the smaller values corresponding to higher power eyepieces; as mentioned in Section III.E, special eyepieces are available for spectacle wearers, e.g., the Bausch and Lomb Hi-Point series. The field stop in Fig. 40 serves to delineate sharply the well-corrected field of view of the eyepiece and it is also a convenient place to put a graticule or scale. The Ramsden eyepiece is well suited for use with a graticule but in the form shown it is not very well color-corrected.

The *Huygens eyepiece* (Fig. 41a) is the simplest and most popular for routine microscopical use; complete transverse color correction is obtained but the primary image from the objective is formed inside the eyepiece, at the position of the field stop; this has the disadvantage that if a graticule is needed, it has to be placed at this point and its image as formed by the eyelens alone will have chromatic aberration.

The *compensating eyepiece* (Fig. 41b) is used to give transverse chromatic aberration to compensate the transverse color inherent in immersion objectives; the eyelens is focused on the field stop plane, so that a graticule can be used there; the distance from field lens to field

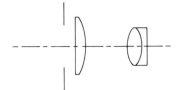

Fig. 42. Kellner or achromatized Ramsden eyepiece.

stop can be variable, as in, e.g., the Watson series, to provide variable transverse color.

The achromatized Ramsden or Kellner eyepiece (Fig. 42) gives good color correction and has an external field stop, so that it is used in filar micrometer eyepieces and similar measuring devices where the scale or graticule must be outside the eyepiece.

These four types of eyepiece serve most ordinary purposes in microscopy; Table V gives typical focal lengths and magnifications.

TABLE V
EYEPIECES

Magnification	Focal length (mm)
6	40
10	25
15	15

Projection eyepieces usually have a triplet eyelens and the exit pupil is closer to the eyelens than in an ordinary eyepiece. However, a compensating eyepiece serves just as well for most projection purposes. Occasionally, negative lenses are used for projection and for microphotography; they have the advantage that some degree of field flattening is introduced.

Some manufacturers now offer zoom microscopes, i.e., systems with magnification continuously variable over a certain range. The zoom optical system is an almost afocal system of variable magnification between the objective and the eyepiece and so it does not strictly come under either classification; however, it is convenient to mention these systems briefly at this point. They have some attractions for low power stereo microscopes, e.g., dissecting microscopes, but the systems incorporated in stands intended for high power work which have been tried by the present writers do not perform very well. The total magnification range is about 2 and both focus and aberration correction wander appreciably through this range. It could not be claimed that zoom systems are adequate for critical work as yet.

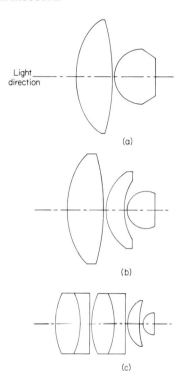

Light direction

(a)

FIG. 43. Condensers: (a) Abbe, (b) aplanatic, (c) achromatic. All are suitable for oil contact with the microscope slide.

(b)

(c)

G. CONDENSERS AND ILLUMINATING SYSTEMS

The condenser has to form an image of the effective light source in the plane of the object and the divergence of the rays forming this image must be enough to fill the aperture of the high NA objectives. Thus, a condenser rather resembles a microscope objective in reverse, but since the aberration correction required is relatively crude the design is correspondingly simple. Figure 43 shows some typical condenser designs, all of which can be used as immersion systems, i.e., the last surface of the condenser is put into contact with the underside of the microscope slide by means of immersion oil; by this means the NA of the illuminating cone can be as large as 1.40. The choice of a condenser depends on practical factors; it is known that the performance of an objective in terms of formal resolution is not affected by the state of correction of the condenser, but on the other hand, a condenser with large aberrations will send light into parts of the object which are not in the field of view and so will produce glare and loss of contrast; again, if the source is small, e.g., a high-pressure mercury arc, a low grade condenser will produce colored patches in the field and will falsify the color rendering. Thus for critical work the more complex condensers can be justified.

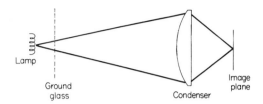

FIG. 44. Critical illumination: the light source is focused on the object plane.

The ordinary method of illuminating a microscope is to use the condenser to focus an image of the source on the object, as in Fig. 44; since most sources have structure which would appear in the image, e.g., the coils of a filament lamp, a ground glass or other diffusing screen is interposed. This system, known for historical reasons as "critical illumination," is wasteful of light because of the diffusing screen and it is inadequate in situations where high light intensity is needed for other reasons, e.g., projection microscopy, deeply absorbing objects, cine work, etc. The system known as Köhler illumination (Fig. 45) is then used. A lens, known as the field lens, lamp lens, or source lens, produces a real image of the light source in the first focal plane of the condenser. Each point in this real image produces a parallel beam passing through the object at an angle depending on the distance of the source point from the axis. The field lens is placed at a distance such that the condenser forms a real image of it in the object plane; the aperture of the field lens is uniformly illuminated by the lamp and thus the object is also uniformly illuminated; also an iris diaphragm placed at the field lens can be used to control the area of the object which is illuminated, thus avoiding glare from unwanted light. The condenser iris is at or near the first focal plane of the condenser, i.e., at the source image, so that it controls the numerical aperture of the illuminating beam. The Köhler system is ideal for exacting microscopical work since it permits precise control over all aspects of the illumination. To cover

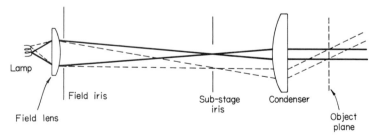

FIG. 45. Köhler illumination: the lamp is imaged at the back focal plane of the condenser.

all possible objective NAs and field diameters, two or three field lenses of different focal lengths are necessary; this is allowed for in modern stands with built-in illumination systems by swing-in lenses or sometimes zoom systems.

H. MECHANICAL PARTS OF THE MICROSCOPE

The mechanical system which holds all the optical components and the specimen is called the microscope stand. Figure 46 shows what might be called the classical form of the microscope stand. The lamp with, if required, field lens, is separate from the stand; the condenser fits into the substage and there is a substage mirror to direct the light into the condenser and a substage iris. The mechanical stage would be almost universally used for biological work. The triple or quadruple nosepiece facilitates rapid interchange of objectives, which is useful in setting up the microscope and finding specific objects. The draw tube permits slight variation of the objective magnification if it is required to set an eyepiece graticule at an exact magnification. The limb is pivoted at the trunnion to permit inclination of the whole optical system to any required angle between horizontal and vertical.

Such an instrument looks almost antique beside a modern stand (Fig. 47) but nevertheless it has advantages for certain purposes, e.g., the choice of optimum tube-lengths. Modern microscope objectives are usually designed for a standard distance of 160 mm between the flange on the screw of the objective and the end of the tube on which the eyepiece rim rests;[1] this distance is known as the mechanical tube-length. Each manufacturer will make his objectives parfocal with each other and likewise his eyepieces, so that the optical components can be interchanged without shifting focus except for a touch of the fine focus adjustment, but different manufacturers naturally have different standards for the relations between these two flanges and the optical components; thus it turns out that an objective made by "A" will not function at the correct magnification in "B's" stand. It also, regrettably, sometimes happens that an objective as actually made is different from the design and needs to be used at some magnification other than the nominal value in order to realize its optimum performance. Thus the discriminating and experienced microscopist will find the draw tube on the classical stand useful on occasion for getting the best out of his optical components.

The substage has a rack and pinion focus movement, a tray for filters,

---

[1] At least one manufacturer, however, corrects the objectives for an infinite tube-length and uses a "tube-length lens" to focus the image at the right distance for the eyepiece.

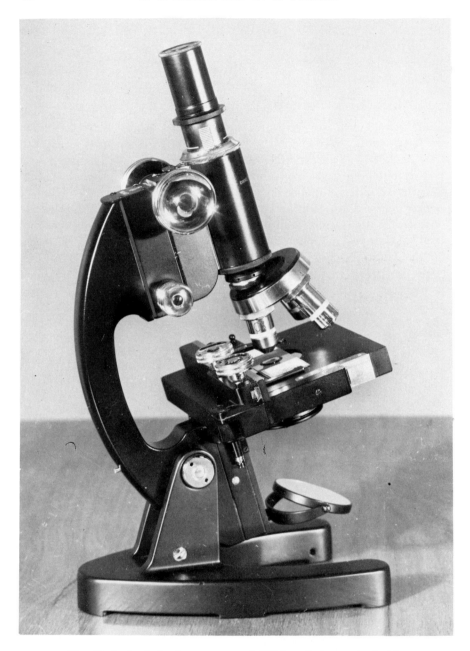

FIG. 46. A classical microscope stand (Vickers Instruments, Ltd.).

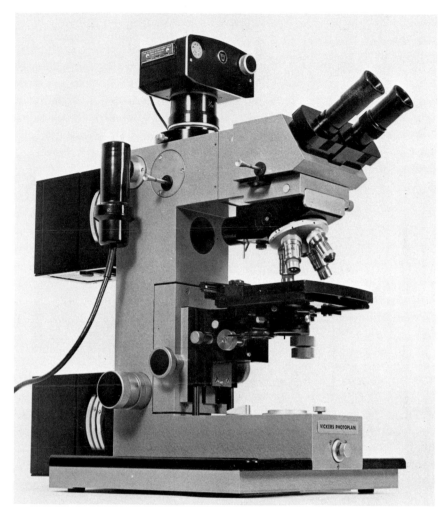

FIG. 47. The Vickers Photoplan research stand, with built-in incident and trans-illumination and camera.

and centering screws for the condenser;[2] it should also have, in a well-designed stand, an adjustable stop to prevent the condenser from being

[2] The condenser centering adjustment is sometimes omitted from modern stands but in a recent survey one of the present writers found no stand without condenser centering in which the illumination could be adjusted really well; centering should be regarded as essential because even if a complete set of microscope equipment leaves the factory well centered, it inevitably happens that in the course of time other components are added and also the fits of the original components become sloppy.

racked above the level of the stage. The mechanical stage should have smoothly operating x-y slides for biological work; the most convenient arrangement is to have the knobs for the rack and pinion movements arranged concentrically on one spindle, so that they can be operated easily by one hand.

The multiple nose-piece is a weak point in many stands because it can become sloppy in angular registration. Attempts have been made to replace it by quick-change devices for individual objectives but these have never become popular, probably because they are less convenient; the objectives which are not in use have to be put somewhere safe.

Fine and coarse focus adjustments are perhaps the most important items; the coarse focus should be smooth and adjustable in stiffness; the fine adjustment should have a range of 1 to 2 mm and a controlled movement of 0.1 μm should be possible. The fine focus control should be calibrated and it is also useful for the limits of the range to be indicated at some convenient point. In modern stands, it is usual for the controls to be concentric and to also be available on either side of the stand.

The distribution of functions and movements over the stand is to some extent a matter of fashion. In modern stands, it is becoming usual to have coarse and fine focus on the stage, so that these movements carry the substage focus and the mechanical stage movements pick-a-back; the upper part then takes interchangeable tube systems for various purposes. These features can be seen in Fig. 47, which shows also a built-in illumination system; the vertical tube can take a 35-mm camera back or a Polaroid camera back, and several alternative systems,

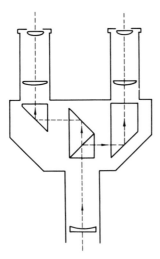

Fig. 48. Binocular head.

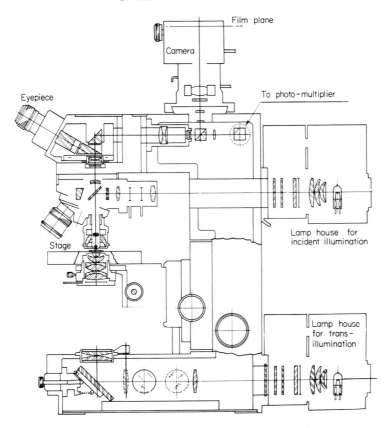

FIG. 49. Optical system of the Vickers Photoplan.

not necessarily all particularly useful for biological work, are available. The binocular attachment would have a negative lens incorporated and prisms for splitting the beam as in, e.g., Fig. 48; also suitable prisms for inclining the beam would be incorporated. Figure 49 shows a detailed outline drawing of the optical systems of the stand in Fig. 47, illustrating these features.

V. Adjustment and Use of the Microscope

A. SETTING UP THE MICROSCOPE

In this section we outline the procedure for setting up a microscope to obtain the best image. We shall describe the procedure with the classical stand in mind (Fig. 46) and it should be noted that some of the steps we mention are either not necessary or, perhaps, not possible with a modern stand. However, a description of the classical procedure serves to emphasize the factors which need attention to obtain a well-aligned system, even if some steps have to be omitted. The technique is in fact similar, *mutatis mutandis,* for adjusting different kinds of optical system and the order of the steps is chosen to ensure good alignment and uniform illumination of the field of view.

(a) The condenser, objective, and eyepiece are removed and the sub-stage plane mirror is adjusted until the lamp shines straight through the tube; preferably at this stage the lamp should have a ground glass diffuser in front of the field lens and behind the field iris (Fig. 45).

(b) The slide is placed on the stage, a $\times 10$ objective and a $\times 5$ or $\times 10$ eyepiece are placed in position and focused on the object by means of the coarse adjustment.

(c) The substage condenser is mounted and racked up until an image of the field iris is seen in focus; this adjustment is facilitated by closing the iris or putting a pointer such as a pencil near it.

(d) The eyepiece is removed and the substage iris is closed down. On looking down the tube at the back of the objective an image of the substage iris is seen; this has to be brought concentric with the objective aperture by means of the condenser centering adjustments and it should then appear to be uniformly illuminated if the mirror is correctly set.

(e) The substage iris is opened until its image fills about two-thirds of the aperture of the objective.

(f) The eyepiece is replaced, the image refocused, and the field iris is set to illuminate the required area of the object; the illuminated area can be centered in the field by a slight adjustment of the substage mirror. A final check should be made on the concentricity of this image of the substage iris and then if necessary the iris can be slightly adjusted to improve contrast.

If a high power objective is needed this should then be inserted and the above procedure from (d) onwards repeated. It is desirable to carry out the preliminary adjustment at low power even when only a high power objective is to be used. If a set of parfocal objectives is in the multiple nosepiece, these can be swung into position in turn and a dab of immersion oil used for the immersion objective. If the objectives are not parfocal, considerable care is needed in setting up higher power objectives on account of the very short working distance; some manufacturers mount the optical components of the objective on a spring plunger system (see Figs. 32 to 35) so that no harm comes of racking it into the coverslip; failing this, it is safest to start with the objective almost in contact and search for the focus by racking away from the slide. Depending on the combination of eyepieces and objectives, it may or may not be necessary to use a special eyepiece to correct for the transverse chromatic aberration of oil immersion objectives; the makers' instructions and the tests described in the next section are the best guide here.

For dark-field illumination, photomicrography, for projection, or for simply observing strongly absorbing objects, the ground-glass must be removed and the complete Köhler system used. The lamp image must be focused in the plane of the substage iris but otherwise the procedure is the same. Very careful attention must be paid to the illumination for color photography and it pays to use an achromatic condenser.

B. TESTING MICROSCOPE OPTICAL COMPONENTS

It is quite easy to learn to judge the quality of a microscope objective and it is well worthwhile to be able to do this, because one is then better able to decide when some just resolvable appearance is an artifact and when it is more probably genuine.

The basic test is known as the *star test*. A microscope slide is covered with an opaque silver film in which there are numerous pinholes of sizes down to about 0.1 μm; the film is protected by a coverslip of correct thickness. Such test slides are commercially available, often with a range of coverslip thicknesses on one slide. The pinholes are viewed with strong illumination (which need not, however, be very carefully aligned) and the aberration correction of the objective is judged by the form of the diffraction image seen. In monochromatic light, the in-focus image should have a light distribution according to Fig. 22 and the diffraction ring systems formed on defocusing should be the same on either side of focus. The different wavelengths in polychromatic light form superposed images scaled according to the wavelength and defocused according to the residual chromatic aberration. The flatness of the field can easily be estimated and the transverse chromatic aberration can be seen and, if necessary, compensated by suitable choice of eyepiece type. Details of the star images to be expected with other aberrations may be found in, e.g., Martin (1961). It should be noted that one is really testing the combined aberrations of the objective and the eyepiece but in practice the latter contributes only to off-axis aberrations. Enough magnification must be used to make the diffraction rings clearly visible and this means using rather more magnification than the values suggested in Section III. Under this condition, it is easy to see when one is looking at a small enough pinhole because there is only diffraction structure and no object shape.

In addition to testing the performance of objectives, the star test can be used to check on centering of components; it can be used to determine the optimum tube length for an objective; and it can be used to set the correction collar on a high NA dry objective. Also, the star test enables one to check on scattered light in the objective from faulty fluorite, poor polish, or other sources.

Test objects such as diatoms are not so suitable for checking objective quality because the details of the image depend strongly on the mode of illumination (see Section VI) whereas this is not so with the star test. Interferometric techniques demand complex equipment not usually available in biological laboratories.

One should not forget to check also for gross defects such as faulty optical cement, loose components, scratched surfaces, and peeling anti-reflection coatings. Regrettably, such things occasionally slip by factory inspection systems.

Very little can be said about testing eyepieces; they contribute chiefly to improvement of astigmatism and transverse chromatic aberration but they are not critical components to manufacture. It may be useful to remark that a Huygens eyepiece shows a blue fringed rim to the field stop when a diffuse source is viewed through it but under similar conditions a compensating eyepiece shows a red or orange rim.

VI. General Theory of Image Formation in the Microscope

A. EXTENDED OBJECTS AND THE MODE OF ILLUMINATION

In Section II we discussed briefly the formation of images of isolated points according to physical optics and we obtained the classical formula [Eq. (23)] for the resolution limit for two bright points. In deriving this expression, it was assumed that the two points were illuminated in such a way that the light beams from them would not interfere, so that we had to add the intensities in the individual Airy patterns in order to obtain the resultant intensity. This is said to be incoherent illumination. However, we might alternatively have assumed that we had to add the complex amplitudes and take the squared modulus to obtain the resultant intensity; this would have been coherent illumination. It is found that either of these cases and also a complete range of intermediate cases (partial coherence) can be obtained by suitable control of the illumination conditions. In practice, the illumination conditions which are actually used generally come in this intermediate range and it is therefore desirable to examine in more detail the effect of the mode of illumination on image formation.

It should not be supposed that variations in the illumination condition cause only minor changes in the image formation. On the contrary, we can see from a simple example that the most profound changes can be produced in the structure of images near the resolution limit. We gave in Fig. 26 the intensity distribution along the center line of two point images separated by the classical resolution limit, 3.83 in z units; this was obtained on the assumption of incoherent illumination, so that we can regard Fig. 26 as the graph of the function.

$$\left(\frac{2J_1(z)}{z}\right)^2 + \left(\frac{2J_1(z-3.83)}{z-3.83}\right)^2 \tag{29}$$

If now we suppose the object to be coherently illuminated, so that both pinholes are illuminated exactly in phase by, e.g., a laser, we should have for the intensity along the center line

$$\left\{\frac{2J_1(z)}{z} + \frac{2J_1(z-3.83)}{z-3.83}\right\}^2 \tag{30}$$

and this function is plotted in Fig. 50. Clearly the two points would not be regarded as resolved by any reasonable criterion and we have to increase the separation to 5 z units, as in Fig. 51, in order to get a clear dip between the maxima. Of course, it should be remembered that we are here considering only the light intensity distribution along

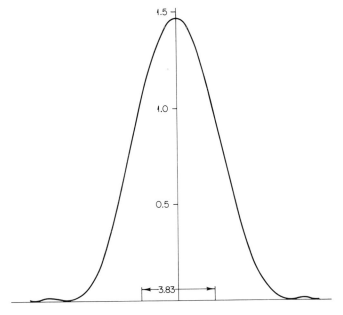

Fig. 50. Image of two coherently illuminated pinholes 3.83 z units apart, illuminated in phase.

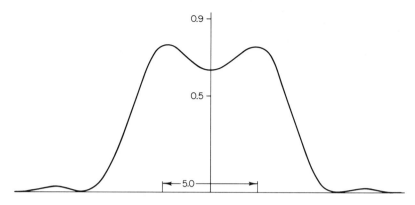

Fɪɢ. 51. Image of two coherently illuminated pinholes 5 z units apart, illuminated in phase.

the center line; other clues to resolution may perhaps be available from parts of the images remote from the center line. Nevertheless, there is a definite difference between the two cases.

Next we note that there was an assumption in Eq. (30) that the light disturbances from the object pinholes were in phase, i.e., they both had maxima at the same instant of time. However, this is only one of many possibilities in coherent illumination; for example, we might assume the disturbances to be in quadriphase, i.e., $\pi/2$ out of phase with each other, for which the expression for the intensity would be

$$\left| \frac{2J_1(z)}{z} + \frac{2iJ_1(z-3.83)}{z-3.83} \right|^2 \tag{31}$$

or they could be in antiphase, giving

$$\left\{ \frac{2J_1(z)}{z} - \frac{2J_1(z-3.83)}{z-3.83} \right\}^2 \tag{32}$$

The result for the latter case is plotted in Fig. 52 and it can be seen that in contrast to Fig. 50 we now have ample resolution for two points 3.83 z units apart.

These results can easily be verified experimentally and they have considerable significance for the experienced microscopist who is used to manipulating illumination conditions. If a collimated beam of laser light is incident normally on the pair of pinholes, the illumination will be coherent and in phase; if the angle of incidence is arc sin $\lambda/2d$, where d is the distance between the pinholes, the illumination will be coherent but in antiphase. The same results are obtainable with a con-

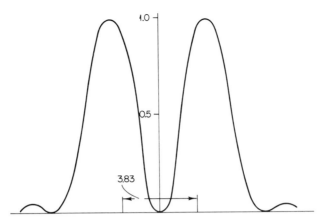

FIG. 52. Image of two coherently illuminated pinholes 3.83 z units apart, illuminated in antiphase.

ventional light source and with small stops suitably placed in the plane of the substage iris.

The above discussion has been restricted to an object consisting of a pair of pinholes in a metalized slide and this is perhaps the simplest form of extended object which may be conceived. However, it is reasonable to suppose that similar effects would be found with any form of object other than a single bright point.

B. COHERENT ILLUMINATION WITH A GENERAL EXTENDED OBJECT

In order to proceed further with a general treatment, we have to introduce the notion of a periodic object, i.e., a regular repeated structure. Abbe did this and arrived at the first systematic description of image formation in a microscope under a reasonable approximation to ordinary conditions of use. In Fig. 53 we show a monochromatic point source of light at the focus of a condenser producing a collimated beam; the object is a periodic structure with basic period of length σ. Under these

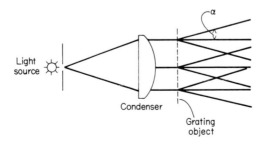

FIG. 53. Diffraction of a coherent illuminating beam by a periodic structure.

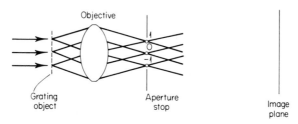

FIG. 54. Formation of the object spectra in the back focal plane of the objective.

conditions the object will be illuminated coherently and the complex amplitudes arriving at all points will be in phase. The object then acts as a diffraction grating, diffracting beams in directions α', given according to the usual theory of the diffraction grating by

$$\sin \alpha' = m\lambda/\sigma \qquad (33)$$

where λ is the wavelength of the light and m is any integer, positive or negative, for which a real value of α' can exist, i.e., we must have $|m\lambda/\sigma| \leq 1$. If now the microscope objective is placed behind the grating, the diffracted beams are brought to foci in the image side focal plane of the objective as in Fig. 54; the three foci shown are numbered -1, 0, 1 to agree with the corresponding values of m in Eq. (33). These beams then go on to the image plane, i.e., the plane conjugate to the object, where they overlap and, since they are coherent, they form an interference pattern. Abbe showed that the image is formed in this way as an interference pattern between the diffraction spectra by simple calculations and by some very beautiful experiments. It is clear from Fig. 54 that in order to get any image at all, i.e., any variation of intensity, at least two spectra must be collected by the objective, which sets a lower limit on the collecting angle or numerical aperture of the lens. It is usually assumed that the effective aperture stop of a microscope objective is at the back focal plane[3] as indicated in Fig. 54, and then it can be clearly seen how a change in the aperture will allow a different number of spectra to be transmitted. It is a reasonable assumption that the image will become a truer representation of the object as more spectra are used in the reconstruction. Also, we could extend these ideas to a nonperiodic object by postulating some kind of Fourier analysis of the object into periodic structures of different spatial frequencies.

The above sketch of the Abbe theory can now be filled in with a

---

[3] This is generally not strictly true; very often several lens rims act jointly as the aperture stop, but it is a convenient simplification for the Abbe theory.

mathematical treatment. We saw in II.A that the complex amplitude distribution in the image of a point object can be expressed as the Fourier transform of the pupil function [Eq. (17)]. More explicitly, the light disturbances from an object point should ideally arrive at the exit pupil with uniform phase relative to the ideal or paraxial image point, i.e., the wavefront arriving at the exit pupil should be a portion of a sphere centered on the image point. If the optical system (microscope objective) has aberrations, this will not be the case and we can denote the distance between the wavefront and the ideal spherical shape by $\varphi(x,y)$ where x and y are coordinates in the pupil (Fig. 55); $\varphi(x,y)$ is the wavefront aberration. The complex amplitude in the pupil is then exp $(i2\pi/\lambda)\,\varphi(x,y)$ and the pupil function $a(x,y)$ of II.A is then defined as exp $(i2\pi/\lambda)\,\varphi(x,y)$ for (x,y) inside the pupil and zero elsewhere. The Fourier transform relationship of Eq. (17) then holds. In the theory of image formation, it is usual to call the image of a point object the point spread function, as mentioned in Section II.B, so that the function A of Eq. (17), regarded as a function of the image plane coordinates, is the amplitude point spread function and $|A|^2$ would be the intensity point spread function.

In order to discuss the formation of images of extended objects, it is convenient to denote positions in the object plane not by their actual coordinates but by the coordinates of the corresponding paraxial image points; thus if an object point is denoted by (ξ,η), we mean that its actual coordinates are $(\xi/m,\ \eta/m)$ where m is the magnification. This convention has the advantage of eliminating the factor m at various points. Suppose now that we have an object which is illuminated coherently as in Fig. 53 and let its amplitude transmission be $K(\xi,\eta)$, ξ and η being as above the corresponding image plane coordinates. The function K would be real if the object consisted only of regions of varying opacity but it could be complex, indicating that the phase of the transmitted wave varied; thus we speak of amplitude objects, of phase objects, and of

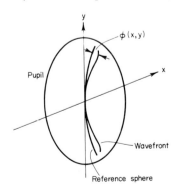

FIG. 55. The wavefront aberration $\phi(x,y)$ in the pupil.

general objects showing both amplitude and phase variations. An element of the object at the point $(\xi'\eta')$ produces in the image the amplitude point spread function centered on (ξ',η') i.e., it produces a complex amplitude proportional to $A\ (\xi - \xi',\ \eta - \eta')$ at some other point (ξ,η). In fact if the element is $d\xi'\ d\eta'$, we can say that the effect at (ξ,η) will actually be

$$K(\xi',\eta')\ A\,(\xi - \xi',\ \eta - \eta')\ d\xi'\ d\eta' \tag{34}$$

The total complex amplitude at (ξ,η) will be the sum of contributions from *all* elements of the object, so that the complex amplitude in the image will be

$$K'(\xi,\eta) = \int_{-\infty}^{\infty} \int_{-\infty}^{\infty} K(\xi',\eta')\ A\,(\xi - \xi',\ \eta - \eta')\ d\xi'\ d\eta' \tag{35}$$

The ranges of integration are set formally from $-\infty$ to $+\infty$ just as in the Fourier transform, but in practice the object transmission K is only nonzero over a finite range. The right-hand side of Eq. (35) is a *convolution integral;* it occurs in many fields and it represents the smearing or blurring of one function K by replacing every point of it by another, A. The convolution theorem (Bracewell, 1965) states that if any function, say K', is the convolution of two others, say K and A, then the Fourier transform of K' is the product of the transforms of K and A. We write these three functions as Fourier integrals:

$$K(\xi,\eta) = \int_{-\infty}^{\infty} \int_{-\infty}^{\infty} k(s,t)\ \exp\left[-2\pi i(s\xi + t\eta)\right] ds\ dt \tag{36}$$

$$K'(\xi,\eta) = \int_{-\infty}^{\infty} \int_{-\infty}^{\infty} k'(s,t)\ \exp\left[-2\pi i(s\xi + t\eta)\right] ds\ dt \tag{37}$$

$$A(\xi,\eta) = \int_{-\infty}^{\infty} \int_{-\infty}^{\infty} \mathcal{L}(s,t)\ \exp\left[-2\pi i(s\xi + t\eta)\right] ds\ dt \tag{38}$$

The variables s and t are in the Fourier transform domain which has, as will be seen below, a simple physical meaning in this connection. The convolution theorem then states that

$$k'(s,t) = k(s,t)\mathcal{L}(s,t) \tag{39}$$

In order to give meaning to this equation, we note first that Eq. (36) expresses the complex amplitude distribution in the object as a sum of complex Fourier components; the Fourier component $\exp\left(-2\pi i(s\xi + t\eta)\right)$ corresponds to s cycles per unit length in the ξ direction and t in the η direction and the magnitude of this component is $k(s,t)\ ds\ dt$. There is a similar meaning to Eq. (37): It gives the Fourier components in the image distribution of complex amplitude. Thus s and t are spatial frequencies, and Eq. (39) states that the magnitude of a given spatial frequency pair (s,t) in the image distribution is equal to that in the object distribution multiplied by a factor $\mathcal{L}$ whose magnitude depends on s and t. Thus $\mathcal{L}$ is similar to what is called in communication theory a transfer function; it has a simple representation in terms of the properties of the objective. Equation (17) stated

$$A(\xi,\eta) = \int_{-\infty}^{\infty} \int_{-\infty}^{\infty} a(x,y) \exp\left[-2\pi i\left(\frac{\xi x}{\lambda R} + \frac{\eta y}{\lambda R}\right)\right] dx\, dy \qquad (17)$$

or, changing the dummy variables of integration to $s = x/\lambda R$, $t = y/\lambda R$,

$$A(\xi,\eta) = \lambda^2 R^2 \int_{-\infty}^{\infty} \int_{-\infty}^{\infty} a(\lambda Rs, \lambda Rt) \exp\left[-2\pi i(s\xi + t\eta)\right] ds\, dt \qquad (40)$$

Comparing this with Eq. (38), we see that

$$\mathcal{L}(s,t) = a(\lambda Rs, \lambda Rt) \qquad (41)$$

i.e., the transfer function is simply the pupil function suitably scaled; the multiplicative factor $\lambda^2 R^2$ is not important since the transfer function has to be multiplied by a suitable normalizing constant to make it equal to unity for zero s and t.

Equation (41) has a very simple significance. If the objective has no aberrations, which is the usual assumption in starting to discuss such matters, we can represent its aperture by a circle of radius b (the exit pupil radius) as in Fig. 56, and Eq. (41) then states that the transfer function is unity for all spatial frequencies s and t such that

$$\lambda^2 R^2(s^2 + t^2) < b^2 \qquad (42)$$

and it is zero for all other frequency pairs. If the objective has aberrations but if there is no absorption, then $|a| = 1$ and, provided the above mentioned normalization has been carried out, the transfer function will be a complex number of unit modulus, i.e., all frequencies up to the limit given by Eq. (42) will be transmitted without attenuation but with varying phase shifts.

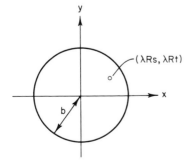

FIG. 56. A point $(\lambda Rs, \lambda Rt)$ in the pupil corresponding to the spatial frequency pair (s,t) in the object amplitude distribution.

We can transfer these results back to the object space very easily.

The spatial frequency s_1 which corresponds in object space to s in image space is given by $s_1 = ms$, where m is the transverse magnification. Also for large convergence angles m is given by $n \sin U/(b/R)$ where $n \sin U$ is the numerical aperture of the objective, so that the limiting spatial frequency in object space is given by

$$(s_1{}^2 + t_1{}^2)^{\frac{1}{2}} < n \sin U/\lambda_0 \qquad (43)$$

where λ_0 is the vacuum wavelength of the light.

To illustrate this, consider first a sinusoidal distribution of complex amplitude in the object,

$$K(\xi,\eta) = c \cos 2\pi s_0 \xi \qquad (44)$$

where as previously s and ξ both refer to the image space; thus the limiting value of s_0 which can be transmitted will be $n \sin U/(\lambda_0 \, m)$. This rather unlikely microscopical object has an amplitude transmission coefficient which varies as $|\cos 2\pi s_0 \xi|$ and at the same time the phase transmission changes stepwise by π each half period. If we use the inverse of Eq. (36) to Fourier analyze K, we find

$$k(s,t) = \tfrac{1}{2}c\delta(t)\{\delta(s - s_0) + \delta(s + s_0)\} \qquad (45)$$

where $\delta(t)$ denotes the Dirac delta function (Bracewell, 1965). The physical meaning of this is that the object represented by Eq. (44) produces just two diffraction spectra in the back focal plane of the objective (Fig. 54) and these appear at points with pupil coordinates $(\lambda R s_0, 0)$ and $(-\lambda R s_0, 0)$. If the aperture stop is large enough to transmit these, then the object will be reconstructed at the image plane as the interference pattern between these two beams; if the aperture stop does not transmit the beams, then nothing is seen.

A more sophisticated example is a square wave object with amplitude transmission as in Fig. 57; it is easily shown that this can be represented as the Fourier series,

$$K(\xi,\eta) = \frac{1}{2} + \frac{2}{\pi}[\cos 2\pi s_0 \xi - \tfrac{1}{3} \cos 6\pi s_0 \xi + \tfrac{1}{5} \cos 10\pi s_0 \xi - \cdots] \qquad (46)$$

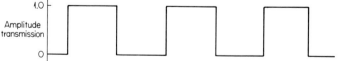

FIG. 57. Square wave amplitude object.

and this will produce spectra in the pupil at the origin and at $(\pm \lambda R s_0, 0)$, $(\pm 3\lambda R s_0, 0)$, $(\pm 5\lambda R s_0, 0)$, etc. If the pupil only transmits the central or zero-order spectrum, i.e., if its radius b is less than $\lambda R s_0$, then the image plane will show only uniform illumination, corresponding to the constant term $\frac{1}{2}$ in Eq. (46). If the pupil is enlarged so that $\lambda R s_0 < b < 3\lambda R s_0$, the first-order spectra are transmitted and the complex amplitude in the image will be

$$K'(\xi,\eta) = \frac{1}{2} + \frac{2}{\pi} \cos 2\pi s_0 \xi \qquad (47)$$

What is actually seen is, of course, the intensity distribution, which has the form

$$|K'(\xi,\eta)|^2 = \frac{1}{4} + \frac{2}{\pi^2} + \frac{2}{\pi} \cos 2\pi s_0 \xi + \frac{2}{\pi^2} \cos 4\pi s_0 \xi \qquad (48)$$

If the aperture stop is further enlarged to include the third-order spectra $(3\lambda R s_0 < b < 5\lambda R s_0)$, the complex amplitude in the image is

$$K'(\xi,\eta) = \frac{1}{2} + \frac{2}{\pi} \cos 2\pi s_0 \xi - \frac{2}{3\pi} \cos 6\pi s_0 \xi \qquad (49)$$

and the intensity is

$$|K'(\xi,\eta)|^2 = \frac{1}{4} + \frac{20}{9\pi^2} + \frac{2}{\pi} \cos 2\pi s_0 \xi + \frac{2}{3\pi^2} \cos 4\pi s_0 \xi$$

$$- \frac{2}{3\pi} \cos 6\pi s_0 \xi + \frac{4}{3\pi^2} \cos 8\pi s_0 \xi + \frac{2}{9\pi^2} \cos 12\pi s_0 \xi \qquad (50)$$

Figure 58 shows the complex amplitudes for these two cases [Eqs.

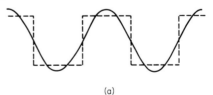

(a)

FIG. 58. Complex amplitude distributions in the images of the square wave object (Fig. 57) with (a) only the zero- and first-order spectra included, and (b) zero, first-, and third-orders.

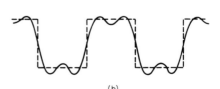

(b)

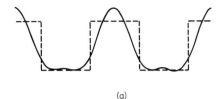

(a)

FIG. 59. Intensity distributions corresponding to (a) and (b) in Fig. 58.

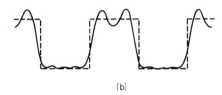

(b)

(47) and (49)] and Fig. 59 shows the intensities [Eqs. (48) and (50)]. It can be seen that the inclusion of the third-order spectra improves the resemblance of the image to the object considerably; however, there is still a great difference between the intensity distributions and Fig. 59 shows the effect of the third-order spectra in creating subsidiary maxima. If we use an objective of the greatest available NA, say 1.3, to look at a grating with light of wavelength 0.55 μm, we see from Eq. (43) that the spatial frequency which is just transmitted is about 2300 cycles per mm. The image of this would appear as in Fig. 59a and since the maximum NA has been used, this is clearly all that can be done. We are left with an image which is very different in form from the original object.

In the above examples, we have assumed that the illuminating coherent wave is incident normally on the object, so that the illuminating phase is constant over the object. If in Fig. 53 the point source were displaced laterally, the illuminating beam would be incident obliquely on the object plane and there would be a linear change of phase across the object. Quite marked changes in image formation occur under such conditions. The simplest example is that of a periodic object which is not resolved in axial coherent illumination, i.e., all spectra except the zero order are blocked by the aperture stop. If the illuminating beam is inclined in the appropriate direction, the effect is to shift all the spectra bodily across the pupil and if the pupil diameter is not less than the spacing between orders, it is then possible to get the zero order and one of the first orders transmitted, as in Fig. 60. Thus if the square wave

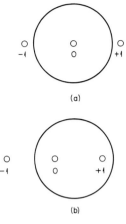

(a)

FIG. 60. The effect of oblique illumination. (a) Axial coherent illumination,
only zero order transmitted; (b) oblique
coherent illumination, zero and $+1$
orders transmitted.

(b)

object of Fig. 57 is illuminated at an angle α in the ξ section, the complex
amplitude transmitted by it becomes.

$$\exp\left(\frac{2\pi i}{\lambda}\alpha\xi\right)\left\{\frac{1}{2} + \frac{2}{\pi}\ (\cos 2\pi s_0\xi - \tfrac{1}{3}\cos 6\pi s_0\xi + \cdot\ \cdot\ \cdot)\right\} \qquad (51)$$

in place of Eq. (46) and if the $+1$ and zero orders are transmitted
by the aperture stop, the intensity in the image is

$$\left|\exp\left(\frac{2\pi}{\lambda}i\alpha\xi\right)\left(\frac{1}{2} + \frac{1}{\pi}\exp(2\pi i s_0\xi)\right)\right|^2 = \frac{1}{4} + \frac{1}{\pi^2} + \frac{1}{\pi}\cos 2\pi s_0\xi \qquad (52)$$

this is shown in Fig. 61. The image is markedly different from Fig.
59a, (the image obtained with axial illumination and the two first-order
spectra), and, of course, even more so from what would be obtained
with axial illumination and only the zero-order spectrum, i.e., uniform
illumination! This effect has been used to gain marginal improvements
in performance near the resolution limit.

C. PARTIALLY COHERENT ILLUMINATION

In general, a microscope is used with the condenser iris opened to
nearly the same NA as the objective and under this condition the illumination cannot be said to be coherent enough to ensure that the effects
seen will be anything like those described in the preceding section. On

FIG. 61. Image of the square wave object (Fig. 57) under conditions of
Fig. 60b.

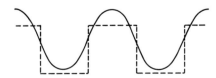

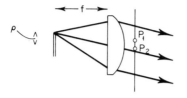

Fig. 62. Partially coherent illumination. The illuminating plane waves are incident on the object over a range of solid angle governed by the angular subtense of the source at the condenser.

the other hand, it is also then not adequate to assume that the Airy patterns formed by different parts of the object combine in the image by addition of intensities as in II.C. The intermediate situation of partial coherence involves rather complicated calculations and we give here only a qualitative and approximate treatment to indicate the general nature of the effects to be expected. A rigorous exposition is given by Born and Wolf (1965).

We can explain the concept of partial coherence most simply in terms of interference effects depending on the angular extent of the illumination. Figure 62 shows schematically Köhler illumination of an object plane by a source (in practice the source is imaged into the substage iris); the source has diameter 2ρ and the condenser has focal length f, so that any point on the object plane is illuminated by a cone of semiangle ρ/f. Let P_1 and P_2 be two such points; we wish to determine how the Airy patterns formed as images of P_1 and P_2 should be combined, whether by adding amplitudes or intensities or perhaps some intermediate method, these three corresponding, respectively, to complete coherence, complete lack of coherence, and partial coherence between the illumination at P_1 and P_2. We could in principle answer this question by sampling the illumination at P_1 and at P_2 and attempting to form interference fringes between the two beams. If the fringes were formed with high contrast, there would be good coherence; if no fringes were formed, the two points would be incoherently illuminated; moderate contrast corresponds to partial coherence. In fact, the contrast defined as follows for a fringe system

$$\text{Contrast} = \frac{I_{\max} - I_{\min}}{I_{\max} + I_{\min}} \tag{53}$$

is the generally adopted measure of partial coherence. Clearly the extreme values of unity and zero correspond to perfect coherence and incoherence, respectively.

Referring again to Fig. 62, if the source were restricted to a single point at the center of the substage iris, then P_1 and P_2 would be illuminated with a single wavefront and in exact phase agreement, so that there would be perfect coherence. If the source is then enlarged to radius

ρ, the points P_1 and P_2 will be illuminated by other wavefronts at angles ranging up to ρ/f and the extreme one will illuminate P_1 and P_2 with an optical path difference equal to $P_1P_2\rho/f$ between them; thus the fringe system formed by this wavefront will be displaced by the fraction $P_1P_2\rho/f\lambda$ of a fringe relative to the fringe system formed by the axial wavefront. Since the two wavefronts in question come from different points in the source, we have to add the intensities in their fringe patterns. Thus if the optical path $P_1P_2\rho/f$ amounts to a substantial fraction of a wavelength or more, the summed fringes formed by all the source points will be correspondingly reduced in contrast. Thus the coherence of the illumination is essentially dependent on the angular subtense of the source[4] and we can say that if ρ is chosen so that

$$(P_1P_2\rho/f) \ll \lambda/4 \tag{54}$$

the points P_1 and P_2 will be coherently illuminated. Thus coherence of illumination is essentially a function of the distance between a pair of points; in microscopy, we naturally choose this distance to be of the order of magnitude of the resolution limit of the objective as given by Eq. (23). But ρ/f is actually the numerical aperture of the condenser since ρ is the radius of the substage iris, so we see that the condition for substantially coherent illumination is, omitting numerical factors of order unity,

$$NA_{condenser} \ll NA_{objective} \tag{55}$$

It can also be shown, although this is beyond the scope of the present chapter, that the condition for substantially incoherent illumination is

$$NA_{condenser} \gg NA_{objective} \tag{56}$$

However, this condition cannot be realized unless the numerical aperture of the objective is rather small, since the practical upper limit for numerical apertures for both condensers and objectives is about 1.3. The ratio, condenser NA divided by objective NA, is denoted by S and it follows from the above that this ratio is a useful measure of coherence conditions in microscope illumination, at any rate when the condenser aperture is a circular, axial iris. Values of S much less than unity correspond to coherent illumination, values very greatly exceeding unity to incoherent illumination, and intermediate values to partially coherent illumination. In this context "greatly exceeding unity" can generally be

---

[4] This applies to illumination of a microscope by conventional sources such as filament and discharge lamps. Laser illumination is essentially coherent for all practical purposes in microscopy and very special measures would have to be adopted to de-cohere it, if this argument were to be applied.

taken to mean greater than about 3 but this depends to a large extent on the form of the object and on the aspects of the image which happen to be of interest.

The details of the methods of calculating images of partially coherently illuminated objects are very complicated but in essence they depend on the following principle. The actual source is assumed to be at the substage iris as in Fig. 62 and each point P of the source then produces an illuminating plane wave at an obliquity to the axis which depends on the distance of P from the axis. Thus P will produce an image in oblique coherent illumination, as explained at the end of Section VI.B for a square grating; all such images are incoherent with each other, since they are formed by different points in the source, so the complete image is found by integrating the intensities of these separate images. The computations are quite lengthy and not many have so far been carried out. Examples are given by Martin (1966).

We show in Fig. 63 calculated images of a circular aperture of diameter almost equal to the diameter of the Airy disc with various values of the parameter S and in Fig. 64 some images of an opaque edge. Several qualitative conclusions can be drawn from these figures and other similar results in the literature. First, as is already known, the physical image is nothing like the image according to geometrical optics when we look

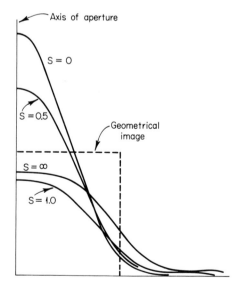

FIG. 63. Images of a circular aperture in partially coherent illumination; the diameter is 8 z units, i.e., just over the diameter of the first dark ring of the Airy pattern [adapted from Slansky (1960)].

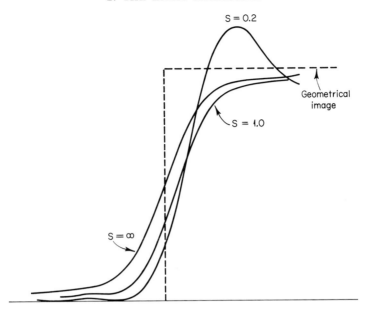

FIG. 64. Images of a straight edge in partially coherent illumination [adapted in part from Watrasiewicz (1965)].

at detail on the scale of the resolution limit. Second, variation of the coherence conditions produces very large changes in the image, but it is not possible to state simple rules to predict these changes. Third, the condition $S = 1$ (equal condenser and objective apertures) does not give substantially the same result as incoherent illumination, from which it follows that approximately incoherent illumination can only be realized for rather small objective apertures. Finally, since sharp-edged objects can have images which are qualitatively very different from the geometrical images, the qualitative differences make it very difficult to carry out accurate estimates of, e.g., size and transmission; there can even be the appearance of nonexistent membranes or other artifacts due to fringes along edges. We may conclude that variations in the mode of illumination can cause major changes in the appearance of detail near the resolution limit. In visual work, there are further changes due to the Mach effect, already discussed in III.D. These matters are discussed in more detail by Martin (1966).

ACKNOWLEDGMENT

The authors are grateful to Mr. B. O. Payne of Vickers Instruments, Ltd. for Figures 30, 32, 33, 35–37, 46, 47, and 49.

REFERENCES

Abbe, E. (1879). *Sitz.-Ber. Jen. ges. Med. Natur.* 129–142.

Born, M., and Wolf, E. (1965). "Principles of Optics," 3rd ed. Pergamon, New York.

Bracewell, R. N. (1965). "The Fourier Transform and its Applications." McGraw-Hill, New York.

Fresnel, A. (1816). *Ann. Chim. Phys.* **1,** 239.

Harris, Jr., F. S. (1964). *Appl. Opt.* **3,** 909.

Huygens, C. (1690). Traité de la Lumière (Leyden).

Martin, L. C. (1955). "Geometrical Optics." Pitman, New York.

Martin, L. C. (1961). "Technical Optics," 2nd ed. Vol. II. Pitman, New York.

Martin, L. C. (1966). "Theory of the Microscope." Blackie, Glasgow and London.

Martin, L. C., and Welford, W. T. (1966). "Technical Optics," 2nd ed., Vol. I. Pitman, New York.

Slansky, S. (1960). *Rev. Opt. Theor. Instrum.* **39,** 555.

Watrasiewicz, B. M. (1965). *Opt. Acta* **12,** 391.

Welford, W. T. (1962). "Geometrical Optics, Optical Instrumentation." North-Holland Publ., Amsterdam.

Welford, W. T. (1968). The Mach effect and the microscope. *Advan. Opt. Electron Microsc.* **11,** 41–76.

CHAPTER 2

Electron Microscopy

V. E. COSSLETT

I. The Limits of Optical Observation

A. Introduction

The electron microscope is primarily useful because it possesses a higher resolving power than that of any system employing light as illuminant. Therefore it is necessary first to discuss the limits imposed on direct observation, either with the unaided eye or through an optical instrument, by the nature of light itself. It will appear that the best resolution attainable is about 2000 Å with visible light and 1000 Å with ultraviolet, assuming an object of high contrast. The theoretical limit with electrons is below $\frac{1}{10}$ Å. Present uncorrected electron lenses have achieved 3–4 Å with high contrast specimens, but only 10–15 Å with biological material. For comparison, the spacing of atoms in molecules or in crystals ranges from 1 to 4 Å. The high resolution of the electron microscope is bought at the cost of certain disabilities concerning the types of specimen which may be examined, in particular, that they must be desiccated and thus dead, or at least in a state of suspended animation. These difficulties will be discussed in later sections of this chapter.

B. RESOLVING POWER

The wave nature of light sets a lower limit to the size of object point that can be resolved, that is, clearly seen as separate from a point of the same size when the two are just in contact. It is clear that contrast must be involved in any such observation, and it is assumed for further discussion that the object points are completely opaque. An exact treatment by diffraction theory shows that the angular width 2α of the beam received by the eye (or other recording system) is also involved (cf., Fig. 1). Regarding the information about the object as being conveyed in a spherical wave originating at the object, it follows that the wider the wavefront gathered by the eye, the greater is the information recorded and hence the easier it is to distinguish the point from a neighboring point. The resolving power as limited by diffraction can then be defined as

$$R_\mathrm{D} = \lambda/N_\mathrm{A} \tag{1}$$

if λ is the wavelength of the illumination and N_A is the numerical aperture of the viewing system. Account must be taken of possible differences between the refractive index of the media in which the object and image are, respectively, situated, since this will influence the velocity of light and hence its wavelength. The full definition of numerical aperture is $N_\mathrm{A} = \mu \sin \alpha$, where μ is the refractive index of the object space relative to the image space. So long as the angle α is small, as it is in present electron lenses, we can approximate its sine to its tangent and write $N_\mathrm{A} = r/f$ for objective lenses (where r is the radius of the lens employed and f its focal length), so long as $\mu = 1$.

Equation (1) is the simplest form for the resolving power, but some expressions current include a multiplying constant, the exact value of

FIG. 1. Angular aperture of lens.

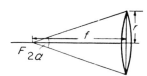

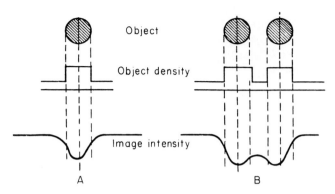

FIG. 2. Intensity distribution across image (a) of single particle, (b) of two
neighboring particles.

which depends on an estimation of the minimum difference in contrast
that is detectable. Figure 2 shows the appearance of the images of two
opaque object discs, with the corresponding density graph. The minimum fall in contrast, along the line joining them, for points to be resolvable will depend on the properties of the recording system and will
vary from person to person when this is the eye. It is conventional,
following Abbe, to take 25% as the minimum discrimination of the
eye and this results in a slight decrease in the value of R_D as given
by Eq. (1). For the discussion of electron microscope resolving power,
it is adequate to take Eq. (1) as it stands for the diffraction limitation,
or rather, since $\alpha = \sin \alpha$, to use the form

$$R_D = \lambda/\alpha \tag{2}$$

In the normal electron microscope, both object and image are at ground
potential and hence in regions of the same refractive index, electrically
interpreted, so that μ does not enter into Eq. (2).

C. The Eye and the Microscope

In the case of optical recording systems, the numerical aperture has
to be evaluated with due regard to the refractive index around object
and image and to the angular aperture. When this is done for the case
of the unaided eye, Eq. (1) gives a resolving power of 0.1 mm, which
agrees well with practical tests under optimum viewing conditions. An
optical microscope, as explained in Chapter 1, is in principle simply
a device for increasing the effective numerical aperture of the viewing
system. The object may be brought close up to the objective lens, giving
a large value of α, and N_A can be further increased by surrounding
the object by oil of high refractive index. In this way the value of

N_A can be made as high as 1.5 in the best case, so that the resolving power from Eq. (1) becomes less than the wavelength of the light used. By using oblique illumination, a further factor of up to 2 can be secured, so that the absolute limit of optical microscopy can be set at $R = \lambda/3$. With visible light of mean wavelength 5400 Å, we have $R = 1800$ Å; with monochromatic uv of 2400 Å, the theoretical limit is 800 Å. Although it is difficult to carry out tests of resolving power that accord precisely with the postulates of the theory, these values agree closely with the best results claimed in visible and uv microscopy, respectively.

It must be stressed that resolving power as thus defined involves the discrimination of two neighboring points; it does not represent the extreme limit of detection of isolated particles. This is a distinction which has to be borne in mind especially when evaluating the resolving power from electron micrographs, where it is rare to find pairs of points sufficiently close to apply the proper measure of R. It is much easier to determine the minimum size of particle that is visible in the image. However, the relation between this limit of detection and R depends on both the shape and the contrast of the object (Section VII.C). For this reason, any claim to extreme resolution on the basis of sharpness of picture has to be accepted with caution.

D. ELECTRON ILLUMINATION

An electron beam may be shown, both by theory and by diffraction experiments, to have wave properties and a relation can be found between its velocity and its effective wavelength. Translated into terms of the voltage (V) employed to accelerate it, the equivalent wavelength of an electron beam (in Å) is given by

$$\lambda = (150/V)^{1/2} \tag{3}$$

neglecting a relativistic correction term which becomes important only at very high voltages. It follows that the wavelength of a 60 kV beam is 0.05 Å, i.e., very much less than atomic diameters. Beam voltages of this order are needed in electron microscopy to give reasonable object penetration. Equation (1) shows that a resolving power 100,000 times better than that of the best optical microscope would be possible with a perfect electron microscope. In fact, as discussed in Section IV, the available aperture of present electron lenses is so small, on account of spherical aberration, that the theoretical limit is about 2 Å, and the best practical value attained is around 3.5 Å. The optical microscope attains its theoretical limit only because means have been found of correcting the aberrations from which a simple glass lens also suffers. It is thus possible to make a compound immersion objective of $N_A = 1.4$, whereas the normal electron lens has to be stopped down to between $\frac{1}{100}$ and $\frac{1}{1000}$ numerical aperture.

Even so, the minuteness of the electron wavelength allows a resolution of a few Angstrom units to be obtained. Hence the electron microscope already largely closes the gap between the limit of optical observation and the size range of X-ray methods. In this region, a great variety of biological detail occurs, whether isolated units such as viruses and macromolecules or structural units of larger organisms and bulk tissue, such as muscle and nerve fibrils, cell membranes, and chromosomes. The electron microscope has already been applied to many of these specimens, and the general principles of preparing such material for inspection will be outlined in later sections. First, the physical principles of the electron microscope and its operation will be described, and in Section III the essential differences between optical and electron microscopy will be further discussed. It may be said at once, however, that the two techniques are complementary and should always be used to reinforce each other. It is obvious that optical control of the reliability of electron micrographs must be used as far as it will go; it is now also beginning to be appreciated that the electron microscope can repay this debt in kind, by providing a check on artifact production in many of the established staining techniques of optical microscopy.

II. Basic Principles of Electron Microscopy

A. The Lens System

In its general lines of construction and operation, the electron microscope closely resembles the optical microscope. Each requires a source of illumination, a condenser lens for controlling it, an objective lens, and one or more further lenses for providing high magnification. The main differences between the two instruments are those of detailed design, arising from the special properties of electron beams as compared with light. Some minor differences of principle arise from the same cause, e.g., the electron image is viewed on a fluorescent screen, not directly, and interchangeable lenses are not required owing to the continuously variable power of magnetic fields. Since practically all commercially available electron microscopes employ magnetic lenses, the discussion will be carried on in terms of these rather than of electrostatic lenses. The differences between the two types are in any case matters of practical details of design and operation rather than of principle or ultimate performance.

A schematic diagram of the imaging process in the electron microscope (Fig. 3), compared with an optical microscope (inverted from its

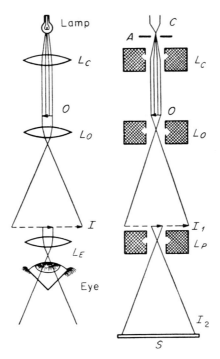

FIG. 3. Electron microscope and optical microscope: schematic diagram of image formation.

normal position), makes clear the essential similarity of the two. The electron illumination is drawn from a white hot tungsten filament C by the strong electric field maintained by a negative high potential applied between C and an anode A, which is grounded. A shield electrode, at a slightly higher negative voltage than C, surrounds it and serves to increase the brilliance of the beam that emerges from the aperture in A. The whole assembly is known as the electron gun and serves to deliver an intense narrow beam into the condenser lens L_c, which transmits part of it through a limiting aperture at its center. By varying the strength of the magnetic field in L_c, the angle of divergence and brightness of the illumination falling on the specimen O are controlled. The whole electron path, and therefore the specimen also, must be in vacuo. O is situated just before the front focal plane of a magnetic objective lens L_o of short focal length, which produces an enlarged image I_1 in the front focal plane of the projector lens L_p, also of high power. L_p throws a further enlarged electron image on to a fluorescent viewing screen S or on to a photographic plate which can be exposed by raising S. Focusing is carried out by varying the magnetic field in L_o, so that I_1 falls in the correct position relative to L_p. Magnification is varied by changing the strength of L_p, which has such a great depth of focus that it is unnecessary to readjust L_o except when the strength of L_p is greatly changed.

It will be seen that the operation of an electron microscope is unlike that of an optical microscope in two essential respects. First, the lenses are continuously variable in strength since the current flowing in the windings of the magnetic lenses can be smoothly controlled with a vari-

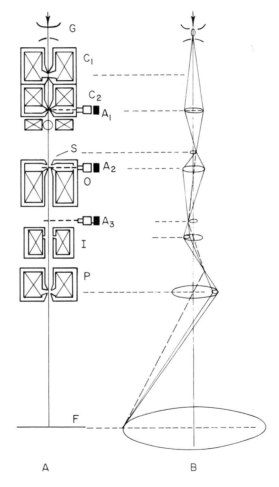

Fig. 4. Three-stage electron microscope (A) in schematic cross section and (B) showing formation of the image. G, electron gun; C_1 and C_2, first and second condenser lenses; O, objective lens; I and P, intermediate and projector lenses; S, specimen; F, fluorescent screen; and A_1, A_2, A_3, beam limiting apertures.

able resistor. Focusing involves no mechanical motion of the lenses but only the variation of a current that changes the focal length of L_0. Second, the magnification is produced in roughly equal proportions by objective and projector, the latter being of high power and controlling the magnification. In practice, it is found that too much distortion of the image is produced if the projector is reduced greatly in strength so that interchangeable projector polepieces are sometimes provided in magnetic microscopes having two stages of magnification. In the optical microscope, magnification is usually altered by changing the objective lens, of course. Most modern electron microscopes incorporate a third stage of magnification (Fig. 4a), by means of a weak intermediate lens I placed between O and P (cf. Section VI.D). In this case, the intermediate lens is varied for magnification change, and O and P require only minor adjustment. As I can be used either to increase or reduce magnification, its combination with P allows a very wide range of magnification (1000 to 250,000) to be covered without changing polepieces. In addition, it is now usual to have two condenser lenses, to permit more flexible control of the illumination (cf. Section V.B). So most electron microscopes have five lenses (Fig. 4b) and some have six, including a special diffraction lens (cf. Section VI.E). The different commercial models of electron microscope on the market are described by Cosslett (1970a).

B. Image Formation

The production of detail in the electron image depends on differential scattering of the illuminating beam in different parts of the specimen. Even with an accelerating voltage of 100 kV, the penetrating power of the beam is so poor that only a very thin specimen—less than 0.25 microns (μm) thick for biological material—can give enough transmission to produce a visible picture. Most objects contain regions of varying thickness and these will transmit electrons to different extents, so that corresponding contrast variations are seen in the image. Scattering rather than simple absorption is the mechanism by which electrons are removed from the beam, each electron usually suffering many collisions before being finally absorbed. In the light microscope, scattering is negligible and contrast is produced almost entirely by differential absorption in the object. In present electron lenses, the permissible aperture is so small that even slightly scattered electrons are removed from the beam and so assist in creating contrast. The mean angle of scattering and hence the contrast will decrease as the beam voltage is raised, whereas penetration will increase (see next section). Most standard electron microscopes provide a moderate range of operating voltage, 40–100 kV for instance, to allow some control over contrast when examining specimens of different thicknesses.

The probability of scattering depends on the amount of matter in the path of the electron, so that the contrast in the image displays directly the distribution of matter in the cross section of the object presented to the beam. "The amount of matter" here means the product of the number of atoms per unit area and their physical density. Hence the electron image can give no information about the chemical constitution of an object unless the local thickness is known and unless it is locally homogeneous. Attempts to identify particular elements by their specific absorption of electrons have so far failed.

The electron microscope image thus gives information about the distribution of matter in the specimen, understood as mass thickness. It is not directly comparable with the optical image of the same region, since local absorption of light depends on chemical constitution rather than the actual amount of matter present. As is well known, the transparency of many materials (especially in biology) prevents their direct observation by normal light microscopy, with the result that various specific staining techniques have been evolved for producing visible contrast in selected constituents of a specimen. It is probable that a similar variety of selective treatments will be discovered for electron microscopy, but so far progress has been slow. The principle here is to absorb heavy elements on to specific biological structures, so that their electron scattering power is increased. So far it appears that the absorption of agents such as lead salts, osmium tetroxide, or phosphotungstic acid is too general, although some success has been achieved in methods for localizing enzymes.

It should be noted that phase contrast microscopy gives a picture that resembles very closely the electron image in its contrast structure. The production of phase delay in a light beam is determined by the optical path length traveled in the specimen, and this by definition is given by the product of thickness and refractive index. Hence the contrast in a phase image is essentially dependent on the same factor as in the electron image—the amount of matter in the path of each component beam. An electron micrograph can thus be more profitably compared with a phase contrast than with a normal optical micrograph. Such comparisons indicate that the degree of distortion in the preparation of material for electron microscopy is much less than originally supposed.

III. Practical Differences between Optical and Electron Microscopy

As indicated in the previous section, the chief operational difference between an electron and a light microscope lies in the continuous variability of the focal length of an electron lens. So the operator looking at the electron image simply turns control knobs, as on any television set, to obtain the clearest picture and also to change its magnification. This advantage in convenience is offset, however, by several limitations that are not encountered in optics. These require separate discussion; they almost all arise from the interaction of the electron beam with the specimen, as discussed above, and from the aberrations of electron lenses. Limits are set to the thickness of specimen that can be examined, conditions of high vacuum and electrical stabilization are imposed that add to the complexity of operating and servicing an electron microscope, small apertures have to be adjusted at the proper locations in the beam path, and precautions have to be taken about electrical and thermal disturbance of the specimen. It is desirable to consider at this stage the nature and operational influence of these special features of electron microscopy; their proper control will be dealt with in Section VIII in connection with the attainment of high performance.

A. Specimen Thickness

The poor penetration of electrons, even at high accelerating voltages, requires that the object examined be extremely thin. The limiting factor is the loss of energy by electrons through inelastic scattering in passing through the object, i.e., by collisions in which energy is transferred to the atoms struck. As a result, a proportion of the electrons is slowed down, and the beam acquires a voltage spread that is equivalent to loss of monochromatism in light. As explained in Section IV.D, electron lenses have severe chromatic aberration which causes blurring of the image unless the beam is uniform in voltage. Hence this chromatic confusion increases with the thickness of object traversed and, in fact, spoils the image well before the object is opaque to electrons.

It is found in practice that biological material can be examined up to a (dry) thickness of about 0.1 μm (1000 Å) at a voltage of 50 kV, so long as a resolution better than 100 Å is not required. At 100 kV, the tolerable thickness is roughly twice this figure. As a general rule, at these voltages the specimen thickness should not be more than 10 times the required resolution; with compact material, such as metal films, it should be thinner. Hence the great depth of focus available in the electron microscope (see Section VI.C) cannot be usefully employed in transmission operation at high resolution. Its main advantage lies in the image space, where it allows great latitude in the positioning of the recording plate.

The extreme thinness thus demanded in the specimen is not such a great limitation as might be expected on the types of material that can be examined. Just because the electron microscope is needed to explore beyond the limits of the optical microscope, the objects to be examined will often themselves be very minute, such as bacteria, viruses, macromolecules, fibrils. The thickness of 1000 Å quoted above refers to dried material, so that bacteria which are up to 1 μm thick in the wet state still show considerable detail at 50 to 70 kV, unless grown on a very rich medium. But the larger microorganisms, such as spermatozoa, protozoa, spirochetes, are already too thick for internal detail to be seen, even at 100 kV. The solution is either to submit them to mild enzymatic digestion or to embed and section them. The high voltage electron microscope, however, now permits appreciably greater thicknesses to be examined, up to 1 μm or more at 1000 kV (see Section XII).

B. Vacuum Operation

The ready scattering of electrons also requires that their path within the microscope be evacuated of air as completely as possible. A residual vacuum poorer than about $5 \cdot 10^{-4}$ mm Hg is liable to give too much background fog in the image, owing to scattering of the beam, as well as leading to electrical breakdown in the gun and shortening of the filament life. High speed pumps have to be provided to maintain the vacuum and the instrument itself has to be constructed in such a way that it is leak tight and yet easily demountable in certain parts for servicing. This is readily achieved with rubber sealing rings (gaskets) but it adds to the complexity of the apparatus and admits more possibility of breakdown. Indeed, the necessity for vacuum operation, along with that for voltage and current stabilization, provides the main reason for the high cost of an electron microscope compared with the optical instrument.

Fortunately the design and construction of vacuum systems is now so well developed that very little trouble is to be expected from an electron microscope in this respect. The parts most likely to develop a leak are naturally those most often opened, such as the object chamber, the electron gun, and the camera. Simple instructions are always provided by the makers about testing for high vacuum and on renewing gaskets which show signs of wear. Similarly, advice is given on the care of pumps and their regular recharging with oil. As regards routine operation, pumps of such speed are now fitted that the delay after changing the specimen or photographic plate (or film) is reduced to the order of a minute. With some emulsions, however, the pumping time may be so long that it is advisable to prepump them in a separate bell jar and transfer them to the microscope as rapidly as possible.

C. Monochromatic Illumination

In optical microscopy, monochromatic illumination is sometimes desirable; in electron microscopy, it is essential. As mentioned above and in detail below (Section IV.D), the lack of correction of electron lenses makes it necessary to ensure high uniformity of velocity in the electron beam. Since the focal length of a lens is determined by the strength of its field as well as the electron velocity, it follows that the electrical supply to the lens must also be highly constant. In a normal magnetic objective lens, for instance, the energizing current must be held constant to within a microamp in a total of 100 mA. A corresponding stability, better than 1 part in 100,000, has to be ensured in the high voltage supply.

In practically all modern instruments, the required stabilization of both high tension and lens supplies is produced by electronic means, operating on the feed-back system by comparison of part of the output voltage with long-life reference batteries. Such equipment is again very complex and the possibility of breakdown is by no means negligible. A more important consideration, however, is that the best resolution cannot be obtained from an electron microscope unless the circuits are all tuned up to their peak of efficiency. Small faults can only too readily occur, leading to the generation of parasitic oscillations or the departure of amplifiers from their optimum conditions, which will impair the stabilization of the supplies for the microscope.

So long as a comparatively poor resolution of the order of 50 Ångstrom units is adequate, as it is for some routine purposes, the electrical equipment can be largely left to itself apart from annual tests of components. But if a resolution better than this is needed, the circuits must be regularly checked, and if 10 Å is sought, then very great care must be given to stabilization. Meters are provided for measuring some of the main currents and voltages in the circuits, and usually more detailed advice is provided in the operating handbooks about the testing of output with an oscillograph. But it has to be recognized that it is impossible to keep an instrument in peak performance without the regular services of an expert in electronics. In many biological laboratories, such specialists are now available. In places where they are not, it is advisable to make arrangements with the makers of the electron microscope to have it serviced regularly. Even if only modest demands are made on it, such attention is desirable; when really high resolution is consistently required, then it becomes indispensable. In optical microscopy there is little to go wrong with the instrument, and the attainment of high performance with a first-class microscope is largely a matter of skilled operation. With the electron microscope, skill in operation plays a minor part, the determining factors being the proper preparation of the specimen on the one hand, and the careful tuning-up of the machine beforehand on the other. Apart from checking of the electrical supplies, the correct alignment of the electron microscope is of the greatest importance—partly again because of the chromatic requirements—and this will be discussed in Section VIII.F.

D. Contrast and Apertures

As mentioned in Section II, contrast in the electron image arises from differential scattering in the specimen, those electrons which are scattered through an angle of more than a few thousandths of a radian being lost from the imaging beam. They may be eliminated most simply by placing a small stop in the objective lens, the size being determined by the value of the spherical and diffraction aberrations for the lens and the resolving power desired. In a normal type of magnetic lens with a focal length of 3 mm and an optimum semiaperture of 8×10^{-3} rad, corresponding to a resolution limit of 3 Å, the aperture in the stop would be 50 μm in diameter. Metal discs with holes in this size range are provided for most electron microscopes, with a device for supporting them in the appropriate position in or near the back focal plane of the lens. The aperture can be centered during observation of the image by means of micrometer controls.

In principle a physical aperture is not essential for high resolution. The severe spherical aberration of the objective will have an aperturing effect and throw scattered electrons out of the imaging beam. But operation without an aperture demands very careful alignment and control of the illuminating conditions and also results in poorer contrast in the image than when an aperture is used. It is therefore best to go to the trouble of centering an aperture, at least wherever the specimen is of poor natural contrast. Many specimens, especially when fixed with osmic acid or other heavy metals, are already of high contrast and operation without an aperture is perfectly satisfactory. The same is true of metal-shadowed preparations, provided that a metal coating tens of Ångstrom units thick can be tolerated. When details finer than 100 Å are sought, however, it is preferable to insert an aperture and use little or no shadowcasting.

Normally the aperture disc will receive heavy electron bombardment, as a result of which a semiconducting organic deposit forms on it. It appears to be impossible to avoid such contamination, which also occurs on the specimen and indeed on any surface reached by electrons (cf. Section VIII.C). It comes from residual vapors in the column of the instrument (Heide, 1958, 1962), which cannot be removed to the necessary degree in a demountable system such as an electron microscope must be. It does not form on regions which are so heavily bombarded that the temperature rises above about 200°C, so that the condenser lens aperture is usually not affected. All other apertures, and especially the objective aperture, have to be cleaned from time to time. The interval varies with the machine and the frequency of its use and may be a few days or a few weeks. The symptoms are drifting of the image and pronounced astigmatism (Section VIII.C). The instructions book should give details of how to remove and clean the aperture. If there is an aperture in the projector, to prevent scattered electrons from reaching the image, this also must be removed and cleaned from time to time.

E. EFFECT OF ELECTRON BEAM ON SPECIMEN

In passing through the specimen, the electron beam conveys charge and thermal energy to it, as well as causing contamination to be deposited. In optical microscopy it is rare for the specimen to be in any way affected by the illumination, but in electron microscopy the first two effects mentioned are inseparable from the imaging process. Contamination of the specimen can now be reduced to a very low level, if not entirely eliminated, by surrounding it with a cold chamber (see Section VIII.D).

Electrostatic charging is rarely troublesome, since few electrons are actually absorbed in thin specimens and such charge as is received leaks away to the metal support grid very rapidly. Difficulty is occasionally found when relatively large pieces of insulating material are examined, such as inorganic crystals or glass (Curtis and Ferrier, 1969). A heavy coating of shadowing metal on the support film often provides sufficient conduction in such cases. An alternative is to employ a low voltage electron beam for decharging, as devised for electron diffraction operation (Trillat and Oketani, 1950; Mahl and Weitsch, 1962).

Heating of the specimen is more difficult to avoid and may be excessive when thick specimens are used. Apart from more frequent inelastic scattering, energy is then more often delivered to the object by electrons which are completely stopped in it. It appears that semi-opaque specimens, under the moderate illumination required for viewing at 5000 to 10,000 magnifications, can reach a temperature of the order of 150 to 200°C. By using small area illumination, dark adapting the eye, or by fitting a small aperture just above the object, it is possible to keep the temperature down to 50 to 60°C. Several methods have been proposed for measuring the local temperature (cf. Reimer, 1967, p. 223).

For sensitive specimens, it is simplest to try the effect of fitting a small metal washer above and in contact with the electron microscope specimen grid, as is possible in several types of specimen holder. Alternatively, it may be necessary to make a special holder of high thermal capacity. Usually it is sufficient to use minimum illumination and dark-adapt the eye. For this reason, it is good practice to operate at low or moderate magnification, requiring low illumination (order of 10^{-4} A/cm^2), and to employ a $\times 10$ optical viewer for focusing the image (cf. Section VI). It is also advantageous to use as high an operating voltage as possible (see Section XII.C).

Not only does the beam heat the specimen, it often produces chemical changes in it. This may simply be a reduction of organic substances to carbon but may be more complex. Little systematic investigation has been made of these effects, but evidence is now coming from the use of high energy electron beams in nuclear physics and radiotherapeutics. It is highly probable that most organic tissues are quickly reduced to carbon under ordinary conditions of illumination and some constituents can be completely evaporated, especially embedding medium. The loss of material is about 50% from methacrylate and 25% for araldite or vestopal, and even greater for support films such as collodion; for Formvar, it is about 30%. The rate of loss can be reduced by lowering the illumination, but at normal levels (10^{-3} A/cm^2) it is complete in less than 1 min (Lippert, 1960). When the object is thick and dense as in the case of inorganic crystals, it often happens that the main bulk evaporates, leaving an empty shell that has been formed by contaminant deposition. For further discussion of the effect of the beam, see Section X.B.

Because of the temperature rise, and also the damaging effect of ionization, it seems improbable that living matter can be examined in the electron microscope. Even if it can survive drying, as is possible with some bacteria, it is highly probable that it will be killed by the amount of exposure to the electron beam needed for focusing and photographing (Cosslett, 1951, p. 261).

IV. Elements of Electron Lenses

A. ELECTROSTATIC LENSES

In order to operate an electron microscope, it is not at all necessary to know how an electron lens focuses the beam. But, as in optical microscopy, an understanding of fundamental principles makes it easier to grasp the need for those small refinements of technique that make all the difference in getting the best possible performance from an instrument. It also helps the nonphysicist to appreciate the limitations of

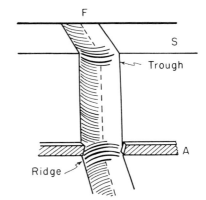

FIG. 5. Gravitational model of electrostatic field between a plane cathode F, bias shield S, and plane anode A.

the electron microscope. As electrostatic lenses are the simplest to understand, and enter into the electron gun if not into the imaging system, they will be described first; the mode of action of magnetic lenses is more complicated, as the image is rotated as well as focused.

The analogy between gravity and electrical potential helps in getting a picture of how electrons are deviated in electrostatic lenses, which in essence are nothing more than properly dimensioned metal tubes or discs (electrodes) maintained at different voltages. Just as material particles move downhill from a higher to a lower level under the attraction of gravity, if unconstrained, so do charged particles move towards an electrode carrying charge of opposite sign. Consider a simplified model of an electron gun. As electrons are negatively charged, the cathode is represented as at high negative potential and the anode at zero (ground). The latter necessarily has an aperture and it is usual to place a rather smaller aperture just before the cathode in the bias shield S (Fig. 5). If a rubber sheet be imagined pinned down along the cathode edge, supposing it to be a flat surface, and also beyond the anode face, the conformation shown in Fig. 5 will result. There will be a trough in the shield aperture and a ridge in the anode aperture. It can be realized at once that particles emitted from a cathode point F on the axis of the system will run together in the trough and be spread out in passing through the anode. In such simple electrode systems, an aperture in an electrode of the same sign as the particle will have a focusing action and one in an electrode of opposite sign will have a defocusing or diverging action. The relative strength of these effects depends on the potential applied to the electrodes and on the speed of the particle passing through them.

In the gun, the electrons are emitted from F at low velocity so that they are sharply focused by quite a small voltage on the shield, or even by zero voltage, into a small spot or "crossover." Thus the bias voltage on an electron microscope gun is a very sensitive control of illumination (cf. Section V.A). At the anode, the particles are at the "foot" of the slope and have such a high velocity that the diverging effect is small. Usually, however, they have been so strongly acted upon at S that they are already diverging when they reach A, i.e., the crossover is between S and A, as indicated in Fig. 3. It is possible to design the electron gun so that the combined action of S and A actually gives a convergent beam at a spot well beyond A, where the specimen will be placed (Steigerwald, 1949). A condenser lens is then not needed for focusing the illumination on to the specimen, as in most electron microscopes (see next section).

It is clear that the combination of S and A will have some sort of focusing action no matter what the point of origin of the electron beam, i.e., F need not be close to S. If it is in fact a gun at some distance delivering a beam into a system of two apertures at different voltages, these will focus the beam to an extent that will depend on their potential difference and spacing, that is, such a pair of apertures forms the simplest type of electron lens. A more common type, as used in television tubes, consists of two coaxial cylinders with a small gap between their ends and at different voltages (Fig. 7). The form of the field is much as in Fig. 5, but the trough now merges rapidly into the ridge within the second tube. As the electrons will still be traveling faster over the latter than through the trough in the first tube, the overall focusing action is always positive. The accelerator of a high voltage electron microscope comprises a series of 10, 15, or even 20 such lenses with electrodes at successively increasing voltage.

The type of lens used in electrostatic electron microscopes is shown in Fig. 6, consisting of three apertures, the outer two being at common voltage (ground) and the center one connected to the cathode supply.

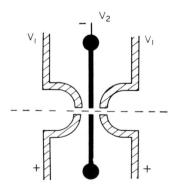

Fig. 6. Electrostatic unipotential lens.

Such a unipotential lens, using only a single voltage, has a very short focal length. It might seem that electrons would not pass the central aperture, as it is at the same potential as the cathode, but in fact there is a "trough" at its center owing to the effect of the neighboring electrodes—just as there is in S in Fig. 5 even if S is at the same voltage as F. Hence there is a slowing up of electrons in the unipotential lens, with consequent strong focusing action. If a still higher negative potential is applied to it, then electrons cannot pass and it functions as a mirror. Such a system can be used as an electron filter or energy analyzer for studying the energy lost by electrons in passing through the specimen (Metherell, 1970).

B. CARDINAL POINTS

These focusing fields have a definite axial extension, even in the case of apertures, and thus correspond to "thick" optical lenses. Their imaging action can be represented by a set of cardinal points, just as in the latter, and the same optical formulas can be used. For weak lenses, it proves to be adequate to use the simple form: $1/f = 1/u + 1/v$, where u is the object distance and v the image distance from a lens of focal length f. The typical arrangement of cardinal points in electron lenses is rather different from that in normal optics, however, as the principal planes are crossed over. Figure 7 shows their order and illustrates their definition. A principal plane is located at the intersection of the direction of incident parallel rays and their emergent direction produced back. As shown, the principal plane P_2 of image space falls actually before the center of the electron lens and lies in the object space. A similar plane P_1 is defined by a pencil of rays imagined as coming from the image space. The focal planes naturally are located where point images F_1 and F_2 are formed by such parallel beams arriving

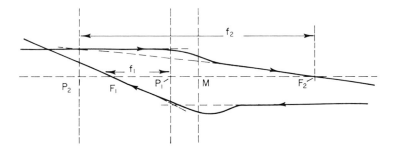

FIG. 7. Electron trajectories and cardinal points in a thick electrostatic lens that has its geometrical middle at M.

from either direction. The distances of these from P_1 and P_2 respectively give the focal lengths f_1 and f_2.

If the distances of object and image from their respective focal planes are now p and q, Newton's relation holds true as in optics:

$$p.q. = f_1 f_2 \tag{4}$$

This is the basic optical formula for calculating image positions in electron lenses for any given object position. Since the lens is not now a solid medium, it follows that the object may be situated actually within the lens, but it can be shown that Eq. (4) still holds. In a normal objective lens, the specimen may be in or very close to the center of the lens.

From experiment, and to a close approximation by calculation, the focal length of an electrostatic lens can be found for any geometrical arrangement of electrodes and its variation with applied voltage followed. Such information will be found in the standard textbooks of electron optics (Glaser, 1952; Grivet, 1965; Hall, 1966). The important point to notice is that the focal length can be varied over a wide range simply by changing the voltages applied. In a strong unipotential lens, the focal length will be roughly equal to the electrode separation (a few millimeters). Danger of electrical breakdown prevents smaller focal lengths being obtained and makes the use of electrostatic lenses less reliable than that of magnetic lenses.

C. Magnetic Lenses

The action of magnetic fields on electrons is more complicated than that of electric fields. The difference may be pictured thus: In an electrostatic field, electrons have a tendency to follow the lines of force, whereas they tend to spiral around magnetic lines of force. The focusing effect of a magnet is roughly speaking that of a potential trough between the north and south poles, if these are hollow and an electron beam enters through one of them (Fig. 8). But the spiraling influence causes the electrons to move around the axis at the same time as they are urged towards it. The result is that the image is rotated by an arbitrary angle with respect to the object, depending on the field strength and the electron speed, compared with the simple inversion of image that occurs in an optical or electrostatic lens. It is also found that a magnetic lens has only "troughs" and no potential "ridges," so that both halves of the lens field have positive focusing action. Hence magnetic lenses are always positive (that is, convex); they also all have aberrations of positive sign, so that correction cannot be carried out by matching positive and negative components, as in glass optics.

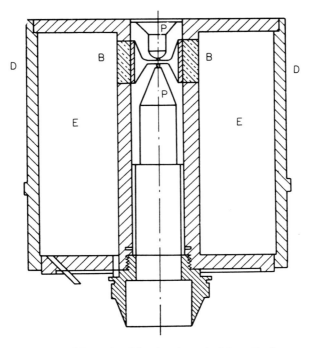

FIG. 8. Magnetic objective lens (axial section).

The typical construction of a magnetic lens is shown in Fig. 8. Polarity is produced in an iron casing D by current flowing in a slab winding E of many thousand turns of copper wire contained within it; permanent magnets can also be used. The magnetic field is concentrated into a small region on the axis by fitting conical polepieces PP of soft iron into the upper and lower parts of the axial casing, separated by a non-magnetic ring B. The focal length in a strong lens is roughly equal to the bore or the separation of the polepieces, which are usually about the same; the minimum value at 100 kV is rather less than 2 mm (Ruska, 1966). The focal strength depends on the geometry of the pole-pieces and on the current I and number of turns N in the winding (ampturns). In a weak lens, the relation is in first approximation:

$$f = KV/I^2 \tag{5}$$

where K is a constant and V is the voltage by which the electron beam is accelerated. In the strong lenses of electron microscopy, K does not remain constant as I increases. The typical variation of f with I is shown in Fig. 9; the focal length at first falls rapidly and then approaches a minimum value. Practical lenses naturally operate at the

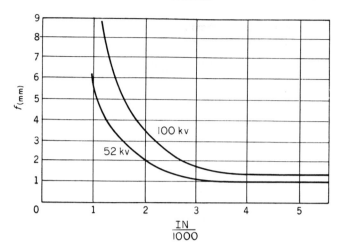

Fig. 9. Variation of focal length of magnetic objective lens with excitation (IN = ampere-turns), for two beam voltages.

highest currents allowable without special cooling, so as to approach their minimum f; several models in fact employ water cooling. With too high an excitation, however, saturation of the iron may occur and the focal length increase again.

The disposition of principal planes in magnetic lenses is the same as in Fig. 7. The normal optical constructions for ray paths may be used and the thick lens formula again holds good [Eq. (4)]; but it must be noted that the effective focal length of a strong lens is different when used as projector from what it is when used as objective. The distinction is illustrated in Fig. 10, where a ray from the right parallel to the axis is bent so strongly as to cross it within the lens, at F_0. A highly magnified image will be so far away that the emergent rays are effectively parallel, and so Fig. 10 with arrows reversed shows the behavior of the lens as objective: An object placed at F_0 will be the origin of a beam passing out to the right. The principal plane for such an objective will be at H_0 and the focal length is f_0. On the other hand, in a projector, the beam which enters almost parallel to the axis, as from the right in Fig. 10, will be bent to cross the axis at F_0, and will emerge at the left strongly divergent. The principal plane of the

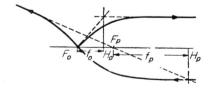

Fig. 10. Ray paths, focal points, and principal planes for objective (left to right) and projector (right to left) operation of a strong magnetic lens.

lens as projector H_p will be where this emergent direction, produced back, cuts the parallel entrant beam; H_p is clearly further back than H_0 and the projector focal length f_p correspondingly greater.

Thus a projector lens so strong that the beam crosses the axis within it will actually have a longer effective f than a weaker lens. If the lens is so strong that the beam crosses in the center of the lens, it will emerge parallel to the axis, being bent outwards in the second half of the lens by exactly the same amount as inwards in the first half; the focal length would then be infinitely great. Hence Fig. 9 shows only the variation of f with I for an objective lens, where the object is always very close to the focal plane. A projector lens shows a similar diminution of f with I at first, but at high excitation it passes through a minimum and rises again. When the final projector is used to vary magnification, this means that for high currents a diminution of the final image size may actually be observed.

D. ABERRATIONS

Magnetic and electrostatic lenses both suffer from the same variety of aberrations as do optical lenses; the magnetic type has in addition three errors due to image rotation. The most serious effects in practice are spherical and chromatic aberrations and astigmatism, all of which are severe by optical standards and rather worse in electrostatic than in magnetic lenses.

1. Spherical Aberration

The cause of this image-marring defect is that the outer zones of the lens are relatively too strong compared with those near the axis, so that the image formed by the former is closer to the lens than that formed by the latter zones. A sharp image can only be obtained by stopping down the lens until a small axial zone alone is used for focusing, with corresponding reduction in intensity of the imaging beam. It is found theoretically that this aberration is inherent in all electron lenses of the simple type now used and could only be eliminated in a much more complicated system.

The practical effect is that what should be a point image is enlarged into a "disc of confusion" of radius R_s given by

$$R_s = \tfrac{1}{4}C_s \cdot \alpha^3 \tag{6}$$

where C_s is a factor depending on the lens design and approximately equal to the focal length; α is the angular semiaperture of the lens (cf. Fig. 1). Hence reducing the aperture rapidly reduces spherical aberration, but as this is done the diffraction effect increases, as indicated by Eq. (2). In practice, a compromise in the size of stop is made such that R_s and R_D are made equal by the value of α chosen. In a strong objective, α will be about 8×10^{-3} and the diameter of the stop 40 to 60 μm. This is the reason for inserting apertures in electron lenses, with the consequent difficulties of centration and regular cleaning. As indicated in Section III.D, it is possible to obtain good resolution without an aperture provided that attention is paid to obtaining illumination conditions that in effect provide a beam of proper angular aperture, but it is easier to use a stop.

2. Astigmatism

Astigmatism arises if the focusing action of the lens is stronger in one direction across the axis than in that at right angles; it is equivalent to the bore being elliptical instead of circular in section. The inherent astigmatism of an electron lens is negligible compared with the spherical error, but a much larger astigmatism is unavoidably introduced by technical limitations in making the lens. In a magnetic lens, this may be due to two causes: The bore may not be truly cylindrical or the magnetic material may not be absolutely homogeneous, owing to blow holes, inclusions, or other faults. The resulting asymmetry of the field gives an astigmatic image: A point object gives two sharp line images, at right angles to each other and to the axis, and separated from each other along the axis. The sharpest representation of the object will be a circular disc of confusion half way between the line foci. The visible effect in the electron microscope will be that the image seems to streak out first in one direction and then in that at right angles as the objective is varied through focus. Even when the astigmatism is insufficient to cause directly visible elongation, it may be enough to be the limiting factor on resolution, rather than spherical aberration.

Fortunately this astigmatism can be corrected by introducing fields that are asymmetrical in the opposite sense to that of the lens. This is done electrostatically by applying variable potentials to a ring of six or eight electrodes mounted in the lower bore of the polepiece, or magnetically by inserting radial iron screws and adjusting their position until correction is obtained, as a result of careful observation of interference fringes in the image. The full procedure is given by Hillier and Ramberg (1947), but microscope makers now supply instructions for their particular type of "stigmator." If astigmatic line foci suddenly appear from a lens that has previously been free of them, it is an indication that an asymmetric field has arisen somewhere along the beam path, most probably by electrostatic charging of an aperture which has become contaminated (cf. Section VIII.G).

3. Chromatic Aberration

From Eq. (5) it is seen that the focal length of the lens depends on the beam voltage as well as on the strength of the magnetic field, as given by the current. Any variation in either V or I will change f and thus confuse the image; this is what is meant by chromatic aberration. Electrons of different voltage have different speed and so will be brought to different focal points; the faster the electron, the "stiffer" the beam and the longer f. The beam voltage must be held constant to such a degree that the disc of confusion arising from the spread of velocities is smaller than that due to spherical aberration and diffraction combined. In practice, this means a degree of stabilization better than 1 part in 100,000 for a resolution of 5 Å. Hence the need for complicated electronic stabilizing circuits—and for ensuring periodically that they are functioning properly. Similarly, the lens current supplies must be highly stabilized, so that the focusing fields do not vary in strength. An accumulator bank is stable enough for resolutions down to 40 Å, but for better performance electronic stabilizing gear is essential.

A small loss of stabilization is difficult to detect from the appearance of the image and can only be found by direct check on the stability of the circuits with a cathode ray oscilloscope. A large chromatic error will cause blurring of the image, which will increase with distance from the axis. It is in practice more important to guard against voltage than current troubles, so that the alignment of the microscope is sometimes carried out while applying variations of voltage; an equivalent test is to reverse the objective current (cf. Section VIII.F).

V. The Illuminating System

A. THE ELECTRON GUN

In practically all electron microscopes, the illumination is provided by a white hot tungsten filament F (Fig. 11), surrounded by a negatively biased shield (or Wehnelt cylinder) S through an aperture in which the beam is drawn off to a positive anode cap A. The exact shape of the electrodes varies, but we need only consider here the essential features of the electron gun. The detailed effect of the electrode parameters and

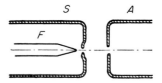

FIG. 11. Electron gun for electron microscope (axial section).

applied voltages on the properties of the beam is given fully by Haine and Einstein (1952) and by Haine *et al.* (1958).

In an electron gun of given design, the filament temperature determines the total electron emission whereas the bias voltage on the shield controls the angular spread of the issuing beam and hence its brightness. The condenser lens concentrates this beam on to the specimen, giving an additional control of illumination. The shape and position of the filament also affects the brightness, so that great care has to be taken when changing it. The gun controls proper will be first discussed and then the optimum use of the condenser.

1. Filament Position

The gun of a given electron microscope is designed for a given position of the filament; it must be on the axis and at the correct distance within the shield aperture. The latter distance ("height") is not highly critical in biased guns, and a simple jig allows it to be checked before inserting the filament in its holder. But the axial position must be more exact than can be preset in this way, and also it may vary when the filament is heated for the first time. An off-axis displacement of 0.001 in. will tilt the beam so much that it may miss the condenser aperture altogether.

When a filament has to be inserted, it should first be adjusted by eye in a jig or dummy shield, then heated in a flame or by passing low current, and readjusted. The makers give instructions as to the correct height. After mounting in the shield, it must be re-examined under a low power lens and set as closely as possible on the axis. Any small residual off-axis error can be corrected by the tilt mechanism of the illuminating system, as described in the alignment procedure (Section VIII.F).

2. Filament Temperature

The current through the filament is controlled by a resistance and is usually indicated on a meter. Increase in current raises the filament temperature and thus the electron emission. The electron illumination increases in brightness up to a limiting maximum if the maximum voltages applied to the anode, the shield, and across the filament itself are fixed. In fact the "ceiling" is reached quite rapidly, since the electron microscope gun is usually self-biased: The beam current flows through a resistor inserted between filament and shield, so that the potential drop across it provides the biasing voltage. The larger the current drawn from the filament, the larger this drop and the higher the bias, so that as the filament temperature is raised a condition is soon reached where the bias throttles the emission. Bias-saturation takes the place of the space-charge saturation of an ordinary triode.

Hence, for a given value of bias resistor, a maximum is fixed to the beam current that can be obtained. Illumination may be controlled during microscopy by varying the filament current; in order to spare the specimen, viewing should be begun with minimum current, which is then increased up to the value needed for observation at the electronic magnification chosen. If the bias is independently variable, its control should also be adjusted for minimum illumination initially. Working with no more than just sufficient brightness for viewing prolongs the life of the filament as well as minimizes damage to the specimen.

3. Bias Voltage

The bias voltage is usually supplied by the potential drop down a fixed resistor, the value of which is so calculated as to provide an adequate range of illumination when the filament current is varied. Alternatively it may be controlled by a simple circuit containing a triode valve (Dolby and Swift, 1960). In some models, a variable bias is provided in the form of a set of resistors which can be switched into circuit in turn. This has the advantage than any small error in filament height can be offset by varying the bias. Also the filament current can now be kept fixed at a moderate value, consistent with long life, and the intensity of illumination varied by selecting different bias resistors. The effect of changing bias is to concentrate the beam emerging from the gun into a wider or narrower solid angle. If the bias is radically changed, it may be found that the filament current also needs slight adjustment. On the other hand, if the gun is not self-biased, but the necessary voltage is supplied from a separate source, the beam will not be automatically throttled as the filament current is raised, and greater care is necessary to avoid specimen damage. But in some electron microscopes the bias is automatically set as the operating voltage is selected, so that only the filament heating current can be varied by the operator.

4. Pointed Filaments

An appreciable increase in brightness can be obtained by using a specially pointed filament instead of the normal bent hairpin. The simplest form is in fact a hairpin with its end ground down, the so-called "lancet" type (Sakaki and Möllenstedt, 1956). Finer tips may be obtained by electrolytic etching (Hibi, 1956), but they are more susceptible to deterioration by adsorbed gases, ion bombardment, and flash-over in the gun. Several manufacturers now supply point-cathodes to fit particular types of electron microscope. These are usually of the order of

1 μm in tip radius, appreciably larger than the points used for field emission microscopy.

There is great difference of opinion as to how much greater brightness a pointed filament gives as compared with a hairpin, in terms of specific intensity (A/cm²/ster) not just in surface brightness (A/cm²) [see Swift and Nixon (1962), Hanszen (1962), Thon (1965), and Pilod and Sonier (1968)]. Part of the divergence in results is probably due to the need for very accurate positioning of the point in relation to the aperture in the shield electrode, if true enhancement by field emission is to be attained. In the best conditions an increase in brightness of four to five times can be expected, at given operating temperature. The life-time of a point is usually shorter than that of a hairpin filament, however, unless extra high vacuum is maintained in the gun. On the other hand, the much smaller effective source size of the point gives the illumination much greater optical coherence, which provides better phase contrast in the image and especially sharpens up Fresnel fringes.

B. CONDENSER SYSTEM

The condenser system is located approximately midway between filament and specimen. It will give brightest illumination, for given gun conditions, when its power is such as to focus an image of the filament on to the specimen ("critical illumination"). This condition is to be avoided, in general, since it is liable to give too great heating of the object and also the illuminated field may be uneven in brightness, owing to variations in emission across the filament ("structure"). In a microscope fitted with a single condenser lens, the only resource is to defocus it so that the illumination is spread over a much larger area; or, of course, the filament current can be reduced. Most electron microscopes now comprise a double condenser system. The first lens is strong and serves to form a highly demagnified image of the cathode (or the crossover) near its back focal plane. This image, which may be only a micron or so in diameter, then serves as a virtual source of illumination for the weak second condenser. The latter throws an image of it onto the specimen, at about unit magnification, or slightly larger. By varying the first condenser, the fraction of the gun current delivered to the second condenser can be varied over a wide range, and varying the latter lens spreads it over a greater or smaller area of the specimen.

So the operator has control of both the current reaching the specimen and its intensity (A/cm²). It is good practice to use just enough illumination to give a readily visible image at the desired overall magnification and to keep the area illuminated to the minimum that conveniently fills the viewing screen, so that the total current falling on the specimen is minimized, to limit the possibility of radiation damage. Except when a wide range of magnification is to be used or a very sensitive specimen is being examined, the first condenser can be kept constant at a high value and the illuminating conditions varied with the second lens only.

For high resolution work or when Fresnel fringes are being observed, the angular aperture of the illumination should be reduced to a value well below that of the objective aperture. This may be done either by defocusing the second condenser, by inserting a smaller physical aperture in it (apertures varying from 100 to 500 μm in diameter are usually available), or by focusing the first condenser more strongly to give a smaller virtual source. It is also desirable to correct the astigmatism of the second condenser, so as to get a circular spot of illumination on the specimen. For this purpose an adjustable electrostatic or magnetic stigmator is built into it, which is varied (following the instructions book) until the illuminated spot is as near circular as possible when the specimen is imaged on the viewing screen. Alternatively the condenser aperture may be withdrawn and the stigmator adjusted until the four-pointed star normally visible becomes a three-pointed star.

VI. The Imaging System

A. SPECIMEN CHAMBER

The specimen to be examined is mounted on a supporting metal gauze (Section IX.B), which is placed in a metal cap or similar holder and inserted firmly into the stage plate of the electron microscope, movable in two mutually perpendicular directions by push rods for searching the specimen. Some instruments contain an air lock by means of which the object holder is transferred to the evacuated interior without admitting air; a separate pumping line is provided for evacuating the lock before opening it up to the rest of the apparatus. Other models have either a pencil type holder that pushes in through a rubber seal with hardly any admission of air, or else have such fast pumps that the whole column can be let down to atmospheric pressure when the object is changed. In the latter case, the prescribed switching routine of the vacuum taps must be followed.

The object holder must always fit firmly in its seating, so as to minimize vibration and to aid thermal contact. The stage plate is moved by controls operated from outside the machine and should travel smoothly and without appreciable backlash even when observed at 100,000 magnifications. Stops are fitted to ensure that the specimen grid cannot be carried right out of the beam, while allowing practically all of it to be viewed. The controls must never be forced in an effort to see a remote corner of the grid, as this may deteriorate the vacuum seals or even damage the stage plate. It is advisable to remove and polish the latter and the surfaces on which it moves at least once a year, or more often if there is any suspicion of sticking.

Most holders now have an attachment for stereophotography, by which the specimen grid is tilted through a small angle between two successive exposures. This is readily operated if so designed that the image can be observed during the tilting process, so that the wanted feature may be kept in the field of view by moving the stage controls. But the two-position type of stereo holder makes necessary a subsequent search to find the wanted field again after tilting; it is best to make a rough drawing of it before doing so.

Holders must always be kept clean and polished, to avoid dirt reaching the specimen and to ensure good thermal and electrical contact between grid and holder. It is desirable to handle the holder (and any item removed from the microscope such as a polepiece or aperture-rod) with gloves or optical tissue, never with the bare fingers, to avoid transferring contaminating greases into the vacuum.

In a run of work where the magnification should be the same, it is essential to employ the same holder right through since a slight difference in axial dimensions will displace the object by that much from the assumed object plane. In the usual type of objective lens, a change of $\frac{1}{1000}$ in. in object distance will make a difference of as much as 1% in magnification.

B. FOCUSING AND MAGNIFICATION CHANGE: TWO-STAGE IMAGING

When the object holder has been placed in position, the image is focused by means of the objective current controls, which usually comprise a coarse, fine, and very fine motion; sometimes an ultra-fine control is also provided. The magnification is varied, in a two-stage instrument, by changing the projector current; a doubling of its object distance ($\fallingdotseq$ focal length) will give a similar change in final magnification, as the projector-image distance is fixed (cf. Fig. 3). The depth of focus of the first image is usually so great that no corresponding change of the objective current is needed, and the variation in the distance OI_1 is relatively so small that the magnification is controlled only by the projector. Hence some microscopes have the projector current meter directly calibrated for magnification. Any such reading is not likely to be accurate to better than 10%, however, unless regularly checked (cf. Section VII.B).

In practice it is usual to start at low magnification in order to survey a wide field and select features of interest in the specimen for greater magnification. As this is done, the intensity of illumination must be increased *pari passu* to keep the image bright enough for ready observation. When a field has been chosen for photographing, it is focused as sharply as possible by using the fine objective control. As already mentioned, it is advantageous to use a low-power viewing telescope for the purpose, and on most instruments they are now standard fitments. A properly designed optical system presents a magnified image without loss of brightness and so permits microscopy at a low electronic magnification and small object loading, while the image is observed at the magnification to which the photograph will be finally enlarged. But there is no point in using an optical magnification greater than makes visible the graininess of the viewing screen, and a ×10 telescope is adequate with normal screens. A monocular will suffice, but a binocular viewer is to be recommended. When high performance is sought, the telescope enables diffraction fringes to be viewed directly, and, when used with high electronic magnification, helps in the detection of minute vibrations or drifting of the image.

C. DEPTH OF FIELD AND DEPTH OF FOCUS

When the image is well focused, it is good practice to change the condenser control so as to diminish the illumination appreciably for photography. The angular aperture is thus reduced, the depth of focus increased, and the resolution improved. Alternatively some microscopes have a means of artificially increasing the value of α during focusing by electrically displacing the beam. In either case, experience will show whether the objective setting is changed in the process, demanding a "blind" readjustment of the fine focus control before exposure.

The meaning of depth of focus is illustrated in Fig. 12, where O is a point in the object. In the image, the resolution may not be sufficient to define O itself, the least observable distance d being larger, owing either to lack of true resolving power in the lens or to graininess in the screen or poor visual acuity. Dividing d by the magnification gives the equivalent detection limit in object space, d/M. Figure 12 shows

FIG. 12. Depth of field within which an object O is imaged with a given degree of sharpness.

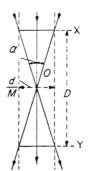

that the objective may be focused anywhere between the planes X and Y, forming an intercept d/M across the illuminating beam, without visible change in the image. The axial separation XY is properly defined as the "depth of field" corresponding to the practical resolution d/M, but it is often called the depth of focus. We have: $d/M = \alpha \cdot XY$. The semiangle α is defined by the objective aperture, but the distribution of electron intensity across it depends on the illuminating aperture as well as on the amount of scattering caused by the object. If the illuminating aperture is reduced by varying the condenser, the image background is minimized. By this method, by use of a viewing telescope, and by bracketing the image by varying the fine focus and setting it at the mean position, it should be possible to get results sharp enough for all normal purposes. For very high resolution work, it may be necessary to take a through-focal series of photographs with blind settings of the fine focus control, progressively increasing in value (Section VIII.G).

The term depth of focus properly refers to the image space, where a similar effect occurs but in a much greater degree. Here the transverse intercept on the beam is greater than in object space by the factor M and the beam angle is smaller by the same factor, since from geometrical optics $I/O = M = \alpha_0/\alpha_i$. Hence the depth of focus in the image is M^2 greater than in the object: With M of the order of 10,000, this gives a factor of 10^8. Consequently the image is in focus practically from the end of the projector lens right away to infinity. Therefore the position of the photographic plate is not at all critical. The camera plane can be well below the viewing screen, which can thus be used as a shutter; or it can be much nearer to the projector, as in the Philips microscope, with smaller picture and shorter exposure.

The advantage of great focal depth follows from the minute apertures imposed on electron microscopy by the aberrations of its lenses. In optical microscopy, the large numerical apertures make the depth of focus extremely small, even at the poorer resolution attainable, so that sections have to be examined plane by plane. In the electron microscope, where the specimen has to be thin in any case, the whole of it is usually in focus at once and discrimination in depth is only possible by stereo-operation. With the much thicker specimens that can be examined in a high voltage microscope, however, a certain degree of focal isolation seems to be possible.

D. THREE-STAGE MICROSCOPES

Since Marton (1945) first constructed an instrument with three stages of magnification, almost all electron microscopes have used this arrangement. By having two independent projector lenses, the magnification can be varied continuously over a wide range and also electron diffraction can be practiced. Figure 13 shows typical modes of operation of the three-stage microscope (Challice, 1950), according to the relative

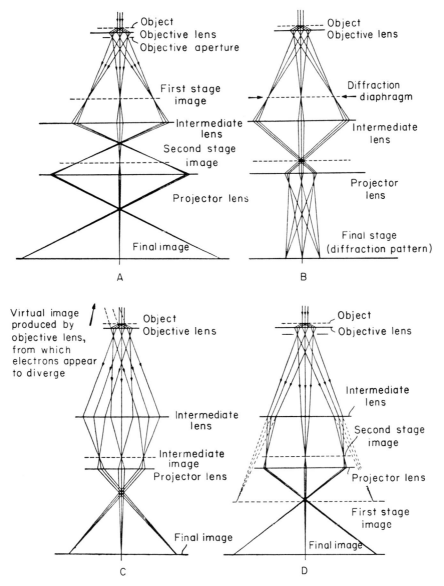

FIG. 13. Three-stage electron microscope: (A) high magnification, (B) diffraction, (C) first low magnification range, (D) second low magnification range.

strengths of the objective and first projector (intermediate lens); the second projector is run at maximum power throughout, to minimize distortion and keep the field of view filled. For high magnification, the objective is operated at high power and the intermediate lens forms an intermediate image (Fig. 13A), so that the total magnification is given by the product of three stages, of the order of 100×, 20×, and 100×, respectively. For diffraction operation, the intermediate lens is reduced in power so as to throw a practically parallel beam into the projector; this is equivalent to forming an image of the electron source on the final screen, so that the diffraction pattern arising from it is also seen (Fig. 13B). For low magnification, the objective power is reduced to give either a first image below the intermediate lens, which the latter then reduces (Fig. 13D), or a virtual first image which is similarly reduced by the intermediate lens (Fig. 13C). In practice, the image below about 500× is distorted so severely that it is useless; but it is possible to cover a range from about 500× to 200,000× simply by moving control knobs. The three-stage instrument can also be used for two-stage operation, of course, by switching out the intermediate lens. Its care and maintenance is essentially the same as that of a two-stage instrument.

E. ELECTRON DIFFRACTION

A powerful auxiliary use of the electron microscope is to produce electron diffraction patterns from crystalline regions of a specimen, from which the chemical nature of a constituent can often be deduced. The simplest way of forming a diffraction pattern is to illuminate the specimen in its usual position with a nearly parallel beam of electrons, i.e., with the second condenser defocused, but with all the imaging lenses switched off and apertures withdrawn. The pattern is then projected onto the viewing screen as a set of bright spots or rings, but unless the bore of the final lens is unusually large or its polepiece can be removed completely, only the central part of the pattern will get through to the screen. To overcome this restriction, a special specimen stage for diffraction can be provided below the final projector lens. Although the whole of the pattern is then visible, it is inconveniently small since the throw of the system ("camera length") is short. However, the compensating advantage that no lenses act to distort the pattern,—it is simply projected on to the screen as from a point source,—makes this method desirable for high resolution diffraction work.

An alternative and operationally very convenient method is shown in Fig. 13B. The diffraction pattern formed by the objective lens is imaged into the projector lens by means of the intermediate lens. To do this, it is only necessary to weaken the latter lens until it is focused on the back focal plane of the objective lens (where the diffraction pattern occurs), instead of on the first stage image. Most electron microscopes have a switch for doing so, which simply changes the excitation of the intermediate lens to the required value. Some instruments now have an extra lens, specially designed for diffraction, fitted between the objective and intermediate lenses. It allows a larger pattern to be formed with less distortion. By varying the relative strengths of the imaging lenses, a wide range of camera lengths can be obtained, which is particularly useful for the long spacings found in biological structures [see Ferrier (1969)]. A small aperture above the intermediate lens (Figure 13B) allows diffraction to be obtained from an area only a few microns in diameter, so that the components of a specimen can be individually examined. This selected area procedure has to be used with great care, however, in order to be sure that the pattern seen actually comes from the area being viewed when the aperture was inserted, owing to the effect of lens aberrations (Agar, 1960; Philips, 1960).

VII. Recording the Image

A. PHOTOGRAPHIC MATERIALS FOR ELECTRONS

The image focused on the viewing screen is photographed on film or plate either by lifting the screen or by turning the camera into the beam (Philips model). With the emulsions generally used, the exposure time is of the order of a second, so that no high speed shutter mechanism is needed and the screen itself can serve as shutter, though it is now usual to have an automatic timer and shutter. Most cameras provide for a small number of successive exposures without reloading, some provide only a single plate, and a few use a length of film on which a large number of exposures may be made. The loading of the camera is usually very simple, a cassette being first charged with plate or film in the dark and then transferred to the microscope in daylight. The camera chamber is isolated from the rest of the apparatus by an air lock, so that reloading can be effected without admitting air to the rest of the column, and often without interrupting microscopy. The large water content of most emulsions makes it necessary to pump for several minutes before the camera chamber can be opened up to the column. This delay can be appreciably shortened if the loaded cassette is prepumped under a bell jar at a rough vacuum for some time before transfer to the microscope.

The photographic material most used for electron photography is of the lantern type, such as the Eastman Kodak Electron Image plate, which has the right speed and contrast characteristics. For specimens lacking in contrast, or when no aperture is used in the microscope, it may be preferable to use an emulsion of the type used for document copying, such as the Photomechanical plate; this has great contrast but is grainy and about five times slower than a high contrast lantern plate. Increasing use is made of cut film, such as Dupont Ortho-litho and Ilford N. 50, since it is easier to handle and store than plates. In roll film, Kodak Microfile Regular and Gevaert C 50 have good photographic properties and pump out relatively rapidly. When ordering, it is desirable to specify nonpanchromatic material since otherwise darkroom work is unnecessarily complicated; the ordinary orthochromatic emulsions are perfectly satisfactory for electrons. Some makers offer for electron microscopy material which is normally panchromatic but from which the sensitizer has been omitted.

Normal development and fixing is employed, but special fine-grain developers can be used when the need arises, as for light images. In general, the graininess to electrons seems to be much the same as to light, and subsequent enlargement of from seven to ten times is practicable with lantern plates. This corresponds to the magnification usually employed in the telescopes used for focusing the image. There is no point in using finer grained material than the accuracy of focusing warrants, unless "blind" photography by a through-focal series is adopted (Section VIII.G). The few investigations so far carried out on the sensitivity of emulsions to electrons show that the range of speed is much less than for light, and also that the exposure–density curve is nearly linear whereas for light the log-exposure versus density curve is linear. For a review of the response of a wide variety of photographic emulsions to electrons, see Valentine (1966).

Nearly all makers of electron microscopes will now supply a closed TV link for remote viewing of the image, frequently combined with an image intensifier system. Remote viewing is especially useful for high voltage microscopy, where the X-radiation level may be above tolerance level when the specimen is being deliberately irradiated with a high beam current in radiation damage experiments. Image intensification, on the other hand, may be indispensable for looking at radiation sensitive specimens such as polymers, or when very thick samples are being examined. Even the best image intensifier is not much more sensitive than the dark-adapted eye, however, though rather lower imaging intensities can be used if a storage tube is available for the display. The main advantage comes from being able to operate in normal low-light conditions, instead of having to be dark-adapted. It is also then quite easy to get a permanent record of a dynamic experiment on film or video-tape.

B. MAGNIFICATION CALIBRATION

A given electron microscope will be provided with a magnification calibration chart, usually in terms of the intermediate lens current. For all normal purposes where an accuracy of no better than 10% is required, this chart will be adequate. The actual magnification will vary according to the position of the specimen, the voltage employed, and possibly magnetic changes in the polepieces. When greater accuracy is demanded, as when investigating particle size distribution, a special magnification calibration must be made in the range of working conditions to be employed. Unless care has been taken to make all object holders of exactly the same dimensions (to better than 0.1 mm), the same holder must be used throughout and the specimen grid always inserted in it in exactly the same way.

Various accurate methods of measuring magnification have been devised, using diffraction grating replicas or quartz fibers of a convenient size that can be checked by optical means (cf. Reimer, 1967, p. 66). Cross-grating replicas of extremely fine spacing (0.5 μm) are now available which allow calibration up to high values of magnification. An accurate procedure is described by Hall (1966, p. 347). For most purposes, however, it is simplest and sufficiently accurate to use standard reference objects of known size, the two most suitable being polystyrene latex spheres and tobacco mosaic virus rods. Polystyrene latex can be made of extraordinarily high uniformity in size and thus is ideal for magnification measurement. It may be added in small amount to many actual specimen preparations, so that reference particles appear on the electron micrographs. Or it can be used in a separate calibration series of photographs taken at various settings of the projector controls. Small samples of the latex (eight sizes) may be obtained from the Dow Chemical Co., Midland, Michigan, U.S.A.

For high magnification work, the latex particles may be found too large; at 60,000 the smallest of them will give an image 5 mm in diameter. The most suitable material is then a crystal of one of the plant viruses or an enzyme such as catalase [see Wrigley (1968)]; the former have spacings of about 250 to 350 Å and the latter about 70 Å. At still higher magnifications, the crystal lattice of an inorganic substance such as platinum phthalocyanine or pyrophyllite may be used, both of which have spacings of the order of 10 Å.

C. RESOLUTION ESTIMATION AND LIMITS OF DETECTION

Nothing is so difficult in electron microscopy as getting a reliable measure of resolution. It is natural for an investigator to be over optimistic about his own results, but even an impartial group of observers will give quite different estimates of the resolution in a particular micrograph. The difficulty is that the proper definition of resolution (Section I.C) can only be applied to two closely neighboring object points of radius equal to the linear resolving power. In most preparations, it will be most unlikely that particles of the right size occur, still less that two will be in contact edge to edge. By using an evaporated film of carbon or carbon-platinum, the resolving power of an electron microscope can be tested, since particles of all sizes down to about 3 Å can be obtained. But it is not practicable to deposit such material on all preparations as a regular method, and other measures of resolution have to be devised for general application. The sharpness of angle where two rods or edges overlap can be employed, or the width of the photometer trace across the image of an opaque edge, the latter giving a direct comparison with the true resolution as previously defined. The applicability of these and other measures is discussed by Cosslett (1951, pp. 142–148).

The most generally useful estimate of resolution, however, is from the size of the smallest particle or other feature that is definitely visible in the image. This limit of detection is best determined by making successively greater enlargements from the negative, until the size is reached at which the image is no longer judged to be sharp when viewed at the distance of distinct vision. Taking the resolving power of the eye as 0.2 mm, the limit of detection is then $0.2/M$ mm, where M is the maximum useful magnification. For instance, if the image is judged to be sharp at 400,000 but not at 500,000, the resolution is 5 Å. This method is essentially the same as that of "quality estimation" proposed by Hillier and Baker (1945). It gives a figure considerably smaller than the resolving limit as properly defined, since the eye can detect an isolated object much more readily than it can separate it from an overlapping neighbor. For opaque disc-like objects, the detection limit should be multiplied by a factor of 3 to get an approximate estimate of the true resolving power. For line objects, the factor is of the order of 10. There is thus an appreciable difference between the limits of resolution and detection, which often leads to exaggerated claims of performance. The same is true for claims based on the resolution of metal lattices, which are much more tests of the thermal, mechanical, and electrical stability of an electron microscope than of its resolving power, apart from the fact that a lattice can be resolved in the presence of appreciable astigmatism. Fresnel fringes (see below) are a better test of stability.

A more objective method of measuring resolving power has been devised by Thon (1967). The test electron micrograph is illuminated with coherent light from a laser on an optical bench, so as to produce a diffraction pattern. The radii of the rings in this pattern are inversely proportional to the spacings between particles in the test specimen. The largest radius recorded thus gives a measure of the resolving power of the electron microscope used.

VIII. Requirements for High Performance

In order to get the very best results from an electron microscope, attention must be paid to a number of disturbing influences which may affect the image. Special care needs to be taken in focusing and photographing, and also in the exact alignment of the lenses. These points of refinement will be discussed in turn [see also Agar (1965), Haine (1961), and Hall (1966, p. 292)].

A. MECHANICAL STABILITY

The need for mechanical stability is obvious if a high resolution image is to be recorded. Vibration of the microscope as a whole is of no account provided that specimen and photographic plate are rigidly connected, so that one does not move relative to the other during exposure. Motion of the specimen is most deleterious, since it is magnified highly in the image. To minimize it, the specimen grid must be firmly held in its holder and the holder firmly fixed in the specimen stage of the microscope. The latter should be designed to ensure this, but small particles of dust or of damaged specimen may hinder proper fit. All the surfaces involved must be kept regularly cleaned and polished.

However, the stage has to be free to move under external control and it is very difficult to combine this requirement with an absence of play. Even if the stage motion and bearing surfaces are kept in good order, there may remain some slight sticking friction. It is good practice to tap the stage controls lightly before making an exposure; if the image is watched during the process, it will soon be found if this is a desirable routine procedure.

In spite of all care, there is bound to remain some degree of play of the specimen with respect to the plate, if only in the order of Ångstrom units, when the microscope is subjected to external vibration. This appears to be one of the major practical limitations on resolution, rather than the theoretical aberrations of the lenses. For high performance, every source of vibration must be kept away from the microscope. The rotary vacuum pump should be as far away as is consistent with good pumping speed; the best solution is to install a vacuum reservoir between it and the diffusion pump, and switch it off during exposure. The diffusion pump heater supply should be checked to ensure that the fluid is not boiling too rapidly, giving undue "bumping."

Any other machinery or apparatus causing vibration should be kept remote from the microscope, which itself should preferably be situated on a firm foundation, e.g., in a basement rather than on a higher floor. Electron microscope manufacturers will usually send experts to make a site test for vibration before installing an instrument. Finally the shutter must be operated as smoothly as possible during exposure, to avoid jarring the column. The exposure should not be less than 1 sec. otherwise jerky action is unavoidable; it should not be greater than 2 or 3 sec., otherwise the probability of external disturbance becomes too high. Photographic material of suitable speed should be chosen to give an exposure of 1 or 2 sec.

B. Electrical Stability

The high voltage supply and the current in the lenses must be adequately stabilized, both as regards ripple and longer term fluctuations. Stability must be of the order of 1 part in 100,000 for a resolution of 5 Å. This requires the services of an electronic specialist to test the circuits under operating conditions; it is not sufficient to test them under no-load conditions. Testing the lens supplies is not difficult, but for the high voltage, a condenser potentiometer must be set up to tap off a suitable fraction of the total voltage. The tuning up of the circuits to this degree of performance is in fact best done by a servicing engineer very familiar with the circuits employed.

Apart from variation in the microscope supplies, the image may be blurred by external alternating fields and distorted by constant fields. Trouble most often arises from stray magnetic fields from transformers, heavy current lines, etc. The microscope column is fitted with protective mu-metal shielding, but this may not be adequate against strong stray fields when high performance is demanded. The magnitude of such effects can be found by exploring the space around the microscope with a coil of many turns (50,000) connected to a sensitive cathode ray tube. The influence of the field on the image can then be estimated by applying a known field in the same region and observing the amount of blurring.

C. Thermal and Electrostatic Stability

In addition to disturbance from external electric and magnetic fields, parts of the interior may become charged and so cause electrostatic deflection of the image. This trouble cannot arise if all surfaces struck by the beam are conducting, but it is found in practice that a nonconducting carbonaceous deposit gradually forms on any such surface, due to residual organic vapors in the column. Thus the apertures and possibly the specimen holder are likely to acquire a deposit, most rapidly where the beam intensity is greatest, i.e., on the objective aperture. The condenser aperture usually receives a load sufficient to heat it above the deposition limit and the specimen itself is in most microscopes protected by a decontaminator.

The visible symptom in the microscope image is flicking of the whole picture, usually repeating at intervals of a second or so, as charge accumulates and leaks away. Before this stage is reached, however, the aperture may already have an asymmetric coating which causes astigmatism; this is only likely to be noticed in high resolution operation, where diffraction fringes can be studied. The remedy is to remove the aperture and clean it according to maker's instructions. If it is of molybdenum, it can be flamed; if of copper, mild attack with formic acid may be adequate. Platinum requires a more complicated chemical treatment. With skill, a ground-down needle may be used to ream out the hole, which is liable to be slightly enlarged in the process; if optical examination shows it to be appreciably noncircular, it should be rejected. If cleaning the aperture has no effect, the object holder should be treated and finally the polepiece itself. If the specimen contains large particles of a nonmetallic nature, it may well be that the trouble arises in the area irradiated. This can easily be tested by repeating observations with a blank specimen grid. Apertures further down the column need attention less often, as the current density is much lower there; however, trouble occasionally arises from small hairs or parts of specimens falling into the projector aperture. It is necessary for all parts that have been cleaned to be washed in ether or other very pure solvent before being replaced in the microscope.

Regular fluttering of the image is probably due to a contaminating deposit, but an irregular effect is most likely due to trouble in the electron gun. It may be due to mechanical instability of the filament or to deposits inside the shield. The latter may be rubbed off with fine emery paper, and the shield aperture similarly treated. From time to time, the outside of the shield and the anode surface and aperture should also be cleaned. All such surfaces should be finished to a high polish to minimize the danger of electrostatic breakdown.

Thermal instability of the specimen as well as electrostatic charging may result from the impact of the electron beam. An immediate break up of the supporting film may occur when the beam comes on, or there may be a gradual warming up of the whole specimen grid and holder, leading to a creep of the image. This proves to be a considerable hindrance to high resolution work, where the specimen should not move more than an Ångstrom unit per second. The most that can be done in a normal microscope is to ensure a close fit of all parts of the object holder to each other and to the support grid, after polishing all such contact surfaces. It is advisable to avoid woven grids, in which the separate strands of wire may move relative to each other under thermal load; plated, flat grids of copper are now generally available. For high resolution work, or if the specimen is very sensitive to electron bombardment, it may be best to use a massive disc as support, bored with a single aperture of 100 μm or so. In such work, it is also desirable to run the lenses for about an hour before starting microscopy, to allow thermal equilibrium to be reached in the objective lens and specimen stage.

D. DECONTAMINATION

It is now usual for a decontaminator to be fitted to all but the simplest electron microscopes. It consists of a metal shield surrounding the specimen grid except for apertures to allow passage to the electron beam. The shield is cooled to a low temperature, usually by a metal rod dipping into liquid air in a container mounted externally to the column; the specimen itself is not cooled, otherwise decomposition may occur (Heide, 1965). Provided that thermal contacts remain good, the rate of contamination should be reduced to negligible proportions. But it is as well to make a check from time to time, by observing over 5 min or so the size of small features in a specimen or better the holes in an openwork support film ("holey" film). If the rate of contamination is not in the order stated in the microscope specification, the manufacturers should be consulted. Most decontaminator systems are delicate mechanisms and should not be interfered with by the ordinary user.

For very high resolution work, or when prolonged observation is desired at normal resolution (as in some dynamic experiments on the specimen), it may be necessary to reduce the partial pressure of residual vapors in the column so as virtually to eliminate the danger of contamination. This can be done by careful attention to degreasing of all components and by replacing pumps using oil by mercury or ion pumps (Hartman and Hartman, 1965). In many cases, it may be sufficient to construct a special specimen chamber, evacuated by its own ion and sorption pumping system [see Forth and Loebe (1968), and Tothill *et al.* (1968)] In all such work, it is most important to avoid introducing any form of grease anywhere into the microscope. Nylon gloves should be worn when handling any component which has to be removed or inserted, especially the specimen holder.

E. VACUUM CONDITIONS

Most commercial electron microscopes are fitted with a valving system such that the machine can only be operated if the vacuum has reached a certain limit, and if the vacuum worsens and causes the beam current to increase, a relay switches off the H.T. and so preserves the filament. It is good policy, however, to aim at getting the best possible vacuum. For one thing, it prolongs filament life; for another, it minimizes beam fluctuations. Slight intermittent leaks of air, especially in the region of the gun, cause the illumination to shift and may spoil an exposure. More severe bursts may cut off the beam altogether, owing to overloading. Whenever such signs are observed, the gaskets and other vacuum seals should be gone over carefully to find the source of trouble. The instructions book should give details of how to go about leak-testing and how to replace gaskets or metal bellows. It is a sound rule to suspect that part of the apparatus that was last interfered with, e.g., when changing a filament. In the worst case, a systematic search must be made for a leak, starting at the top of the column, removing each unit in turn, and sealing a glass or metal blank across the exposed opening. The rate of leak after pumping down may be observed each time on the built-in vacuum gauge; in this way, the faulty component can at length be traced. It should be possible to get the system vacuum tight to the degree that, after shutting off the pumps, it does not rise above the working limit of vacuum for several hours; in some instruments, an overnight "black out" can be attained.

F. ALIGNMENT

The permissible aperture of the electron microscope is so small (10^{-2} to 10^{-3}) that a high degree of precision is demanded in the alignment of the several lenses and the electron gun. The mechanical accuracy of construction can be made such as to ensure the correct parallel alignment of components in a two-stage microscope. Adjustments then have to be provided to bring the axis of each lens into line with that of the next and to ensure that the gun delivers a beam correctly into the condenser. As mentioned, a tilt of the gun is needed since the filament tip cannot be accurately enough positioned in its aperture.

The problem sounds simple, so stated, and it might be thought that the whole column could be set up and clamped by the makers, even if the machining tolerances cannot be made fine enough to ensure exact fit on assembly. The complicating factor is that the magnetic axis of a lens may not coincide precisely with its geometric axis and may vary with operating conditions or with time owing to hysteresis and saturation in the iron. It is usually necessary to realign after changing the filament and advisable to check alignment after any radical change in operating conditions (voltage or lens power).

Two different principles may be used in aligning the column: that the image stays stationary when the lens power is varied or when the beam voltage changes. Current centration has been most used in the past, but it is now appreciated that voltage alignment is more satisfactory for high resolution work, since slight variations are more likely to occur in voltage than in current. If the current and voltage axes coincided, no choice would be needed, but this is hardly ever the case. Current alignment will first be described for a two-stage instrument (objective and one projector) and a single condenser; the same procedure holds for most double condenser systems, but additional adjustments are needed in microscopes having the two condensers separately movable.

The criterion of centration is simple in a magnetic lens, since the image rotates when lens power or voltage changes. This makes centration in some ways easier than in electrostatic lenses, in other ways more difficult as the image may also be displaced sideways. This combination of effects hinders a straightforward deduction of remedial measures from the observed behavior of the image, as the objective current is varied about its normal value; it is preferable to isolate the rotational effect by reversing the current through the objective lens (Haine, 1961). An image point then jumps to a new spot on the screen, and the desired axis will lie half way between the two positions, but such a reversing switch is not provided on most microscopes.

The first step in alignment is to remove the objective aperture and operate at low magnification, to get bright illumination on the screen. The condenser and gun traverse is used to keep this in view as the magnification is raised to a few thousand times; the image should expand and contract roughly symmetrically about the center of the screen as the condenser current is varied. Now change the objective lens current and identify a point in the image which does not move on doing so. Move this point, which is on the axis, to the center of the field by shifting the projector (or objective, if the projector is fixed). The image should now expand about the center when the projector current is varied and rotate about the same point when the objective current is varied. This procedure is best carried out in the diffraction mode of operation, using the central spot of the diffraction pattern as reference point.

If the image is found to shift sideways as the objective current is slightly varied, without reversing, it is a sign that the beam from the condenser is not entering along the objective axis. Use the gun tilt controls to return a chosen image spot to the center of the screen, after it has moved away on reducing the objective current. After this has been repeated several times, the image should rotate about the center of the screen on varying the objective. This centration of the illuminating beam is most important for high resolution work, since an additional rotatory aberration of the image is caused if it is not correctly done.

Finally check that the brightest illumination is obtained in the center of the screen; if not, adjust the gun traverse controls until it is so. Now insert the objective aperture and align it according to the mechanism provided, using a low magnification image. Once properly aligned, only minor adjustments should be needed in day-to-day operation.

The alignment procedure is more complicated in three-stage microscopes, except when the intermediate lens is firmly locked to the projector. When the intermediate and projector lenses are separately adjustable, as is usual in high performance microscopes, the detailed procedure set out in the instructions manual should be consulted; different models differ so much that a general account cannot be given.

For voltage alignment, the same procedure may be followed, except for rapidly switching the H.T. on and off (or applying an extra voltage) instead of reversing the objective. Hillier and Ramberg (1947) give in detail the procedure for an RCA model, in which the projector lens was fixed. Voltage centration is essential if resolution better than 40 Å is sought, in conjunction with correction of the objective polepiece for astigmatism.

G. ASTIGMATISM CORRECTION; THROUGH-FOCAL SERIES

The polepieces of any objective lens will show a degree of astigmatism, no matter how well machined, owing to minute irregularities of bore and nonuniformity of the iron. The result is to limit the resolution to about 50 Å, even with an aperture of 10^{-3}. It is possible to correct this aberration by means of a stigmator, which introduces counterbalancing electrostatic or magnetic fields in the back polepiece of the objective. By careful adjustment of the strength and orientation of the field, astigmatism can be reduced to a negligible amount. The method is tedious, demanding careful observation of the Fresnel fringes formed around opaque particles or around holes in a carbon film, when the beam aperture is very small. Uniformity in width of the fringes is the criterion of correction. By systematic work, the astigmatism, as measured by the separation of the two focal lines (cf. Section IV.D.b), can be reduced to about 0.1 μm, compared with the 5 μm of an average untreated polepiece; this would be equivalent to a resolution of a few Ångstroms, if no other aberrations were present.

As the different types of stigmator differ considerably in design and operation, the instructions manual for a given microscope should be consulted for the detailed correction procedure. It needs to be repeated regularly, because of hysteresis effects in the polepieces or even a slight influence from the specimen holder or the specimen itself. The most important part of the procedure is the evaluation of the astigmatism from the Fresnel fringe width. The picture must be very close to exact focus for this purpose. If the ordinary fine focus control is too rough for such accurate setting, a ten-times finer rheostat should be fitted in the objective lens circuit. By observing the ratio of each control to the next finer, and reading with a meter the current change corresponding to one division on the coarsest, the value of a division on the other controls can be found; for the finest, it must be of the order of a few microamps.

Even so it may not be possible to work accurately enough by eye for high resolution purposes and a through-focal series has to be taken. The true in-focus setting can be found very closely from the circumstance that image contrast is then a minimum; on either side of focus the Fresnel fringes form around the image, thus increasing contrast. After setting the finest control just short of this point, a series of five or six exposures is taken of the same particles, moving the knob by one or two divisions each time, depending on its range. In the negative, it should be possible to see that the fringe width first decreases and then increases again; it is only by luck that an absolutely in-focus picture will be obtained, showing negligible fringes. Typical pictures are usually given in the instructions manual. It is often found at first attempt that one has either started too far off-focus or changed the control in too large jumps. If too far off-focus, the fringes may be so big as to hide the astigmatic effect and lead to the false conclusion that the lens is perfect. Once the near-focus image has been obtained, a systematic procedure of measuring the fringes and deducing which controls to adjust leads to correction by successive approximation [see Chapman (1962)]. In very high resolution work, the presence of residual astigmatism can be detected by optical diffractometry of the electron micrograph (Thon, 1967).

If an objective aperture is used with a corrected lens, even greater care than usual must be taken in cleaning it. A very small unsymmetrical deposit on it will introduce more astigmatism than the correction has taken out. It may be preferable to correct the lens without an aperture, and to use it subsequently in the same way, in spite of the poor contrast. The condenser must then be well defocused to give a small angular aperture and the alignment of the illuminating beam becomes very critical. If the beam makes an angle with the objective axis appreciable greater than its own angular aperture, the aberration for off-axis points may be considerable (Glaser and Grümm, 1952).

H. Focusing the Image

Although a through-focal series is worthwhile when correcting astigmatism, it is too time consuming to be used for every micrograph taken.

The operator soon gets used to judging the near-focus image, however, to within the finest focusing step on the objective lens controls. At high magnifications, it is usually possible to see the Fresnel fringe around some well-defined feature in the specimen, unless it is too thick; a hole gives a much better indication of the minimum contrast position.

At low magnifications, the viewing telescope will not show the fringes and some other means must be adopted to get best focus. If a moderately large aperture is being used, it may be enough to swing the condenser control rather widely, thus varying the angular aperture of the illumination, at the same time slowly changing the objective finer control until minimum contrast is obtained. Exact focus is easier to hit, however, if the objective aperture is withdrawn during this operation, but it must be reinserted before taking a micrograph. Some instruments have a "wobbler" button, which automatically deflects the beam over a large angle for fine focusing, or the usual condenser aperture can be replaced by a disc with multiple apertures which fulfills the same function, but it seems that these methods do not give such good results as removing the objective aperture.

It is frequently suggested that the best image is obtained by a slight degree of defocus from the minimum contrast condition [e.g., Haydon and Taylor (1966)]. By doing so there is certainly an appreciable improvement in the overall contrast of the image, due to the Fresnel fringes (phase contrast), but this may be accompanied by a slight loss of resolution (Sjöstrand, 1955). Best resolution is obtained within the "parafocal" region, where only amplitude contrast is effective (Van Dorsten and Premsela, 1966; Bremmer and Van Dorsten, 1969). For the great majority of biological applications, however, the loss of resolution is less important than the gain in contrast. For very high resolution work, with an ultra-fine objective focus control, the interpretation of the image becomes in any event very complicated (Thon, 1967). To avoid the production of spurious image details (optical artifacts), the operator should take a series of micrographs in differing conditions of operation and assess them with a critical understanding of out-of-focus effects. The misinterpretation of the structure of ferritin is a classical example of the difficulties involved (Chescoe and Agar, 1966; Haydon, 1969; Towe, 1969; Haggis, 1970).

I. SPECIMEN TILTING; STEREOPHOTOGRAPHY

Most electron microscopes, particularly those designed for use by metallurgists, have available a goniometer stage that allows the specimen to be tilted through a wide angle (usually $\pm 30°$) about an axis normal to the electron beam. This accessory is often useful to the biologist also. A tissue section is likely to be cut at random through an extended feature such as a double membrane, so that the observer is left in doubt as to its true nature. By tilting the specimen, it may be possible to bring its extension into line with the beam, giving an end-on view, when a double membrane should clearly appear as such. A goniometer frequently provides rotation as well as tilt, thus allowing a specimen feature to be lined up in the right direction for tilted observation.

Stereophotography is an equally valuable technique whenever a specimen is rich in detail, and especially for the relatively thick sections which can be examined in a high voltage microscope. A simple tilting rod is supplied for many microscopes, with a variation of $\pm 10°$ or so, but a full goniometer gives much greater flexibility of viewing and makes it possible to bring any selected part of a specimen onto the axis, so that it does not run out of focus on tilting. A tilt of 5° to 10° is suitable for most purposes, but a rather larger value may be desirable for fine structural studies. It is quite straightforward to take a stereo-pair of micrographs, because of the great depth of field of the electron microscope, but care is needed in mounting the final prints if the proper 3-D effect is to be obtained, since the electron image has been rotated with respect to the specimen. If a stereo-comparator is available, the images can be evaluated quantitatively, giving the coordinates and so the dimensions of specimen features. The angle of tilt must then be known with sufficient accuracy (see Hudson and Makin, 1970).

Since stereology is being increasingly used in optical microscopy also, there is a considerable amount of information on the technique in the literature. Reviews of different aspects have recently been published by Reimer (1967, Chapter 11.2), Underwood (1969), and Weibel (1969). For a stroboscopic method of three-dimensional reconstruction from ultra-thin serial sections, see Kölbel (1968).

IX. Specimen Preparation

A. PREPARATION FACILITIES; LAY-OUT OF AN ELECTRON MICROSCOPY LABORATORY

With the increasing complexity and refinement of electron microscopical techniques, the facilities to be provided have become correspondingly extensive. Apart from a room for the electron microscope itself (or for several of them) and a dark-room, wet and dry preparation rooms, perhaps separate microtome rooms and a dust-free room, an equipment room and photographic finishing space are also desirable. An allowance of between 750 and 1000 ft^2 per microscope seems to be usual. Most often the best has to be done with such rooms as are made available for the electron microscopist, but anyone fortunate enough to be given a clean start will find much useful information on designing a laboratory for biological electron microscopy in articles by Harrison and Philpott (1968) and Willis (1969).

In view of the great variety of preparation techniques now at our command, only a brief outline is given below. Detailed information will be found in the standard texts (Kay, 1965; Pease, 1964; Reimer, 1967; Sjöstrand, 1968).

B. SUPPORT GRIDS

The standard specimen support is an electrodeposited metal grid, usually in the form of a disc 3 mm in diameter, and obtainable in a variety of mesh sizes. Copper grids with 200 meshes per inch are most commonly used, but other metals available for special needs (e.g., resistance to chemical action) are stainless steel, nickel, platinum, or gold. Special bar layouts are made, giving larger free areas when examining sections and for autoradiography, or much finer mesh for supporting very thin specimens. Grids with micro-numbers or micro-letters make it possible to log and relocate features of interest in a specimen.

Grids should be stored in tubes or pillboxes so that they do not accumulate dust or get corroded. If there is any doubt about their cleanliness, they should be rinsed in ether or redistilled carbon tetrachloride. It is possible to reclaim grids after use, but great care has to be taken to remove all traces of specimen and support film; this is particularly difficult because the electron beam reduces the normal support film to an insoluble carbon or graphite. Usually it will not be found worth the trouble of reclaiming grids of copper or stainless steel. Those of platinum (Siemens) are more resistant and may be cleaned with strong acid or by flaming.

C. SUPPORT FILMS

Some specimens, such as thick sections or surface replicas, can be collected directly on the support grid. But usually a supporting film is necessary, which may be of collodion (nitrocellulose), Formvar (polyvinyl formaldehyde), or carbon. For special purposes, films of beryllium, aluminum oxide, or silicon monoxide have been used; details are given by Kay (1965) and Reimer (1967). Here only the preparation of the three most generally used films will be described: collodion, Formvar, and evaporated carbon. For most purposes carbon is preferred, as it can be prepared thinner and is more stable in the electron beam.

1. Collodion Films

The support film must be very thin, in order not to obscure detail in the specimen to be placed on it. A thickness of 150 to 200 Å is found to provide an adequate combination of transparency with mechanical strength for many purposes; it should also be reasonably uniform and free from holes. Nitrocellulose films have the desired characteristics and are easily prepared. The simplest method is to take up in a fine pipette a quantity of a 2% solution of nitrocellulose in amyl acetate and deposit a drop on the surface of clean distilled water contained in a deep Petri dish or glass bowl. The solution spreads rapidly across the water surface, the solvent evaporates, and a thin collodion film remains. The first film prepared should be treated as a broom and discarded with the surface dust it collects. The second film is prepared for microscopy by dropping onto it the metal support grids, cutting round them with a fine needle, and lifting them off plus film on a washer fixed at the end of a metal rod. Alternatively, an area of film may be removed with such a washer and transferred to a grid supported on a metal pedestal.

Calculation will show what thickness of film is to be expected from a single drop of 2% nitrocellulose spread uniformly over the available water surface; a 200 Å film would need an area of about 300 cm^2. In fact, it is found that the film is thinner near its periphery and rarely reaches to the wall of the container, unless the surface is exceptionally clean. It is desirable to choose the strength of collodion solution in relation to the surface area so that a single drop gives a film of the required thickness. It is difficult to get anything like a uniform film from several drops, since they successively travel less far from the center. A little experience will show the correct strength of solution to give a single drop film of the required thickness. It will also be found that some grades of nitrocellulose yield better films than others.

2. Formvar Films

Formvar is insoluble in solvents that spread well on water, so that films are made from it by stripping from glass slides. A 1% solution in ethylene dichloride or chloroform is poured over the slide, or alternatively the slide is dipped into the solution, withdrawn, and stood on end to drain. When the solvent has evaporated, the slide is slowly immersed in water at an angle of about 30° to the surface, when the film should come away and float on the surface. If stripping does not start automatically on immersion, the edge of the film may be teased off the slide with a needle. It usually helps if the edge of the slide (not the flat) is first scraped with a knife to remove adhering Formvar.

The floating film is then cut and picked up as in the case of collodion. The film will not be uniform, but considerably thicker at the edge to which the solution drained. Interference colors will show the thinner (darker) region. If the film is found on examination to be full of holes, the indication is either that the slide was not dry before spreading the solution on it or that it was breathed upon before it dried. On occasion, a network film may be useful as a support, especially for high resolution work, and the best means of making it have been discussed by Sjöstrand (1957) and Harris (1962); see also Bradley (1965a).

If the slide is new, no cleaning other than a wash in dioxane or a detergent is needed before applying the Formvar solution. If it has been used before, it should first be cleaned with dioxane, preferably in a fume cupboard (owing to the toxic character of dioxane), then flamed, and finally washed in a pure solvent. Formvar films are distinctly tougher than those of collodion and are recommended whenever heavy particles are present in the specimen or when subsequent washing or staining procedures have to be carried out. Collodion films can be stripped from glass in the same way and are also found to be stronger than when cast on water.

It needs to be emphasized that the support films should be free of extraneous particles or bacteria which may be confused with the specimen itself. Solvents should be as pure as possible and preferably redistilled. All containers, droppers, and tools for handling films should be well washed beforehand. The support grids should be kept in tubes or otherwise protected from dust and the water surface or glass slide for film preparation should be kept covered until actually used. Whenever there is doubt, blank films should be examined in the electron microscope for cleanliness. In industrial districts, it may be necessary to arrange a dustproof chamber for film preparation, or better still an air-conditioned room. Bacteria are liable to turn up even in distilled water, so that special care is necessary in preparing films for bacteriological work; specially redistilled water must be used and frequently renewed, since bacteria may be introduced from tools or grids or drop from the air onto the surface.

3. Carbon Films

Evaporated carbon films can be prepared much thinner than surface-spread plastic films (down to 20 Å with care), and so are normally used for high resolution microscopy. They are relatively strong and are more stable, whereas plastic films are partially decomposed in the electron beam, resulting in movement or splitting.

The films are formed by evaporation in the type of vacuum apparatus described later (Section IX.E) for shadowcasting. Instead of evaporating from a boat or strip, as for metals, the carbon is vaporized by passing a heavy current between two pointed graphite rods, just in contact.

When moderately thin films are required, 100 Å or more thick, the carbon is deposited on a clean glass slide and stripped off by immersion in water, as when making Formvar films. For very thin films, 20–100 Å, it is better to deposit carbon directly onto a collodion or Formvar film on individual specimen grids. The plastic is subsequently removed by floating on a suitable solvent or in solvent vapor. The procedure is described in detail by Bradley (1954, 1965a). He prefers to use Bedacryl as film substrate, which is soluble in chloroform or xylene.

Carbon nets or "holey films" can be made by evaporating carbon on to perforated Formvar films prepared as described in the previous section, and then dissolving away the Formvar with chloroform.

D. MICROCHAMBERS

For observing specimens in the wet state, or for carrying out reactions in the gaseous state, increasing use is being made of microchambers. In these, the specimen is isolated from the vacuum by being placed between two plastic films, thick enough to withstand a moderate pressure difference (not necessarily 760 mm Hg), but thin enough to permit the electron beam to pass through without too much scattering. The "windows" for the beam are limited by electron microscope apertures placed on each side of a washer of larger bore which carries the plastic films; its thickness defines the depth available for the specimen.

Detailed descriptions of the construction and operation of microchambers for 100 kV microscopes have been published by Heide (1958, 1962), Hiziya and Ito (1958), Stoyanova (1958), Nekrassova and Stoyanova (1960), Escaig (1966), and Escaig and Sella (1969), apart from early attempts to make them. It should be emphasized that considerable skill is required both in their construction and operation. A controlled supply of an inert gas such as nitrogen has to be fed continuously to the microchamber to make up for the leakage of gas through the plastic films into the vacuum. The technique is easier to use in a high voltage microscope, since much thicker films can be used. The microchamber used by Dupouy et al. (1960) for studying bacteria in the wet state has parlodion-carbon windows 600 Å thick [see also Dupouy (1968)].

E. SHADOWCASTING

The method of depositing metal atoms on the surface of a specimen, from a directed beam produced by evaporation, is an invaluable means of enchancing the contrast of surface details. It was first perfected by Williams and Wyckoff (1944) using gold, but most other metals and many alloys have since been tried. Palladium, platinum-carbon, or a

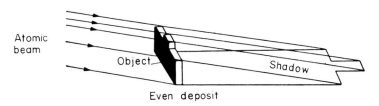

FIG. 14. Schematic diagram of metal shadowcasting technique.

gold-palladium alloy are most favored, but uranium, tungsten, and other high-melting metals have also been used.

The principle is simply that the beam of metal atoms is caught by any projection on the surface, causing a drift on one side and a shadow on the other (Fig. 14). On viewing vertically in transmission, a great accentuation of contrast is obtained in the surface relief, as in the case of a landscape seen from the air in the rays of the setting sun. Heavy elements should be used, so that a very thin layer gives enough contrast differentiation, since electron scattering depends on atomic number and density.

Several types of shadow-casting apparatus are now on the market, but are mostly expensive. As all that is required is a bell jar connected to a vacuum gauge, a diffusion pump, and a rough backing pump, it is not difficult to make a suitable evaporating plant with quite modest workshop facilities. A suitable design, constructed in this laboratory, is shown in Fig. 15. The diffusion pump should be of high capacity

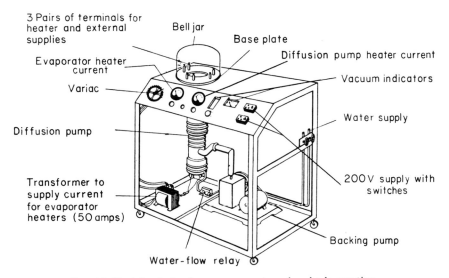

FIG. 15. Sketch of simple vacuum system for shadowcasting.

($\sim$100 liters/sec) and the bell jar no larger than absolutely necessary, so that pumping speed is as high as possible; a vacuum desiccator serves very well. The chief difficulty lies in sealing the current leads through the metal plate carrying the bell jar, but several methods are described in the literature or ready-made units can be bought. It is advisable to have two or three pairs of leads, so that successive evaporation of different metals or of the same metal from several directions may be carried out. It is in any case convenient to reserve different pairs for evaporating different metals, since some require a boat or basket and others simply a V-shaped filament.

Most of the useful metals have high evaporating temperatures so that an even higher melting metal must be used for heating it. A V-shaped tungsten filament is most generally useful, made of wire 0.2–0.5 mm in diameter. It should be kept as short as possible between terminals and will require 15–30 A from a step-down transformer, according to the metal to be evaporated. A small boat of tantalum is suitable for powders such as beryllium. It may be made very simply by punching an indentation in a 2-mm wide strip of fine gauge or by making a longitudinal crease in the strip. For chromium and other metals in the form of chips, and also silica for replica work, a small basket of tungsten wire is made by winding it around the conical end of a large wood screw and filling the gaps with alumina paste.

The simplest geometrical arrangement of source and specimen is that in which the prepared grids lie horizontally on the base plate of the bell jar at a distance of 10 to 15 cm from the filament, which is raised to a height appropriate to the shadowing angle needed. Alternatively, and especially when powders are evaporated from a boat, the grids may be held on a glass slide with spring clips of brass or phosphor bronze, and this in turn clamped above or opposite to the filament on an arm, the angle of which can be varied. A separation of at least 10 cm is desirable to protect the specimen from the effects of heat and to give a sufficiently uniform distribution of deposit over a row of grids.

An angle of 20 to 30° is suitable for the ordinary run of specimens, often approximately specified as $\tan^{-1} \frac{1}{3}$ or $\frac{1}{2}$, respectively. Shallower angles give long shadows which may blanket important detail and also pile up too much metal on the "windward" side of projections. When the features of interest are all very small, as with plant viruses or macromolecules, it may be useful to use an angle of 10 or 15°, but care must then be exercised in calculating particle size since the deposit may be of appreciable relative thickness. For all ordinary purposes, a deposit of thickness 10–20 Å only should be used. Assuming spherical symmetry of evaporation, which is roughly true for a deposit at some distance from a filament, the formula for calculating the thickness t over a surface inclined at an angle θ to the atom beam is

$$t = (m \sin \theta)/(4\pi l^2 d) \tag{7}$$

The thickness of a projection that is normal to the beam is given by omitting the term $\sin \theta$. Here d is the density of the metal used, l the distance of specimen from filament, and m the mass of metal evaporated. A 20 Å layer on a surface at 20° and at a distance of 10 cm would need 15 mg of platinum (density 21.5) or 11.5 mg of 60% gold–40% palladium (density 16.5). In practice, evaporation is not symmetrical nor completely efficient, and the layer may be thinner than calculated. To avoid troublesome weighings, it is convenient to obtain the metal in the form of wire of given mass per unit length so that a few millimeters may be cut off as required. Wire is also easily wrapped around the heating filament, which should be slowly raised in temperature so that the wire does not melt too rapidly and partly fall off.

The chief difficulty in shadowcasting is the tendency of the deposited metal to aggregate into small crystallites, which may be mistaken for particles in the specimen. Even if this is avoided during evaporation, it may occur in the electron microscope if the illuminating beam is too intense. The remedy is to choose a metal that shows minimum granulation and to take precautions in shadowing by using only a thin layer and depositing it in the highest vacuum conveniently obtainable. Alloys or mixtures naturally show less tendency than pure metals to form crystals. Platinum-carbon is most widely preferred, prepared by evaporation as when making carbon films, but using rod electrodes of carbon impregnated with platinum (Bradley, 1959). However, Abermann and Bachmann (1969) claim that an alloy of tungsten and tantalum gives a still finer layer (see also Zingsheim, Abermann and Bachmann, 1970).

The degree of vacuum is of great importance. At least 10^{-4} mm Hg should be aimed at and better for high resolution work. It is not sufficient to pump until this vacuum is nominally obtained according to the gauge, since residual oil and other vapors are desorbed only very slowly from metal surfaces. It is advisable to pump for at least half-an-hour, and improvement in uniformity of deposit is usually found up to an hour or more. It also seems to be of value to install a shutter between filament and specimen, so that the latter is exposed only for the minimum time, which can be controlled.

It is usually sufficient to shadow once only at a given angle and direction, but Drummond (1950) has recommended using two successive beams from different directions to show the shape of particles, and this method has been developed into a method of "portrait" shadowing by

Philpott (1951). Heinmets (1949) went so far to rotate the specimen during shadowing to increase contrast all round. This method has been used by Kleinschmidt *et al.* (1962) in the microscopy of nucleic acid molecules.

The chief reservation about shadowing is that it should not be used when an indication of body distribution of matter in the specimen is sought. It should really be reserved for cases where the primary interest is in surface relief, in spite of the great temptation to get show pictures by its use. In many cases where it might add slightly to the pictorial value, the time consumed in properly shadowing makes it uneconomic, especially in routine runs. There are cases in replica work where it is of value, but if the replicas are properly prepared the contrast should already be adequate for most purposes. Usually the resolution is limited by the replica itself and is not improved by shadowing; the main exception is the stripped replica, in which shadowing is an integral part of the process. Shadowing can also be used for electrostatically stabilizing sections and other specimens [see Boult and Brabazon (1968)], which may have thick regions that absorb electrons from the beam, but evaporated carbon is better for this purpose.

F. PARTICLES IN SUSPENSION

Suspensions of viruses or bacteria in water may be deposited as a droplet directly on to a filmed grid by means of a micropipette or a platinum loop. After about a minute the droplet is drained off. Alternatively bacteria can be collected directly from a culture by forming a local suspension: A drop of water is placed on the surface and the grid touched on to it (Houwink and van Iterson, 1950). These methods have the disadvantages that large surface tension forces occur during drying, and also that aggregation may be great enough to make observation difficult, especially for particle counting.

Better results can usually be obtained by spraying the suspension on to filmed grids by means of an air jet or nebulizer, operated inside a glass vessel. Several models have been described which are readily constructed by a good glass blower or a commercial type can be bought. A detailed description of the various aspects of the technique, including quantitative assay of particles, has been given by Horne (1965a).

G. THIN SECTIONS

For electron microscopy at the usual voltages, sections have to be cut nearly 100 times thinner than for optical microscopy. Instead of a few microns, they must be only a few hundred Ångstrom units in thickness. The reason is that electrons lose so much energy in even such a thin layer that image resolution is impaired by chromatic aberration (Section IV.D.3). As a rough rule, a section should not be thicker than ten times the resolution desired when working at 50 KV, and at 100 kV not more than 40 times the resolution (Cosslett, 1956). These figures assume negligible loss of material from the section in the electron beam. As in fact up to 50% or more may be lost unless very low illumination is used (Section III.E), an additional factor of two can be allowed on thickness. At very high voltages, the thickness:resolution ratio improves greatly, to values of several hundred (Dupouy, 1968; Cosslett, 1969a), so that a 1 μm section can be observed with relatively high resolution at 1 MV (see Section XII.B).

Ultramicrotomes are now commercially available in a variety of models for cutting sections down to a thickness of 100 Å routinely. Special hard embedding materials are needed, to give adequate mechanical support to the tissue during the cutting process; polymerized plastics or chemically setting substances such as araldite are used. Instead of a metal knife, a sharp glass edge cut at a particular angle from sheet-glass is satisfactory. A diamond or sapphire has the advantage of longer life, but at more than proportionately greater cost.

The variety of fixation, dehydration, embedding, and sectioning techniques developed for ultramicrotomy is too great for even an outline to be given here. Comprehensive surveys have been published by Glauert (1965a), Pease (1964), Sjöstrand (1968), and Wachtel et al. (1966), including accounts of freeze-drying procedures. Juniper et al. (1969) deal with the special problems of preparing plant materials. Millonig and Marinozzi (1968) have discussed fixation and embedding methods in great detail. To cope with the load of work entailed in serial section observation, special attachments are now made for the electron microscope in the form of a multiple section adapter for the specimen stage and a roll-film camera of large capacity (Lucas, 1968; Dougall, 1969).

H. STAINING

In place of the colored reagents familiar as stains in optical microscopy, we have to use compounds of heavy metals for electron microscopical staining; that is, substances which have high scattering power for electrons and thus increase the local contrast of the specimen where they are selectively absorbed by particular cell constituents. Staining may be carried out either before sectioning or on the cut thin sections. Salts of molybdenum, tungsten, lead, and uranium are particularly useful. Other substances such as osmium tetroxide are effective for giving an overall increase in contrast, combined in this case with fixation. The negative staining method, in which phosphotungstic or phosphomolybdic acid is applied to a specimen already mounted on a grid, has proved even more effective in showing up ultrastructural details, particularly of viruses; for a review, see Horne (1965b).

In the past 10 to 15 years, systematic studies have been made to discover specific stains for different substances or tissue components. A great deal of experience has accumulated and is reviewed by Glauert (1965b), Horne (1965a), Symposium on Cytochemical Progress in Electron Microscopy (1962), and Sjöstrand (1968). Considerable progress has been made in the localization of enzymes, in regard to antibody–antigen reactions, and in some applications of autoradiography at the electron microscopical level. Compared with optical cytochemistry, however, a great deal remains to be done both in establishing quantitative methods for the larger cell components and in devising any adequate methods at all for locating and labeling macromolecular substructure [see Beer (1965)].

I. Replicas

A large number of replica techniques for examining surfaces of various types was developed before ways were found of thinning metals and cutting ultra-thin sections so that materials could be examined in transmission. Basically the procedure was to coat the specimen surface with a thin layer of a plastic from solution, or of carbon by evaporation, and then remove it by one of the methods used for preparing support films or by backing it with a thicker layer and then stripping both off together. The initial replica film could then be mounted for examination in the electron microscope, or if it was too thick to do so (as with polystyrene or methacrylate replicas), it was used as template to form a second thinner replica.

Replicas are still useful for some special purposes. For instance, Labaw and Rossmann (1969) have recently used the platinum-carbon technique for obtaining high resolution images of L-lactate dehydrogenase crystals, which are sensitive to the electron beam. Henderson (1969) has used a two-stage replica method (polyvinyl alcohol and preshadowed carbon) for studying the surface of sections of a variety of tissues. A platinum-carbon replica is also the final essential stage in the freeze-etching procedure developed by Moor and Mühlethaler (1963) for studying the fine structure of tissues with a minimum risk of artefact.

Full details of these techniques, and of others for rough, fibrous, wet, or porous surfaces, are described by Bradley (1965b) and Reimer (1967, Chapter 16).

X. Artifacts

Since the first edition of this volume (1955), the initial burst of enthusiasm for ultrastructural research, opened up by thin sectioning, has given way to a more cautious and critical attitude. All the same, it may not be out of place to repeat some of the earlier warnings against accepting micrographs at their face value.

Electron microscopy usually involves even greater interference with the specimen than in the normal mounting and staining methods of optical microscopy. So it is important to beware of the risk of artifact production at the various stages of fixation, dehydration, sectioning, staining (and possibly extraction), and exposure to the electron beam. On the whole, experience has shown that the dangers are not so great as were at first thought. The best precaution is to employ several different methods of preparation whenever possible and to compare the results with one another and also with optical observations made with the phase contrast and ultraviolet microscopes.

A. ARTIFACTS OF PREPARATION

1. Fixation

To some extent the lessons learned with the light microscope on the effect of fixatives can be carried over into electron microscopy, but it is necessary to remember that the details sought with the latter are so very much finer, and the best optical fixative may not be best for macromolecular structures. In optical work, fixation has to make features visible within the resolution limit of light, and this must often involve agglomeration or thickening of thinner features in the specimen. In the electron microscope, we wish fixation to leave undisturbed detailed structures of the order of 20 Å or less. One cannot take over automatically the fixation techniques of normal microscopy without critical examination of their results at high magnification.

The most comprehensive study of fixation problems is that of Millonig and Marinozzi (1968), who compared the effects of osmium tetroxide, glutaraldehyde, and potassium permanganate on three typical specimens: serum albumin, sea urchin eggs, and Ascites tumor cells. See also the detailed exposition of ultramicrotomy techniques by Wachtel et al. (1966). Both articles give many references. In general, the conclusion has to be that there is no perfect fixative. Each causes some type of artifact, while giving good preservation of other structures. So the most appropriate fixative must be selected for the features of immediate interest in a tissue. There is obviously great scope for difference of opinion over what is "real" and what is artifact in any set of results.

2. Dehydration

The most likely source of artifacts in electron microscopy is the desiccation essential to vacuum working. Since all biological material has a high water content, the shrinkage on drying is bound to be considerable. It will have two consequences: an overall reduction in dimensions of the specimen, and a disturbance of its internal structure, by displacement where there is differential shrinkage and possibly also by aggregation of material. The gross shrinkage can be estimated by comparison with ultraviolet and phase contrast pictures, although only on a statistical basis, since organisms usually vary considerably in size in a given population. When they are settled onto a supporting film for electron microscopy, there is little shrinkage of the area of contact but great collapse in the third dimension. Even when fixed, bacteria will normally dry down to half their initial height, and without fixation to much less. It is probable that considerable reorganization of the interior structure takes place at the same time, leading to aggregation in some parts and the formation of relative vacuoles in others. Some check on the larger changes are available optically; for instance, the ordinary light microscope, and even better the phase contrast method, shows clearly the dense regions in mycobacteria which appear as granules in electron micrographs.

The small scale desiccation changes can be checked by thin section techniques, in which it can be assumed that the relative motion of material is much less than in simple, dried suspensions. Little comparative work on these lines has been done so far. Alternatively, special methods of fixation and drying may be used, which are likely to lead to less disturbance in the material. Freeze-drying offers a method of both fixing and drying which should leave the natural structures largely unchanged, except perhaps at the molecular level. The results so far of its application in electron microscopy are to some extent contradictory, probably due to the fact that tissues are extremely fragile after such drying. But irradiation with a low voltage electron beam after freeze-drying stabilizes the preparation against the effect of the microscope beam.[1] The alternative method of critical point desiccation devised by Anderson (1951) has similar advantages and defects. The specimen is passed through a series of alcohol-carbon dioxide mixtures until it is perfused with the latter substance under high pressure. Conditions are then raised to the critical point, and the CO_2 allowed to evolve. In this way, no phase boundary passes through the material, as it does even in freeze-drying, so that the disturbance should be minimal.

The temptation with electron microscopy, as with any other new and brilliant technique, has been to use it in every possible direction without

---

[1] A review of the problems connected with freeze-drying and freeze-substitution will be found in *Federation Proc.* (1965); see also, Millonig and Marinozzi (1968, p. 312).

waiting for the careful evaluation of pitfalls and accumulation of critical experience that logic would have demanded. At the moment, the available evidence indicates that the degree of distortion on drying, even in the simplest procedure, is much less than originally feared; the fixation artifacts are equally, and perhaps more, dangerous, and careful attention needs to be paid to the special requirements of electron microscopy in this direction.

3. Embedding and Sectioning

The additional hazards introduced in microtomy are those involved in embedding, in cutting the section, and (when practiced) removing the embedding material. Knife marks are usually immediately obvious but may be mistaken for directional effects or even striations in the tissue unless carefully checked. Displacement of material and vacuole formation is more difficult to identify; such artifacts may be produced either by the knife or during removal of the embedding substance. The latter is simply tested by observations on unextracted sections, in which most features should still be visible except the very finest. Local knife effects, such as shredding of membranes, can to some extent be checked by comparing similar regions of tissues which have met the knife at different angles. But it is impossible to be sure whether an occasional vacuole or break in a membrane is real or not, unless it is large enough to be seen by optical microscopy. One can only be guided by the frequency with which such features are observed.

A detailed description of the techniques of ultramicrotomy has appeared in an earlier volume of this work (Wachtel et al., 1966), including a discussion of the difficulties and pitfalls.

B. Artifacts of Microscopy

1. Effect of Electron Beam

The energy density in the electron beam is enormous and the thicker the specimen the more energy is absorbed by it, with consequent rise in temperature and possibly disruption of chemical bonds in some of its component substances. Even with the minimum beam intensity needed for viewing, it is almost impossible to prevent very thick specimens from moving so much as to disrupt the supporting film. In such cases, ebullition is sometimes observed, as though tightly bound water was being driven off. But with specimens of more normal thickness, such as bacteria, viruses, or thin sections, one rarely sees any change produced by the electron beam, apart from an overall increase in transparency, which may be marked as sections lose embedding material (see Section

III.E). Occasionally, a dense particle, such as occurs in some bacteria, will be seen to clear when the beam intensity is increased, as if some of its contents were being evaporated. But in the normal conditions of viewing, it is unusual to find any artifacts of this type. Whenever a new variety of specimen is to be examined, however, it is always preferable to begin with minimum beam intensity and to watch the image carefully as the intensity is increased in order to spot any possible changes. Dense material, such as crystals of inorganic salts, can often be troublesome and frequently volatilize completely in a strong electron beam. It is also advisable to look for signs of granulation if an unduly thick layer of shadowing metal has been deposited or when a new shadowing metal is being tried.

Although visible artifacts are rarely produced by the electron beam, it should be remembered that molecular changes are usually if not always produced. Biological material is reduced entirely to a form of graphite in an intense beam, and milder irradiation probably causes cross-linkage in long chain compounds, leading to radical changes in chemical constitution. For a discussion of this type of radiation damage, see Reimer (1967, Chapter 9). It appears that, in consequence, most staining reactions cannot be carried out in a specimen that has been exposed to the electron beam. On the other hand, there is an appreciable reduction in the rate of damage (in a specimen of given thickness) at very high operating voltages (see Section XII.C).

2. Artifacts of Imaging

As already pointed out (Section I.C), objects that are so small as to be close to the resolution limit of the electron microscope may not be imaged in their true shape. Polygonal particles will have their corners rounded, and beyond a certain minimum size will all look circular in section. The appearance of a particle of a size approaching the resolution limit cannot, therefore, be any guide to its true shape. Two particles separated by less than the resolved distance will give even more misleading images, appearing as an hour-glass figure of increasing waist the closer they are together. It needs mentioning that these artifacts may appear in any micrograph and are possibly more of a pitfall in routine work at moderate resolution than in special investigations at very high resolving power, when the operator would in any case be examining his pictures with care.

It should be hardly necessary to say that misleading appearances may also arise from experimental accidents, such as shrinkage of the support film during or before shadowing, giving rise to ridges that might be mistaken for filaments. Granulation of metal in shadowing, or in the electron beam, contamination of solvents or distilled water, and settling of airborne particles on specimens must always be looked for; one case is on record of carbon black from a shadowing filament being mistaken for chromosomes.

C. OPTICAL CHECKS

In conclusion, it must be emphasized that electron microscopy is in no sense a method to be relied upon to give the answer to any morphological problem in microbiology without further question. It should always be employed as one research tool among many that can be brought to bear upon the problem. The evidence of the others must also be collected and compared with that of electron microscopy. Optical observations especially should be carried out, to give an independent check on artifact formation as far as their resolution allows. Ordinary light and ultraviolet methods can be very useful, but experience has shown that phase contrast microscopy gives a better, because a visibly similar, comparison. Its contrast scheme is basically the same as that of an electron image, so that the two pictures give very similar information about a biological specimen. Indeed, at the same magnification (say, $\times 1500$), it is often difficult to distinguish one from the other, if the phase contrast "halos" are faint. The information which this technique gives on the living organism makes it an invaluable supplement to electron microscopy.

It is also necessary to plead for open-mindedness in comparing electron micrographs with the results of the established optical methods. Long familiarity with them unconsciously instills the feeling that the latter show exactly "what the specimen really looks like." In the case of stained specimens obviously, but to some extent also in nonstained preparations, this cannot be true, since there is always the possibility of artifact even in optical preparations. The greater danger of artifact formation in electron microscopy should not blind one to the fact that it gives pictures that are valid within their own limitations. It is only necessary to realize that these are different limitations from those of optical methods. In short, each technique provides evidence of value, each gives a partial view of the constitution and structure of the specimen under examination, and it is the task of the research worker to correlate critically all the clues and deduce from them a coherent answer to the puzzle set by Nature.

XI. The Scanning Electron Microscope

The conventional electron microscope is a transmission instrument and can only be used for studying surface structure indirectly, via a replica. Although reflection microscopes have been built, in which an electron beam is incident at grazing angle on a solid specimen and the "reflected" beam is imaged by the train of lenses of a conventional instrument, they have had only a limited use. For biological applications, they are ruled out because of the very high illuminating intensity needed to give a bright enough image at adequate magnification, with consequent rise in specimen temperature. A much more useful technique is that of scanning microscopy, in which a finely focused electron beam (or "probe") is scanned across the surface of the specimen in a regular raster, as for television, and an image is obtained by collecting the backscattered electrons, amplifying the signal, and feeding it to a cathode ray display tube.

This type of sequential imaging had been attempted in the 1940's in several laboratories but failed to produce clear images because of lack of sensitivity of the detecting system. The signal-to-noise ratio was too poor. The effective development of the instrument began about 1950 in the Engineering Laboratory of Cambridge University, under the direction of Oatley, out of which came the prototype of the first commercial scanning microscope, the "Stereoscan." The chief features of the instrument will be outlined here, with some account of its biological applications, which have recently begun to rival the inorganic applications in volume and interest. A fuller review of the subject has been published by Oatley et al. (1965) and a comprehensive bibliography by Wells (1969).

A. Physical Principles

The scanning microscope, in its reflection mode, forms an image by means of the electrons scattered back from the surface of a specimen on which a fine electron probe is incident. The probe is typically a few hundred Ångstrom units in diameter and carries a current of between 10^{-10} and 10^{-12} A at a voltage of 10 to 40 kV. The principle of the method is shown in Fig. 16. The beam emerging from the triode electron gun is focused into a minute probe on the specimen by means of three magnetic electron lenses. It is deflected in a regular raster by scanning coils situated before the final lens, which is also fitted with a stigmator to correct its astigmatism. The specimen surface is arranged at an angle to the beam axis, to ensure a high yield of scattered electrons into the collector. The collector consists of a scintillator, to which a bias voltage may be applied, connected by a light pipe to a photomultiplier and video amplifier. The output from the latter is used to modulate a cathode ray tube in which the beam is scanned in synchronism with the raster on the specimen. For visual display a tube with a long persistence screen

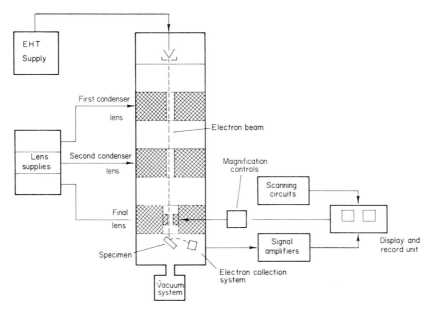

FIG. 16. Scanning electron microscope principle.

must be used, operating at a frame repetition frequency of the order of one per second owing to the weakness of the signal. For photographic recording, a short persistence, high-definition tube is used, the exposure being made in the form of a single frame scan over a time of 1 to 10 min. The magnification obtained is given by the ratio of the width of scan on the tube to the width of scan on the specimen, and is conveniently controlled by varying the latter, as indicated in Fig. 16. Since the specimen is inclined at an angle to the axis, the magnification will be different in the directions across and along its surface.

In the commercial form of the scanning microscope produced by the Cambridge Scientific Instrument Company under the name of the "Stereoscan" (Fig. 17), the three magnetic lenses produce an electron probe of minimum diameter about 200 Å. A column designed for a still higher performance, in which the first two lenses are both of short focal length, has been described by Pease and Nixon (1965). It forms a probe of minimum diameter 50–60 Å.

The resolving power of the instrument is governed by a number of factors: the shot noise in the electron image, the current density which can be focused into the probe, the properties of the electron gun, the aberrations of the electron lenses (the final lens in particular), and the properties of the collector system. The instrumental requirements are to some extent determined by the allowable scanning time, which may

be as long as 10 min. The circuits supplying the high tension to the
gun and the currents to the lenses must be highly stable over this period,
to a few parts in 10^5. The theoretical limitations and practical details
of all these electronic and electron optical factors are fully discussed
by Oatley *et al.* (1965). However, there is a further factor which finally
limits the resolution in the scanning image in practice: the penetration

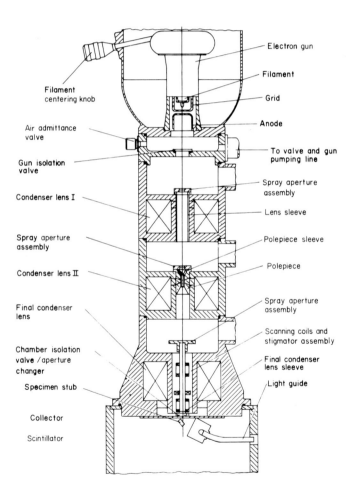

Fig. 17. Scanning electron microscope: cross section of "Stereoscan" instrument
(courtesy of Cambridge Scientific Instrument Company).

of the beam into the specimen. Scattering of the primary electrons entering it and of the secondary electrons leaving it have an important effect in broadening the area from which electrons are received by the collector; the smaller the probe diameter, the greater is the relative influence on the practical resolution. Some improvement can be obtained by reducing the beam voltage and thus the electron range in the specimen, but the signal-to-noise ratio falls at the same time. It is concluded, on present evidence, that a better resolution that 50 Å is unlikely to be achieved in the reflection mode of operation. In transmission, and using a very thin specimen, a much better ultimate resolution can be obtained (Crewe *et al.*, 1968), approaching that of the conventional electron microscope.

B. IMAGE FORMATION

The scope of practical applications of the scanning microscope is determined by the way in which it forms an image and by the factors which determine the appearance of the image, i.e., by the contrast mechanism. Its primary asset is the ability to image surfaces, and this is made particularly useful by the large depth of focus of the probe, which is much greater than that of an optical microscope at the same magnification. Even if the scanning electron microscope could not give a better resolution than the light microscope, it would still have great value because of this large depth of focus. An additional advantage is provided by the fact that the collector can receive electrons from "around a corner," owing to the action of its biasing field. In this way, images can be formed from a larger fraction of the surface of a fiber than can be directly seen, for example, and even from well down into a hole in the surface of a specimen. An important feature of the scanning microscope image is therefore the considerable impression of surface relief it presents, giving almost a three-dimensional effect.

The mechanism of contrast formation in the image is complicated, depending on the physical and electrical conditions over the surface and on whether the collector receives only the true secondary electrons (of low energy) or the reflected primary electrons (of high energy) or both. Normally, the secondary electrons are much the more numerous and also convey more information, but the reflected electrons give greater contrast.

Three factors chiefly determine the image contrast, their relative influence depending on the nature of the specimen examined.

1. Surface Topography

Variation in the local angle presented by the surface to the incident beam profoundly affects the fraction of the electrons, primary or secondary, which is collected. A difference in angle of only 1 or 2° can cause a detectable change in brightness of the image. So scanning images are similar in appearance to those seen in an optical microscope from a surface under oblique illumination. If the surface is very rough, it may be desirable to alter its aspect to the incident beam either by changing its inclination or by rotating it in azimuth.

2. Chemical Constitution

The yield of secondary electrons from a surface at constant inclination depends on the local value of the secondary emission coefficient. For most materials, however, the coefficient differs very little from unity and only rarely does this factor contribute appreciably to contrast. The yield of reflected electrons, on the other hand, depends on the atomic number of the element encountered. The back scattering coefficient varies by a factor of about 6 over the periodic table from carbon to gold, so that contrast differences will be observable whenever the specimen surface contains elements of widely differing atomic number. Aluminum and brass, for instance, show a readily visible difference in contrast.

3. Potential Variations

Since the secondary electrons leave the specimen with very low energies, their trajectories are strongly influenced by the local potential on the surface, so that the proportion reaching the collector will vary. Trajectory plots have been confirmed by experiments, which show that potential differences well below 1 V can be detected without loss of resolution. The corresponding effect of magnetic fields, in domain structures for instance, are more difficult to observe because the fields are too weak.

C. Applications

The various areas of application of the scanning electron microscope follow from the factors determining image contrast. In general, any specimen, except one that is perfectly smooth and composed of a single element, will show detail due to one or other of these factors. Since the conventional electron microscope is of little value for examining surfaces, except by means of replicas, it is clear that the scanning instrument has a wide field of use open to it, provided that a sample of the specimen can be prepared small enough and resistant enough to be mounted in the appropriate position, in vacuo. This proviso makes it difficult, but not impossible, to examine biological material. Very rugged surfaces, such as some fractures, also cause difficulty because the high regions mask the lower ones as seen from the collector. But apart from these special cases, almost any type of specimen can be examined in the scanning microscope. Insulating materials may cause trouble owing to electrostatic charging, but this can be overcome by coating them with a thin layer of aluminum or other metal by evaporation [see Boult and Brabazon (1968)]. An alternative solution is to operate at a very low beam voltage, as low as 1–2 kV, as shown by Thornley (1960). The reason appears to be that the secondary emission coefficient increases as the voltage is lowered, and a beam voltage can be found at which there is no net charging of the specimen. At such a low voltage, the image resolution deteriorates, but this may be acceptable when it is feared that coating with metal would obscure important details in the specimen.

The scanning technique has found several uses in examining transistors, fabricated microcircuits, and other semiconductor devices. It also lends itself to the study of dynamic processes such as thermal decomposition, the activation procedure of a dispenser cathode, and the sputtering of a metal surface by a beam of positive ions. Descriptions of such applications are given by Oatley et al. (1965). Details which are actually below the surface of a specimen may also be revealed by the method of specimen–current recording, in which the output signal is obtained from an electrode in contact with the base of the specimen (otherwise insulated) instead of by collecting the scattered electrons in the usual way (Czaja and Patel, 1965).

The scanning microscope is now being increasingly used for biological investigations, especially where the specimen is resistant to vacuum exposure [see the biological entries in the bibliography by Wells (1969)]. In entomology, it allows the whole of an insect, or a considerable part of it, to be imaged in a single micrograph (Fig. 18) instead of in a series of sections as in the optical microscope. It is also proving valuable for characterizing the seeds and spores of botanical species, as described by Echlin (1969). Figure 19 illustrates the amount of detail which is revealed, in this case in a seed of the Morning Glory flower. A variety of other studies of plant materials and wood are described by Heywood (1969). Other biological applications have been for studying the layer structure of the domestic hen's egg and the heart valve surface in the pig (Little, 1969). These and a number of other applications in biology are described in a review paper by Pease and Hayes (1966). Clarke and Salsbury (1967) have reported marked differences in the appearance of human red blood cells (Fig. 20) in various pathological conditions [see also Salsbury and Clarke (1967); Salsbury et al. (1968)].

A quite different biological field in which the scanning microscope has proved useful is the study of fibers. On the one hand, the structure

Fig. 18. Scanning electron micrograph of eyes of a fly (coated with 800 Å of gold). Magnification ×210 (Cambridge Scientific Instrument Company).

of paper has been the subject of a detailed investigation (Smith, 1959), and on the other hand, both manmade and natural individual fibers have been examined, as well as the spinnerets through which the former are drawn. Boyde and Barber (1968) have been concerned with the study of ciliated and other soft tissues, the preparation of which for scanning microscopy has been discussed in some detail by Boyde and Wood (1969).

The scanning microscope is also useful for examining sections of tissues in transmission. Since no electron lens is needed between specimen and detector, the image resolution is not limited by chromatic aberration as it is in the conventional electron microscope. With a special adaptor fitted to a standard scanning microscope, Swift and Brown (1970) have examined sections 0.5 μm thick at a resolution of 200–300 Å and 1 μm sections at about 500 Å. When working at magnifications more than

20,000×, however, care has to be taken not to burn up the specimen. In a specially designed scanning transmission microscope, Crewe has been able to attain a resolution better than 10 Å on very thin sections (Crewe and Wall, 1970), and to identify uranium atoms by means of an image processing system (Crewe *et al.* 1970).

The observation of biological material in the wet state is obviously beset with much greater difficulties. Attempts have been made by Thornley (1960) and by Pease and Hayes (1966) to do so, either by covering the specimen with a cellulose membrane or by enclosing it in a specially designed cell. Some images were obtained at low resolution, but the possibility of further improvement appears doubtful.

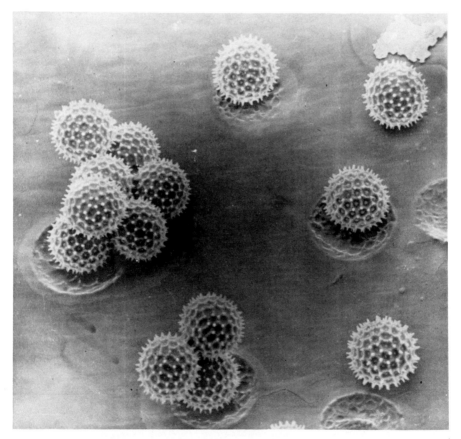

FIG. 19. Scanning electron micrograph of pollen grains of *Ipomoea Purpurea* (coated with gold-palladium). Magnification ×25. (Dr. P. Echlin, Botany School, University of Cambridge.)

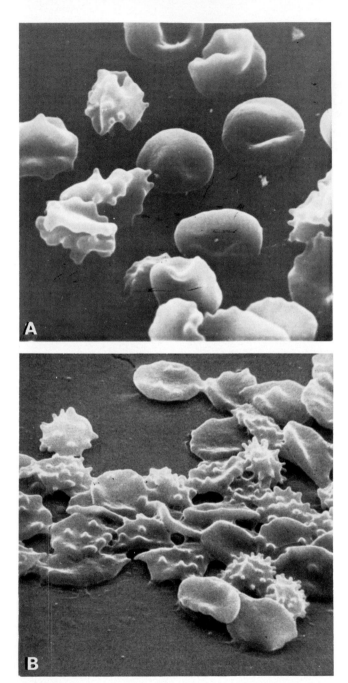

Fig. 20. Scanning electron micrograph of human red blood cells treated with cold agglutinin. (A) After 2 min at 4°C; ×3500. (B) After 30 min at 4°C; ×2400. (Drs. A. J. Salsbury, J. A. Clarke, and W. S. Shand, St. Bartholomew's Hospital, London.)

The prospects for scanning microscopy in biology and medicine are authoritatively discussed in a recent survey article by Hayes and Pease (1968). For other biological applications, see the symposium proceedings edited by Pfefferkorn (1968).

XII. The High Voltage Electron Microscope

A variety of electron microscopes are now available with maximum operating voltages above 100 kV; for instance, 200 kV, 650 kV, or 1 MV. Those in the lower range (up to 200 kV) are scaled up versions of the conventional instrument, the main difference being the use of a two-stage electron gun instead of a single stage of acceleration. The very high voltage microscopes, however, are radically different in construction. Although they were originally produced for use in metallurgical research, they are now finding increasing applications in biology, particularly for examining thick sections.

The main advantages of operating at such high voltages are as follows:

(a) Increased specimen thickness, for given image resolution, roughly in proportion to voltage (with constant angular aperture).

(b) Higher resolution from specimens of near normal thickness, i.e., which give an image of poor resolution at 100 kV.

(c) Less damage due to heat and ionization in a specimen of given thickness.

(d) Higher beam brightness.

(e) Potentially higher ultimate resolution on a very thin specimen.

There are also some disadvantages:

(f) Lower contrast in specimens of normal thickness.

(g) Greater radiation damage at the atomic and molecular level (bond breakage, cross-linking).

(h) Poorer response of viewing screens and photographic emulsions.

(i) Greater X-radiation hazards.

(j) Greater cost, approximately linear with maximum voltage.

Detailed discussion of these factors will be found in review articles by Dupouy (1968) and Cosslett (1969b). The two points of major importance, penetration and energy loss, require some further consideration here.

A. SPECIMEN THICKNESS: CONTRAST AND ULTIMATE RESOLUTION

The image in an electron microscope is formed by those electrons which penetrate the specimen and are collected in the aperture in the back focal plane of the objective. Since the interaction of electrons with a specimen gets less as the voltage is increased (i.e., the scattering cross-section of atoms decreases), not only do more electrons penetrate the specimen but a greater proportion of these are funneled into an aperture of given size. So the image contrast falls with voltage, for a specimen of given thickness. To regain it, one must stain the material more heavily or increase the thickness, provided that the feature of interest is thicker than the section first studied.

The variation of usable thickness with voltage depends strongly on the nature of the specimen and on the angular aperture of the objective lens, which is assumed to remain constant. For constant fractional transmission through a carbonaceous specimen, i.e., for constant contrast, the specimen thickness increases almost linearly with operating voltage when an aperture of usual size is used (5×10^{-3}). With a smaller aperture or for elements of high atomic number, the increase is less marked. For carbon, on the basis of the best available treatment of plural scattering, it can be calculated that, at 500 kV, a 2% contrast would be given by a thickness of 2.5 μm when using an aperture of 5×10^{-3}, but by only 1.2 μm with $\alpha_0 = 1 \times 10^{-3}$ (Cosslett, 1969b). Comparison of this prediction with experiment has shown close agreement (Curtis, 1968), but it has meaning for practical microscopy only if image resolution is also taken into account, which involves a consideration of energy loss and its effect on chromatic aberration.

Before doing so, it should be emphasized again that, whereas practical resolution is limited by chromatic aberration in specimens of average thickness, the limit to the ultimate resolving power of an electron microscope is set by a combination of spherical aberration and diffraction (see Section IV.D). So, for a very thin test specimen and in the absence of disturbing extraneous circumstances, the resolving power will improve as operating voltage is raised because of the shorter electron wavelength. This increase is very slow, however, and amounts only to a factor of two between 100 kV and 1 MV so long as conventional lenses are used. Since it is increasingly difficult to exclude electrical and mechanical disturbances at high voltages, very high resolution projects are for the most part aiming to use only conventional (100 kV) or somewhat higher (200–300 kV) voltages, not the very high values (up to 3 MV) now becoming available for studying penetration.

B. Energy Loss in the Specimen: Chromatic Aberration and Practical Resolution

The energy lost by the electron beam (ΔV) in penetrating the specimen determines the amount of chromatic aberration in the image (Section III.A). The loss depends on the nature of the specimen (atomic number and density), on its thickness, and on the speed of the electrons, i.e., the beam voltage. The thicker the specimen, the greater the energy loss, in direct proportion over the range of thicknesses involved in electron microscopy. But the faster the electrons, the less chance they have of losing energy in a layer of given thickness. So there is less energy loss per unit mass-thickness at high than at low voltage, by a factor of about 2.5 between 100 kV and 1 MV. The relevant quantity for chromatic aberration, however, is the fractional energy loss $\Delta V/V_0$, which decreases even more because of the increase in V_0.

The other factors involved (the chromatic coefficient and the aperture of the objective lens) change very little. Taking relativistic effects into account, the chromatic aberration decreases by a factor of 10 to 15 between 100 kV and 1 MV, *with a specimen of given thickness*. Hence the great improvement in image definition, as distinct from contrast, when ultratome sections of normal thickness are viewed at high voltage. Correspondingly, the old rule that resolution is about $\frac{1}{40}$ the specimen thickness at 100 kV (Cosslett, 1956) becomes something like $\frac{1}{400}$ to $\frac{1}{600}$ of the thickness at 1 MV.

Since energy loss is directly proportional to thickness, this means that specimen thickness can be increased considerably at higher voltages, for a *given image resolution*. So a section 1 μm thick should give nearly as good a resolution at 1 MV as a 500 Å section at 100 kV. The limited experience so far available suggests that the resolution is in fact even better at the higher voltage. Comparison with the figures given for contrast in the previous sections shows that a 1 μm section would give much less than 1% contrast at 1 MV. The need for heavy staining even of thick sections in high voltage microscopy is thereby underlined.

C. RADIATION DAMAGE

If in fact the attempt is made to look at specimens much thicker than 1 μm at 1 MV, it is found difficult to avoid severe damage and even complete disruption of a section. This is a result to be expected: Although the energy loss *per unit mass thickness* is 2.5 times less at 1 MV than at 100 kV, the section itself is now 25 to 50 times thicker than would be used for 100 kV. So the absolute amount of energy loss is now 10 to 20 times greater. The situation can be improved by sandwiching the section between two support grids, to provide better thermal conductivity, or by using an image intensifier. But experience so far indicates that 2 μm is about the maximum specimen thickness that can be continuously viewed at 1 MV in the absence of these measures, and then only at moderate magnification.

Even so, the reduction in radiation damage in given thickness proves to be most valuable in polymer research, where the material is often extremely sensitive to the electron beam. It has been shown by

Kobayashi and Sakaoku (1965) that the dose required (intensity $\times$ time) to destroy crystallinity in a number of different polymers is almost directly proportional to operating voltage. So it was possible to observe samples for 5 times longer at 500 kV than at 100 kV before degradation was complete, as judged from the disappearance of the diffraction pattern.

In addition to this type of damage, which is mainly due to ionization and possibly also to temperature rise, ejection of atoms can also occur when an electron makes a direct hit on an atom and transfers to it enough energy to overcome the bonds binding it in the molecule (or in a lattice, in the case of a metal). Since an atom is so much heavier than an electron, the latter must have a high velocity for enough energy to be transferred in the collision. This threshold velocity corresponds to a beam voltage of about 500 kV in the case of copper, as has now been experimentally confirmed by Makin (1968). For carbon, the threshold is of the order of 30 kV so that damage of this nature is not negligible in the conventional electron microscope. At higher voltages, the cross-section for the process increases rapidly, and considerable numbers of carbon atoms are likely to be knocked out of any organic preparation during prolonged observation, especially with the beam currents required for high magnifications. Direct measurement of the rate of damage is very difficult, in the absence of atomic resolving power, since the vacant sites do not aggregate into larger units as in a metal lattice. However, it may be possible to distinguish atomic ejection from other forms of specimen degradation by a careful series of experiments at different voltages. It is already clear, unfortunately, that above the threshold voltage this effect may make it impossible to image the details of molecular structure with any certainty, if and when an electron microscope capable of resolving individual atoms becomes available (Cosslett, 1970b).

D. CONSTRUCTION AND OPERATION

A high voltage microscope is a scaled-up version of the conventional model, having the same controls and being operated for the most part in an identical manner. An obvious difference is the more massive construction of the column, to provide stronger magnetic lenses but also to give adequate protection from X-rays. A more basic difference is that an accelerator now has to be installed between the electron gun and the microscope proper (Fig. 21). The maximum voltage at which a single stage gun can be operated is about 200 kV, so that a 1 MV microscope must have 7 or 8 stages, for safety. Commercial high voltage microscopes mostly have 10, 15, or even 25 stages to reduce the risk of electrical breakdown.

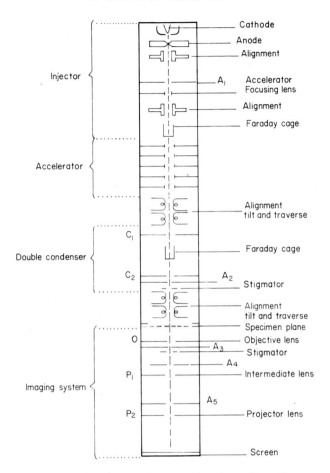

FIG. 21. High voltage microscope in schematic section.

For the operator, the only complication is that such a system cannot be switched on directly at full voltage but has to be "conditioned." This is a procedure by which the voltage is raised stepwise to the full operating value over a period of tens of minutes at the beginning of the day's work. One can then carry out microscopy at any lower voltage for several hours. Otherwise the high voltage microscope is entirely similar in operation to any conventional instrument, with double condenser, three-stage imaging, and usually a separate diffraction lens as well.

As a consequence of the reduced rate of energy loss at high voltages, the response of viewing screens and of photographic emulsions falls with voltage. Screen brightness can be restored by having a thicker layer of phosphor, at the cost of some loss in resolution. So it is best to have

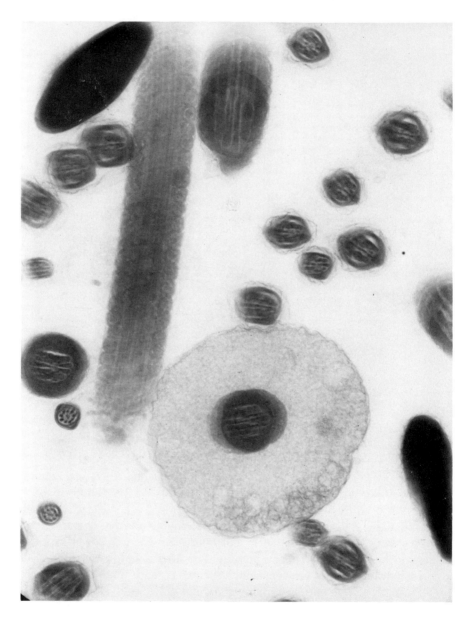

FIG. 22. Section about 1 μm thick through spermatozoa in epidydimis of the hamster. One of a stereopair taken with $\pm15°$ tilt from horizontal at 1000 kV. Magnification $\times20,000$. (Prof. K. R. Porter, Harvard University, and A.E.I. Scientific Instruments, Harlow.)

a thick screen for routine observation and a thin screen for focusing the image. Photographic emulsions are slower by a factor of 2 to 4 between 100 and 600 kV, according to the type of plate or film, from experiments in the author's laboratory (Jones and Cosslett, 1970), but it is likely that special emulsions for high voltage microscopy can be developed.

E. APPLICATIONS IN BIOLOGY AND MEDICINE

The value of high voltages for viewing thick sections and isolated tissue components or microorganisms is now being actively explored. A review of some of the first results has been published by Cosslett

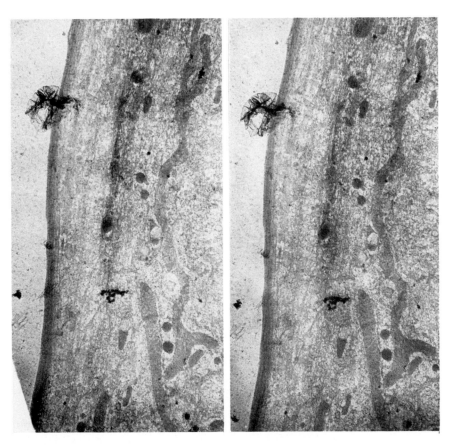

FIG. 23. Human skin fibroblast (intact and unsectioned); grown in culture on filmed grid, fixed Karnovsky and 2% O_sO_4 in phosphate buffer, stained uranyl-magnesium-acetate and lead citrate, and dried by Anderson critical point method. Stereopair taken with ±5° tilt at 1000 kV. Magnification ×15,000. (Prof. H. Ris, Wisconsin University, and Prof. G. Dupouy, Toulouse).

(1969b), and recently some more systematic observations have been reported by Hama and Porter (1968), Nagata *et al.* (1969), and Ris (1969).

The initial conclusions are that sections of thickness up to about 1 μm can be readily imaged at 500 kV and up to 2 μm at 1 MV (Fig. 22). Even greater thicknesses will give informative images, but at lower resolution and with greater background. For such specimens, the darkfield technique devised by Dupouy (1968) and Dupouy *et al.* (1969) gives better results (see also Dubochet *et al.*, 1970). Clearly the interpretation of detail in such thick sections is greatly facilitated by stereo-viewing (Fig. 23). Open-work structures, as in the retina or chromosomes, are immediate targets for high voltage microscopy, to avoid the labor of cutting large numbers of thin sections and the difficulty of reconstructing their three-dimensional architecture by superimposing them.

As appropriate preparation and staining techniques are developed for thick sections, as they had to be for ultra-thin sections, and as the art of stereogrammetry becomes more familiar, there can be little doubt that the high voltage electron microscope will prove to be as valuable to the biologist as it already is to the metallurgist. A particularly inviting prospect is the ability to look at sections a micron or so thick, as prepared by standard cytological methods, alternatively in the optical and the electron microscope.

XIII. Other Types of Electron Microscope

A great variety of electron microscopes working on entirely different principles from those considered above have been designed and built experimentally. One broad category involves the imaging of a surface in terms of the electrons emitted from it by bombardment with ions, electrons, or light (visible or ultraviolet), or by thermionic emission at high temperatures. Into a quite different category fall instruments which have a conventional electron source, but make use of special means of imaging such as an electron mirror or prism. Energy analyzing microscopes, which form images from electrons that have lost a definite amount of energy (or none) in the specimen, come into this second category.

It would be impracticable even to outline the mode of operation of all these types here, and for the most part they are not available to the ordinary microscopist. It may be more useful to mention briefly three instruments which are now beginning to become available commercially, and to which access may therefore be obtained for trying out biological applications: the analyzing microscope, the emission microscope, and the field electron (and field ion) microscope.

A. The Analyzing Electron Microscope

Instruments are now appearing that combine the capabilities of a normal electron microscope with those of the electron probe microanalyzer, in which point analysis of chemical elements is carried out by X-ray spectrometry. In some an analyzing system is fitted to an existing type of electron microscope (Siemens, JEOL, Hitachi), in others an electron microscope is attached to an existing type of microanalyzer (Cameca, ARL, MAC). In either case one or another capability is appreciably below optimum performance. At least one completely new design has now appeared (AEI) which attempts to combine both high microscopical resolution with high analytical accuracy.

These hybrid instruments have so far been used mainly for metallurgical problems on thin films or extraction replicas, but they have obvious applications in biology and medicine, especially for the quantitative estimation of mineralization or the detection of foreign bodies. The present state of their performance, and examples of their usefulness in biology, are described in Chapter 1 of this volume (Hall, 1970).

B. The Emission Microscope

Several types of emission microscope have been devised, but the only type commercially available is the "Metioscope" (Balzers A. G., Liechtenstein), and a similar instrument made by JEOL. In these microscopes, the specimen surface is imaged by means of the secondary electrons emitted from it when bombarded by an ion beam. The imaging system is almost conventional, consisting of an electrostatic or magnetic immersion objective, followed by two stages of magnification by magnetic projectors.

From the nature of its electron source, this type of microscope is primarily of value for studying metallic and other inorganic specimens. It might find some biological applications, however, perhaps in combination with microincineration techniques. A review of the principles and practical use of the various types of emission microscope has been published by Möllenstedt and Lenz (1963).

C. The Field Electron and Field Ion Microscopes

Field emission microscopes also produce images of the emitting surface, which in this case is an extremely fine metal tip with a radius in the order of 1000 Å. In the field electron microscope, the image is formed by electrons extracted from the tip by a very strong positive electrostatic field. It has a resolution of about 15 Å and requires an ultra-high vacuum. The field ion microscope has the polarity reversed and contains a rare gas (helium, neon) or hydrogen at moderately low pressure. The image is formed by gas molecules which become ionized on the tip surface and are then repelled in almost straight lines on to the fluorescent viewing screen. So a projection image is formed of the emitting sites on the tip at a magnification given by the ratio of the tip-screen distance to the tip radius, which can be of the order of 1 million times.

The field ion microscope has a resolution of a few Ångstrom units, so that individual atoms on a tungsten or platinum tip are readily resolved. It has been used for adsorption and desorption studies, in which other elements or compounds are deposited onto a tip by evaporation. Observations have also been made of defects in metals and of radiation damage. Several attempts have been made to image molecules, but none has yet been successful. If an adequate technique could be found, the resolution should certainly be adequate for showing the atomic structure of macromolecules, unless they are disrupted in the strong electrostatic field at the tip.

Although the principle of the instrument is simple and the images easy to record, the technique as a whole is somewhat demanding because the tip has to be kept at liquid hydrogen temperature. A full account of the subject has recently been published by Müller and Tsong (1969) and a shorter review by Brandon (1968).

REFERENCES

Abermann, R., and Bachmann, L. (1969). *Naturwissenschaften* **56**, 324.
Agar, A. W. (1960). *Brit. J. Appl. Phys.* **11**, 185.
Agar, A. W. (1965). *In* "Techniques for Electron Microscopy" (D. H. Kay, ed.), 2nd ed., pp. 1–42. Blackwell, Oxford.
Anderson, T. F. (1951). *Trans. N.Y. Acad. Sci.* **13**, 130.
Beer, M. (1965). *In* "Quantitative Electron Microscopy" (G. F. Bahr and E. H. Zeitler, eds.), pp. 282–287. Williams & Wilkins, Baltimore, Maryland.
Boult, E. H., and Brabazon, E. J. (1968). *J. Sci. Instrum.* **1**, 565.
Boyde, A., and Barber, V. C. (1968). Electron Microscopy 1968, *Proc. 4th Euro. Reg. Conf. Elect. Microsc., 1968*, **II**, p. 51. Tipografia Poliglotta Vaticana, Rome.
Boyde, A. and Wood, C. (1969). *J. Microsc. (London)* **90**, 221.
Bradley, D. E. (1954). *Brit. J. Appl. Phys.* **5**, 65, 96.
Bradley, D. E. (1959). *Brit. J. Appl. Phys.* **10**, 198.
Bradley, D. E. (1965a). *In* "Techniques for Electron Microscopy" (D. H. Kay, ed.), 2nd ed., pp. 58–74. Blackwell, Oxford.
Bradley, D. E. (1965b). *In* "Techniques for Electron Microscopy" (D. H. Kay, ed.), 2nd ed., pp. 96–152. Blackwell, Oxford.
Brandon, D. G. (1968). *Advan. Opt. Electron Microsc.* **2**, 343–404.
Bremmer, H., and Van Dorsten, A. C. (1969). *Z. Angew. Phys.* **27**, 219.
Challice, C. E. (1950). *Proc. Phys. Soc. London Sect. B* **63**, 59.

Chapman, J. (1962). *J. Sci. Instrum.* **39**, 273.

Chescoe, D., and Agar, A. W. (1966). *J. Microsc. (Paris)* **5**, 91.

Clarke, J. A., and Salsbury, A. J. (1967). *Nature (London)* **215**, 402.

Cosslett, V. E. (1951). "Practical Electron Microscopy." Butterworths, London.

Cosslett, V. E. (1956). *Brit. J. Appl. Phys.* **7**, 10.

Cosslett, V. E. (1969a). *Z. Angew. Phys.* **27**, 138.

Cosslett, V. E. (1969b). *Quart. Rev. Biophys.* **2**, 95.

Cosslett, V. E. (1970a). *Proc. I.E.E.* **117**, 1489.

Cosslett, V. E. (1970b). Berichte der Bunsen-gesellschaft (in press).

Crewe, A. V. and Wall, J. (1970). *J. Mol. Biol.* **48**, 375.

Crewe, A. V., Wall, J., and Langmore, J. (1970). *Science* **168**, 1338.

Crewe, A. V., Wall, J., and Welter, L. M. (1968). *J. Appl. Phys.* **39**, 5861.

Curtis, G. H. (1968). Thesis, Cambridge University.

Curtis, G. H., and Ferrier, R. P. (1969). *Brit. J. Appl. Phys.* **2**, 1035.

Czaya, W., and Patel, J. R. (1965). *J. Appl. Phys.* **36**, 1476.

Dolby, R. M., and Swift, D. W. (1960). *Proc. 2nd Eur. Reg. Conf. Electron Microsc., 1960,* **1**, p. 114. Ned. Vereniging voor Electronenmicroscopie, Delft.

Dougall, J. (1969). *J. Microsc. (Paris)* **8**, 10a (abstract).

Drummond, D. G. (1950). "The Practice of Electron Microscopy." Royal Microscopical Soc., London.

Dubochet, J., Ducommun, M., Zollinger, M., and Kellenberger, E. (1970). *J. Ultrastruct. Res.* (in press).

Dupouy, G. (1968). *Advan. Opt. Electron Microsc.* **2**, 167–250.

Dupouy, G., Durrieu, L., and Perrier, F. (1960). *C. R. Acad. Sci.* **251**, 2836.

Dupouy, G., Perrier, F., Enjalbert, L., Lapchine, L., and Verdier, P. (1969). *C. R. Acad. Sci. Ser. B.* **268**, 1341.

Echlin, P. (1969). *J. Microsc. (London)* **88**, 407.

Escaig, J. (1966). *C. R. Acad. Sci.* **262**, 538.

Escaig, J., and Sella, C. (1969). *C. R. Acad. Sci. Ser. B* **268**, 532.

Federation Proc. (1965). *Proc. Fed. Amer. Soc. Exp. Biol.* **24**, 2.

Ferrier, R. P. (1969). *Advan. Opt. Electron Microsc.* **3**, 155–218.

Forth, H. J., and Loebe, W. (1968). Electron Microscopy 1968, *Proc. 4th Eur. Reg. Conf. Elect. Microsc., 1968,* **1**, p. 231. Tipografia Poliglotta Vaticana, Rome.

Glaser, W. (1952). "Grundlagen der Elektronenoptik." Springer, Vienna.

Glaser, W., and Grümm, H. (1952). *Oesterr. Ing. Arch.* **6**, 360.

Glauert, A. M. (1965a). *In* "Techniques for Electron Microscopy" (D. H. Kay, ed.), 2nd ed., pp. 166–253. Blackwell, Oxford.

Glauert, A. M. (1965b). *In* "Techniques for Electron Microscopy" (D. H. Kay, ed.), 2nd ed., pp. 254–310. Blackwell, Oxford.

Grivet, P. (1965). "Electron Optics." Pergamon, New York.

Haggis, G. H. (1970). *J. Microsc. (London)* **91**, 221.

Haine, M. E. (1961). "The Electron Microscope." Spon, London.

Haine, M. E., and Einstein, P. A. (1952). *Brit. J. Appl. Phys.* **3**, 40.

Haine, M. E., Einstein, P. A., and Borcherds, P. H. (1958). *Brit. J. Appl. Phys.* **9**, 482.

Hall, C. E. (1966). "Introduction to Electron Microscopy." McGraw-Hill, New York.

Hall, T. A. (1970). This volume, Chapter 3.

Hama, K., and Porter, K. R. (1968). *J. Microsc. (Paris)* **8**, 149.

Hanszen, K. J. (1962). *Proc. 5th Int. Congr. Electron Microsc., Philadelphia, 1962,* **1**, p. KK-11. Academic Press, New York.

Harris, W. J. (1962). *Nature (London)* **196,** 499.

Harrison, G., and Philpott, D. (1968). *RCA Sci. Instrum. News* **13,** 6.

Hartman, R. E., and Hartman, R. S. (1965). *In* "Quantitative Electron Microscopy" (G. F. Bahr and E. H. Zeitler, eds.), pp. 409–416. Williams & Wilkins, Baltimore, Maryland.

Haydon, G. B. (1969). *J. Microsc. (London)* **89,** 251.

Haydon, G. B., and Taylor, D. A. (1966). *J. Roy. Microsc. Soc.* **85,** 305.

Hayes, T. L., and Pease, R. F. W. (1968). *Advan. Biol. Med. Phys.* **12,** 85–137.

Heide, H. G. (1958). *Proc. 4th Int. Congr. Electron Microsc., Berlin, 1958,* **1,** p. 87. Springer, Berlin.

Heide, H. G. (1962). *J. Cell Biol.* **13,** 147.

Heide, H. G. (1965). *In* "Quantitative Electron Microscopy" (G. F. Bahr and E. H. Zeitler, eds.), pp. 396–408. Williams & Wilkins, Baltimore, Maryland

Heinmets, F. (1949). *J. Appl. Phys.* **20,** 384.

Henderson, W. J. (1969). *J. Microsc. (London)* **89,** 369.

Heywood, V. H. (1969). *Micron* **1,** 1.

Hibi, T. (1956). *J. Electronmicrosc. (Japan)* **4,** 10.

Hillier, J., and Baker, R. F. (1945). *J. Appl. Phys.* **16,** 469.

Hillier, J., and Ramberg, E. G. (1947). *J. Appl. Phys.* **18,** 48.

Hiziya, K., and Ito, T. (1958). *J. Electronmicrosc. (Japan)* **6,** 4.

Horne, R. W. (1965a). *In* "Techniques for Electron Microscopy" (D. H. Kay, ed.), 2nd ed., pp. 311–327. Blackwell, Oxford.

Horne, R. W. (1965b). *In* "Techniques for Electron Microscopy" (D. H. Kay, ed.), 2nd ed., pp. 328–355. Blackwell, Oxford.

Houwink, A. L., and van Iterson, W. (1950). *Biochim. Biophys. Acta* **5,** 10.

Hudson, B. and Makin, M. J. (1970). *J. Sci. Instrum.* **3,** 311.

Jones, G. L. and Cosslett, V. E. (1970). *Proc. 7th Int. Congr. Elect. Microsc., Grenoble, 1970,* **1,** p. 349.

Juniper, B. E., Cox, G. C., Gilchrist, A. J., and Williams, P. K. (1969). "Techniques for Plant Electron Microscopy." Blackwell, Oxford.

Kay, D. H., ed. (1965). "Techniques for Electron Microscopy." Blackwell, Oxford.

Kleinschmidt, A. K., Lang, D., Jacherts, D., and Zahn, R. K. (1962). *Biochim. Biophys. Acta* **61,** 857.

Kobayashi, K., and Sakaoku, K. (1965). *In* "Quantitative Electron Microscopy" (G. F. Bahr and E. H. Zeitler, eds.), pp. 359–376. Williams & Wilkins, Baltimore, Maryland.

Kölbel, H. (1968). *Z. Wiss. Mikrosk.* **69,** 1.

Labaw, L. W., and Rossmann, M. (1969). *J. Ultrastruct. Res.* **27,** 105.

Lippert, W. (1960). *Proc. 2nd Eur. Reg. Conf. Electron Microsc. 1960,* **2,** p. 682. Ned. Vereniging voor Electronenmicroscopie, Delft.

Little, K. (1969). Unpublished work.

Lucas, J. H. (1968). Electron Microscopy 1968, *Proc. 4th Eur. Reg. Conf. Elect. Microsc., 1968,* **1,** p. 247. Tipografia Poliglotta Vaticana, Rome.

Mahl, H., and Weitsch, W. (1962). *Z. Naturforsch. A* **17,** 146.

Makin, M. J. (1968). *Phil. Mag.* **18,** 637.

Marton, L. (1945). *J. Appl. Phys.* **16,** 131.

Metherell, A. J. F. (1970). *Advan. Opt. Electron Microsc.* **4,** 263.

Millonig, G., and Marinozzi, V. (1968). *Advan. Opt. Electron Microsc.* **2,** 251–342.

Möllenstedt, G., and Lenz, F. (1963). *Advan. Electron. Electron Phys.* **18,** 251–329.

Moor, H., and Mühlethaler, K. (1963). *J. Cell Biol.* **17,** 609.

Müller, E. W., and Tsong, T. T. (1969). "Field Ion Microscopy." Amer. Elsevier, New York.

Nagata, F., Hama, K., and Porter, K. R. (1969). *J. Electronmicrosc.* (*Japan*) **18**, 106.

Nekrassova, T. A., and Stoyanova, I. G. (1960). *Dokl. Akad. Nauk SSSR* **134**, 467.

Oatley, C. W., Nixon, W. C., and Pease, R. F. W. (1965). *Advan. Electron. Electron Phys.* **21**, 181–247.

Pease, D. C. (1964). "Histological Techniques for Electron Microscopy." Academic Press, New York.

Pease, R. F. W., and Hayes, T. L. (1966). *Proc. 6th Int. Congr. Electron Microsc., Kyoto, 1966*, **2**, p. 19. Maruzen, Tokyo.

Pease, R. F. W., and Nixon, W. C. (1965). *J. Sci. Instrum.* **42**, 81.

Pfefferkorn, G., ed. (1968). "Beiträge zur Elektronenmikroskopischen Direktab-bildung von Oberflächen," Vol. I. Remy, Münster.

Philips, R. (1960). *Brit. J. Appl. Phys.* **11**, 504.

Philpott, D. E. (1951). *J. Appl. Phys.* **22**, 982.

Pilod, P. and Sonier, F. (1968). *J. Microsc.* (*Paris*) **7**, 313.

Reimer, L. (1967). "Elektronenmikroskopische Untersuchungs-und Präparations-methoden." Springer, Berlin.

Ris, H. (1969). *J. Microsc.* (*Paris*) **8**, 761.

Ruska, E. (1966). *Advan. Opt. Electron Microsc.* **1**, 116–179.

Sakaki, Y., and Möllenstedt, G. (1956). *Optik* (*Stuttgart*) **13**, 193.

Salsbury, A. J., and Clarke, J. A. (1967). *J. Clin. Pathol.* **20**, 603.

Salsbury, A. J., Clarke, J. A., and Shand, W. S. (1968). *Clin. Exper. Immunol.* **3**, 313.

Sjöstrand, F. S. (1955). *Sci. Tools* **2**, 25.

Sjöstrand, F. S. (1957). *Proc. 1st Eur. Reg. Conf. Electron Microsc. Stockholm 1956*, p. 120. Almqvist and Wiksell, Stockholm.

Sjöstrand, F. S. (1968). "Electron Microscopy of Cells and Tissues." Academic Press, New York.

Smith, K. C. A. (1959). Pulp and Paper Mag. Canada **60**, T366.

Steigerwald, K. H. (1949). *Optik* (*Stuttgart*) **5**, 469.

Stoyanova, I. G. (1958). *Proc. 4th Int. Congr. Electron Microsc. Berlin, 1958*, **1**, p. 82. Springer, Berlin.

Swift, D. W., and Nixon, W. C. (1962). *Brit. J. Appl. Phys.* **13**, 288.

Swift, J. A., and Brown, A. C. (1970). *Proc. Scanning Elect. Microsc. Sympos. Chicago*, p. 113. Illinois Inst. Technology, Chicago.

Symp. Cytochem. Progr. Electron Microsc., 1962. Roy. Microscopical Soc., London.

Thon, F. (1965). Siemens Eg. Bericht 1/7 (quoted by Reimer (1967), p. 11).

Thon, F. (1967). *Siemens Rev.* **34**, 13, 1967.

Thornley, R. F. M. (1960). *Proc. 2nd Eur. Reg. Conf., Electron Microsc., 1960*, **1**, p. 173. Ned. Vereniging voor Electronenmicroscopie, Delft.

Tothill, F. C. S. M., Nixon, W. C., and Grigson, C. W. B. (1968). Electron Microscopy 1968, *Proc. 4th Eur. Reg. Conf. Elect. Microsc. 1968*, **1**, p. 229. Tipografia Poliglotta Vaticana, Rome.

Towe, K. M. (1969). *J. Microsc.* (*London*) **90**, 279.

Trillat, J. J., and Oketani, S. (1950) *J. Rech. Cent. Nat. Rech. Sci. Lab. Bellevue* (*Paris*) **2**, 189.

Underwood, E. E. (1969). *J. Microsc.* (*London*) **89**, 161.

Valentine, R. C. (1966). *Advan. Opt. Electron Microsc.* **1**, 180–203.

Van Dorsten, A. C., and Premsela, H. F. (1966). *Proc. 6th Int. Congr. Electron Microsc., Kyoto, 1966,* **1,** p. 21. Maruzen, Tokyo.

Wachtel, A. W., Gettner, M. E., and Ornstein, L. (1966). *In* "Physical Techniques in Biological Research" (A. W. Pollister, ed.), Vol. IIIA, pp. 173–250. Academic Press, New York.

Weibel, E. R. (1969). *Int. Rev. Cytol.* **26,** 235.

Wells, O. C. (1969). *Symp. Electron, Ion and Laser Beam Technol. 10th, Gaithersburg,* pp. 509–541. San Francisco Press, Inc., San Francisco.

Williams, R. C., and Wyckoff, R. W. G. (1944). *J. Appl. Phys.* **15,** 712.

Willis, R. A. (1969). *Med. Biol. Illustration* **19,** 82.

Wrigley, N. G. (1968). *J. Ultrastruct. Res.* **24,** 454.

Zingsheim, H. P., Abermann, R., and Bachmann, L. (1970). *J. Sci. Instrum.* **3,** 39.

CHAPTER 3

The Microprobe Assay of Chemical Elements

T. A. HALL

I. Introduction

The phrase *microprobe analysis* is usually understood as a short name for *electron-probe X-ray microanalysis,* and it is this technique which is almost exclusively the subject of the present chapter, although in Section IX we shall also consider briefly the potential of certain other microprobe techniques which have not yet been applied so extensively in biology, namely X-ray probe analysis, electron-probe electron-energy-loss analysis, ion-probe analysis, laser-probe analysis, and proton-probe X-ray microanalysis.

Electron-probe X-ray microanalysis is a means of identifying chemical elements and measuring their amounts *in situ* within microvolumes at the surface of thick specimens or within thin specimens. The specimen is exposed to a beam of electrons focused into a small spot, usually 1 μm or less in diameter at the surface (see Fig. 1). Characteristic X rays are generated within the volume reached by the electrons, usually limited to a few cubic microns or less. The constituent chemical elements within the microvolume can be identified by spectroscopic analysis of the X rays, usually by means of diffracting crystals which can be

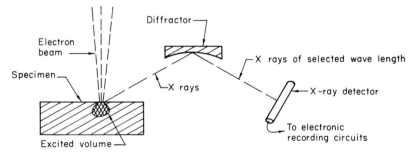

Fig. 1. The principle of electron-probe X-ray microanalysis.

set to the suitable angle to reflect a given characteristic rediation while discriminating quite effectively against all other X rays. The X-ray detector in Fig. 1 is usually a gas-filled proportional counter.

The salient features of the method are:

(1) The specimen must be analyzed in the vacuum of the electron-optical column.

(2) Analysis is conducted *in situ* with a spatial resolution of the order of 1 μm.

(3) The X-ray spectra of the elements are much simpler than the optical spectra, and the characteristic wavelengths depend on atomic number in a simple and orderly way. Consequently, when diffracting crystals are used, interference between characteristic X-ray lines hardly ever occurs.

(4) The chemical elements are analyzed as such, since the characteristic X-ray wavelengths and excitation probabilities are virtually independent of the state of chemical binding. (However, at the long-wavelength end of the X-ray spectrum, wavelengths $\lambda > 5$ Å approximately, slight "chemical shifts" are detectable with high quality spectrometers, so that the diffracting crystal should be peaked separately for each specimen or standard.)

(5) No chemical reactions or stains are necessary for the sake of the elemental analysis itself (although stains may sometimes be needed in order to *see* the important features in the specimen, so as to place the probe where desired).

(6) The method is "nondestructive," at least on a 1-μm scale. (See Section VIII for qualifications.)

(7) Suitable diffractors and detectors are now routinely available for the assay of all elements of atomic number $Z > 5$. (Diffractors and detectors for the range $2 < Z < 6$ have been developed but are not yet routinely available.) Sensitivity, very roughly speaking, is uniform for the entire range $Z > 8$, but not as good for $Z \leq 8$.

(8) The method can be used for quantitative analysis (Section VI). In thick specimens, one can measure local elemental weight-fractions. In thin specimens, one can measure local elemental weight-fractions, local spatial concentrations (i.e., amount of element per unit volume), and local total mass per unit area.

(9) For elements of atomic number $Z > 8$ in biological tissues, the minimum measurable amount of an element is commonly of the order of 10^{-16} g, and the minimum measurable *local* weight-fraction is commonly of the order of 10^{-4}, i.e., 100 ppm. Of course, even in a specimen where the mean bulk weight-fraction of an element is far below the limit of detectability, it may still be possible to assay spots where the

concentration is locally elevated above the limit. The weight-fraction limit refers to the specimen in the form presented to the probe; thus it is a dry-weight value for unembedded dried tissue and must refer to the totality of tissue plus embedding medium if embedding medium is present in the assayed microvolume.

II. Components and Modes of Operation of the Standard Microprobe

A. COMPONENTS

In recent years most of the commercial microprobes have evolved to a design to which we shall refer as the "standard" modern probe. The basic components, indicated in Fig. 2, are:

1, 2. An electron gun and electron lenses, quite similar to those in electron microscopes, but used here to produce a beam of electrons focused to a microspot at the specimen surface. Electrons in the range of energies 2–50 keV are commonly available.

3. Coils or plates to deflect the electron beam, either for fine adjust-

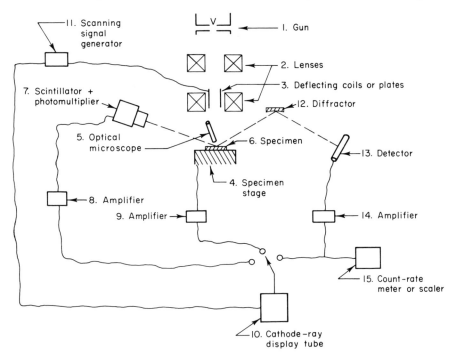

FIG. 2. The chief components of standard microprobes.

ment of the position of the probe or else to sweep the probe over the specimen in a scanning pattern.

4. A specimen stage which can be translated mechanically to position the specimen. Automatic drives are usually available for alternating between specimen and standards and for performing continuous or stepwise linear scans of the specimen.

5. An optical microscope for looking at the specimen while it is in position for analysis and even while it is under the probe.

6. A specimen (not really part of the instrument).

7–11. Equipment for producing scanning electron images according to the principle of the scanning electron microscope (see Chapter 2). The probe is scanned over a rectangular area of the specimen and the image is produced on a synchronously scanned cathode-ray display tube with a persistent screen (component 10), the brightness of the tube being modulated at each instant by an electron signal from the specimen. The available electron signals include probe electrons scattered back or "reflected" from the specimen, low-energy "secondary" electrons knocked out of the specimen, and the "specimen current" which flows from the specimen to ground. (The specimen current is equal to the total current of the incident probe minus the sum of backscattered and secondary electron fluxes.) The intensity of backscattered or secondary electrons or both may be monitored by a Faraday cup or by a scintillator and photomultiplier plus associated electronics (components 7 and 8), while the specimen current can be utilized through the amplifier 9. All standard modern probes can provide some kind of scanning electron image, but most are not equipped to provide all three types.

12–15. One or more X-ray spectroscopes, consisting of diffractors, detectors, pulse amplifiers, and associated electrical circuits including count-rate meters and scalers to record the X-ray counts. (Each detected X-ray quantum produces one discrete electrical pulse in the detector.)

B. Modes of Operation

1. Scanning X-ray Images

An X-ray emission image is produced when an X-ray signal, rather than an electron signal, is used to modulate the brightness of the display tube during rectangular scans. If the X-ray signal comes from a spectroscope set to a particular characteristic line, the image will be a map of the distribution of the corresponding element. It is also possible in thin specimens to map the distribution of mass per unit area by modulating according to the intensity of the X-ray continuum (see Section VI.B.2).

The rectangular scanned area is adjustable, the maximum scan being of the order of 1 mm on a side and the minimum of the order of a few μm. However, the X-ray diffractors are cylindrically curved crystals which focus, at any setting, onto a line at the specimen surface, and high-performance crystals are substantially out-of-focus and less sensitive to radiation originating approx. 20 μm or more from this line. Therefore with most modern probes there are facilities for performing rectangular scans by scanning the probe along the crystal's line of focus while moving the *specimen* at right angles, so that the radiation is always generated on the crystal's focal line.

X-ray images are very useful when the interesting elements are present in high concentrations (perhaps 10% or more). Indeed when there are only a few such elements, a set of X-ray images may give all the information desired, without need to refer to any optical or electron image of the specimen. This is often the case in mineralogy and metallurgy. However the X-ray image is much less satisfactory when elemental concentrations are low, for two reasons: First, the X-ray counting rates become so low (especially in thin specimens like tissue sections) that long "time" exposures are needed to photograph worthwhile images from the display tube. Exposure times for a single image may range from 10 min. up to a few hours. Second, a large part of the X-ray count may be due not to the characteristic X rays for which the spectrometer is set but to background, mainly continuum X radiation from all of the constituent chemical elements of the specimen and from extraneous

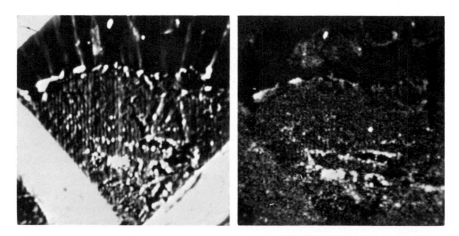

FIG. 3. Scanning images of a section of an artery containing mineralized plaques. Left: Scanning electron image. Right: Scanning image based on calcium X rays. [From Hall *et al.* (1966), courtesy of Wiley, New York.]

sources. Then the X-ray image is no longer a legitimate picture of the distribution of one element.

In biology, the concentrations of the interesting elements are usually low and X-ray images are accordingly rarely valuable. Instructive images have been produced in the study of the distribution of iron in iron-storage diseases [hemochromatosis and hemosiderosis, Marshall (1967)], and most often in the study of the distributions of calcium and phosphorus in mineralizing tissues (Fig. 3, for example).

2. Linear Scans

In the linear-scan mode, usually the spectrometer is set for a particular element and the counting rate is recorded while the probe is made to scan along a line on the specimen surface. The resulting graph of elemental distribution along the line is particularly useful when significant variation is expected only along one dimension in the specimen. An outstanding example is dental tissue studied from the outer surface in-

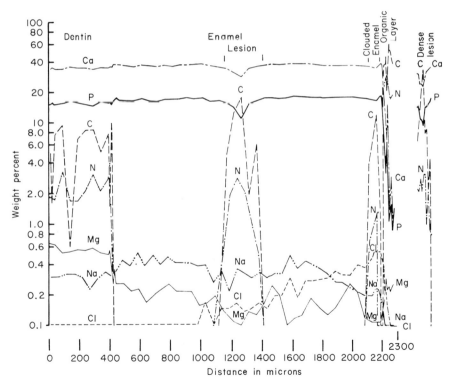

Fig. 4. One-dimensional chemical profile through a section of human tooth. [From Andersen (1967a), courtesy of Wiley (Interscience) Publishers, New York.]

wards to the enamel-dentin junction [cf. Fig. 4, from Andersen (1967a)].

In thin specimens, one can also obtain a record of the distribution of total mass per unit area along the line of scan by recording the intensity of the X-ray continuum generated in the specimen (Section VI.B.2).

The X-ray signal can be displayed on the cathode-ray tube as well as on a chart recorder. Especially useful is a triply exposed photograph of the display tube, superposing a scanning electron image of the specimen, a montage of the line of scan, and the graph of X-ray intensity. Such a record shows precisely where the line ran along the specimen. The triple exposure, and the use of the continuum X-ray signal as well, are exemplified in Fig. 5 which shows the variation of sulfur and total-mass signals along a single line of scan going from dermis through epidermis and corneum in a thin section of skin [cf. Sims and Hall (1968), also Section VI.B for full interpretation].

Sometimes relevant histological structures are not apparent in the scanning electron image but can be seen in the optical microscope. In this case, it may be possible to establish the position of the line of scan in histological terms by a faint luminescence produced as the probe strikes the specimen. Otherwise, the line of scan can usually be discerned in the optical microscope *after* the scan is completed, in the form of a track of "contamination," i.e., carbon deposited on the specimen surface where the probe has "cracked" organic vapor from vacuum pump oil and other sources.

In the study of low elemental concentrations, it is important again to realize that much of the signal may consist of background, which is quite variable locally especially in linear scans across thin specimens. Hence it may be necessary to perform additional scans along the same line with the spectrometer set off the peak angle so as to record and subtract the background, or else to use a special technique for automatic subtraction of the background (Hall and Werba, 1969).

3. Static Probe Analysis

In static probe analysis, which is the most common mode of study of biological specimens, the interesting microareas are analyzed one at a time, the probe remaining fixed on each spot while X-ray data are accumulated.

The first problem in this mode of operation is to put the probe in the right place. If the relevant histological structure is apparent in the scanning electron image, placement is very convenient: One forms a scanning image on the persistent screen of the display tube, stops the scan, and sets the beam of the display tube onto the desired point in

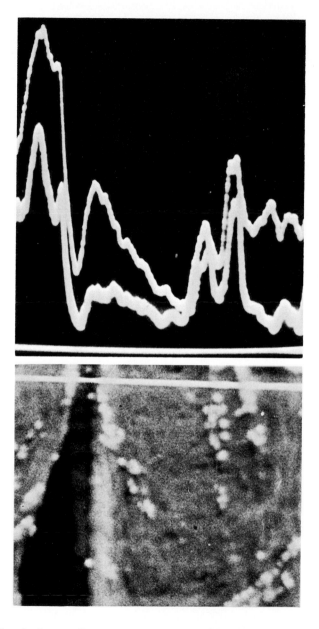

FIG. 5. Signals from a linear scan across a section of skin, correlated with the image of the section. The two traces show the sulfur X-ray and the continuum X-ray intensities.

the image; the probe itself should then simultaneously be fixed onto the corresponding point in the specimen. (It must be noted here however that a slight misregistration between scanning image and static probe can occur in maladjusted instruments, as described elsewhere (Hall *et al.*, in press), so that the registration should always be checked in advance on some simple object like a fine metallic grid or around the edges of the specimen.)

If the relevant histological detail cannot be seen in the scanning electron image but is visible in the optical microscope, the position of the probe may be established as described earlier—by luminescence during the analysis, or by subsequent observation of the contamination spot.

Once the probe is set on the chosen microarea, if the constituent elements have to be identified, the diffracting crystals may be rotated manually or automatically through a wide range of angles while the X-ray signal is recorded from a count-rate meter onto a chart recorder. The constituent elements appear as peaks in the resulting graph, as exemplified by Figure 6. The appearance of a given peak is unambiguous evi-

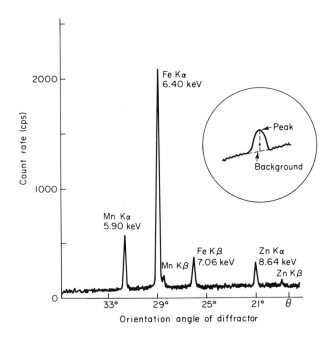

FIG. 6. Line spectra obtained during a static-probe analysis. Inset: Determination of the background which is to be subtracted from the peak to establish the genuine characteristic X-ray signal. [From Hall *et al.* (in press), courtesy of Charles C Thomas, Publisher, Fort Lauderdale, Florida, to be published.]

dence of the presence of the corresponding element. However, low peaks are occasionally due to the presence of the element *outside* of the nominally analyzed volume, either because of stray electrons striking other places or because of X-ray fluorescence, as discussed in Section IV.A.

Most often, the analyst is not interested in many elements but wants to determine the local amounts of a single or a few particular ones. Most modern probes have two or more spectrometers, so that two or more elements may be assayed simultaneously. For any one element, an X-ray spectrometer is set to the peak of the chosen characteristic line with the probe on the specimen, and one then records counts with the probe static for a suitable time over each selected microarea. At high elemental concentrations, the background correction may be negligible but at low concentrations the background also must be measured by counting with the diffractor slightly offset. When the peak/background ratio is poor (for example, the inset in Fig. 6), it may even be necessary to record at several points around the peak in order to establish the correct background value B, so as to arrive at a valid result $P - B$ for the characteristic X-ray count. [For thin specimens, an alternative method for background correction, requiring no counts off the peak angle, has been described by Hall and Werba (1969).]

For the *quantitative* analysis of one element, the only additional required datum is a similar count obtained from one standard of known composition under similar operating conditions. In Section VI, we describe how one deduces elemental weight-fractions from these data.

Static probe analysis is by far the most sensitive mode of operation, since at low counting rates it is necessary to dwell on each spot for a relatively long time, usually in the range 10–100 sec., in order to reduce statistical error to an acceptable level.

III. Types of Instrumentation and Their Performance

Besides the "standard" form, microprobe instruments exist now in many configurations with widely differing capabilities. A central purpose of this article is to enable the reader to judge which types are suited to his specimens.

One important feature is the type of X-ray spectrometer, since the sensitivity depends obviously on the efficiency of detection of the X rays and also, as we shall see, on the X-ray wavelength discrimination. Another feature which may be crucial in practice is the mode of viewing the specimen, since even the best X-ray sensitivity and analytical spatial resolution may be futile if one cannot see well enough to put the probe on the structures or regions he wants to analyze.

A. Modes of Viewing the Specimen

1. Optical Microscopy

Standard probes are usually equipped with optical microscopes providing magnifications up to approx. 600×, with objective-lens numerical apertures up to 0.4 or 0.5. This corresponds to a spatial resolution considerably better than is usual for the X-ray analysis itself, but still does not always reveal all the relevant histological detail. (For example, one may want to analyze microareas free of certain cytoplasmic organelles which may not be visible at 600×.) Usually microscopy by transmitted and by reflected light are both available. Both of these modes and all magnifications up to maximum are commonly available while the specimen is in position for analysis, but in some instruments there is some restriction on mode or magnification of optical viewing while the probe is actually impinging.

Polarizing optics are usually available or easily fitted. Phase or interference optics are not provided and would rarely be useful, since the specimens are dry and generally too thick for these techniques to work well.

The microscopy by reflected light is quite useful for opaque specimens, for example, sections of teeth and bone. Microscopy by transmitted light is generally preferable with transparent specimens, for example, stained sections mounted on transparent slides. In such specimens, the histologist can have just the view to which he is accustomed, except for the relatively low maximum magnification.

While optical microscopy in the probe is an important asset, two limitations should be noted:

a. The position of the probe can usually be established in the optical microscope as mentioned earlier, by faint luminescence or subsequent observation of contamination, but these methods are somewhat marginal at best and are sometimes inadequate.

b. A much more serious limitation is that in many studies stains should be avoided because of the risk of removing or displacing the elements to be analyzed, and adequate images of unstained tissues are often unobtainable by optical microscopy because of the low contrast inherent in this mode of viewing.

2. Scanning Electron Image

The signal for the scanning image in the standard instruments most often comes from a scintillator which responds to both backscattered and secondary electrons. If the scintillator is electrically biased slightly negatively, the secondary electrons can be excluded by virtue of their low energy and one can have an image based solely on backscattered electrons. If the scintillator or another electron detector is offset from the path of the backscattered electrons and is somewhat positively biased, the backscattered electrons will have too much energy to be deflected to the detector and the image can be based solely on secondary electrons. However, although backscattered electron, secondary electron, and specimen–current images differ in certain ways which make one or another advantageous in some cases, none of these modes seems to have any decisive advantage with the flat specimens commonly studied in the microprobe.

In the standard probe, the best spatial resolution of the scanning electron image is in the range 0.1–0.3 μm; the contrast in the image, while quite different from that of optical images, is generally no more revealing in biological specimens; and in comparison with the optical

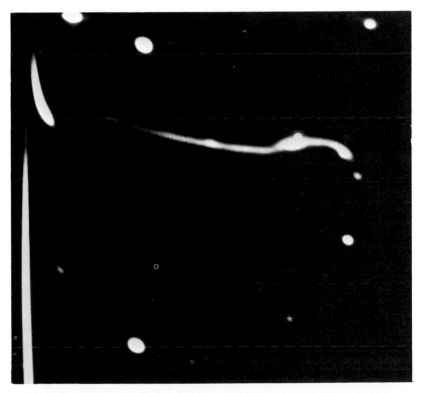

FIG. 7. Scanning image of a rat sperm cell. A dot was written onto the image on the cathode-ray display tube at the point where a high zinc X-ray signal was observed. [From Hall *et al.* (1966), courtesy of Wiley, New York.]

mode, the main advantage of the scanning image is in the ease of positioning the probe for static-probe analysis or linear scans (cf. Fig. 7).

However, X-ray spectrometric attachments are now available for microprobe analysis in commercial scanning electron microscopes (SEM's), in which the image can be much finer. The SEM usually has three lenses rather than two, to produce a smaller electron spot, so that the best spatial resolution of the scanning image is generally of the order of 100 Å (0.01 μm). In addition, with the rough or irregular surfaces often studied in the SEM, another advantage of the scanning image comes to the fore: Because the electron beam is much less divergent than optical illumination (half-angle perhaps 0.02 rad.), the electron images have great depth of focus, ranging up to 1 mm or more at low magnification. In the secondary-electron image, it is even possible to see features which are deep down inside cavities. Thus, in comparison with optical images, the image conventionally available in the commercial SEM is far superior in both resolution and depth of focus, and is uniquely suited to the study of thick specimens with irregular surfaces. Nevertheless, histological features such as cell and nuclear boundaries and cytoplasmic organelles are not usually visible in conventional scanning electron images of unstained specimens.

3. Conventional Transmission Electron Image

Microprobe analysis can now be performed in several types of instrument in which the specimen can be viewed by conventional transmission electron microscopy (Fig. 8). One first looks at an entire field on a fluorescent screen in the usual way, then selects the microarea to be analyzed and narrows down and centers the electron illumination onto this area, and finally does a static probe analysis by means of an attached X-ray spectrograph.

The chief limitation to this mode of viewing is that the specimen must be thin enough for transmission electron microscopy—no more than 1–2000 Å thick at most for the electron-beam energies generally available. A further consequence is that the amount of material under the probe and hence the X-ray intensity are reduced together with the thickness. Specimen preparation is also a rather demanding procedure, especially when there is a danger of loss or displacement of the assayed elements (see Section VII). But there are three major advantages:

a. Since the probe is positioned by a gradual reduction of the illuminated field, which is continuously visible, there is no doubt at all about the position of the probe in relation to the image.

b. The spatial resolution in the image is quite good—in some instru-

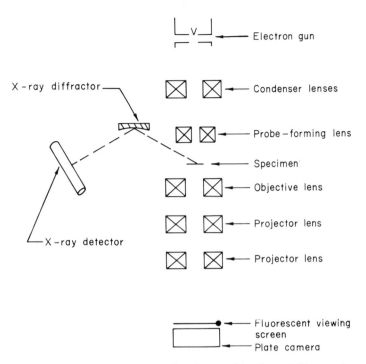

Electron gun

X-ray diffractor

Condenser lenses

Probe-forming lens

Specimen

Objective lens

Projector lens

Projector lens

X-ray detector

Fluorescent viewing screen

Plate camera

Fig. 8. Configuration of components for the combination of X-ray microanalysis with conventional transmission electron microscopy.

ments, fully up to the performance of high quality electron microscopes, i.e., 5–10 Å point-to-point. (One must not confuse the spatial resolution in the image and the spatial resolution of the X-ray analysis. The latter, which is determined by separate factors discussed in Section IV.C, cannot be nearly as good.)

c. The contrast in transmission electron microscopy is intrinsically very high, so that cell outlines and some cellular features can readily be seen even in unstained sections.

4. Scanning Transmission Electron Image

Transmission electron microscopy can also be performed in a scanning mode. As indicated in Fig. 9, this mode is simply one more method of scanning electron microscopy, in which the signal is taken from a detector placed to receive some of the electrons transmitted through the specimen. In comparison with the conventional transmission mode, there is the advantage of compactness and ready addition to a pre-existing scanning instrument, since no electron lenses are required on the exit side of the specimen.

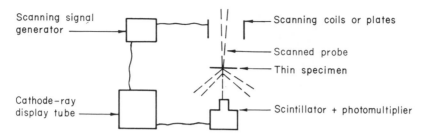

FIG. 9. Configuration of components for transmission scanning electron microscopy.

In this mode, the point-to-point spatial resolution, or the resolution in amorphous specimens, cannot be finer than the diameter of the electron beam. While sufficiently intense beams with diameters down to approx. 20 Å have been produced in laboratory instruments with field-emission "point" cathodes in the electron gun (Crewe *et al.*, 1968), in commercial instruments with conventional hairpin filaments the best beam diameters and spatial resolutions have been in the range 50–100 Å. Although this resolution is not as good as achieved in the conventional transmission mode, the *contrast* and hence the appearance of the images is quite similar, as might indeed be expected since the contrast depends essentially on the aperture-angle of the accepted electrons and not on whether the specimen is illuminated all at once or sequentially.

B. MODES OF X-RAY SPECTROSCOPY

1. Fully Focusing Diffracting Crystals

Radiation of a given wavelength is efficiently reflected from a diffracting crystal only when the angle of incidence is within a very narrow range, at the so-called "Bragg angle." Since a microprobe generates X rays virtually from a point in a specimen, it is obvious that the angular condition cannot be satisfied over any substantial area of a flat crystal, and therefore curved ones are used. The best diffractors fitted to microprobe instruments are crystals cylindrically bent to a radius $2R$ and ground to conform to the radius R of a Rowland circle ("Johansson geometry"). Such crystals are called fully focusing because radiation originating from a line (a point in the two-dimensional diagram of Fig. 10) may be focused precisely into a conjugate line (where one may place a detector entrance slit). The range of wavelengths which may be analyzed by a given crystal is limited according to the well-known Bragg equation, which gives the relation between the diffracted wave-

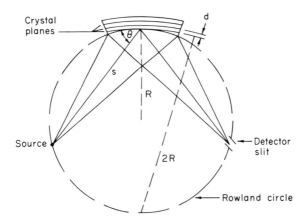

FIG. 10. Geometry of a fully focusing crystal diffractor.

length λ, angle of incidence θ, and the spacing d between the relevant planes of atoms in the crystal:

$$n\lambda = 2d \sin \theta \qquad (1)$$

(Here n is the order of diffraction, a small positive integer. First order, $n = 1$, is used predominantly.) In any case $\sin \theta$ in Eq. (1) cannot exceed unity, and, in fact, in practice the available angular range of the spectrometer is usually something like $15° < \theta < 70°$, so that one must have a group of crystals with different spacings d in order to cover a wide range of wavelengths. However a fully focusing crystal can perform well over the entire range of mechanically accessible angles.

One measure of spectrometer performance is "peak-to-background ratio." This is the ratio obtained when the probe is set onto a pure element and the counting rate at the peak of the main characteristic line is divided by the background rate, determined as indicated in the inset of Fig. 6. The ratio varies from element to element and from one diffractor to another, but is generally in the range 500–1000 with fully focusing spectrometers. In some instruments, even better ratios, up to 5000:1, have been obtained for a few elements.

There are five main sources of the background:

a. In collisions with the atoms of the specimen, the probe generates not only characteristic radiation but also a continuum of nonspecific X radiation, some of which is so close in wavelength to the selected characteristic line that it is diffracted simultaneously.

b. Imperfections in the diffracting crystal (chiefly disoriented zones) lead to the simultaneous diffraction of other wavelengths.

c. Some radiation of other wavelengths is deflected from diffractor to detector not by diffraction but by simple X-ray scattering.

d. The X radiation which strikes the diffractor generates some secondary ("fluorescent") radiation consisting of the characteristic X rays of the chemical elements making up the diffractor.

e. Scattered or stray electrons may strike some surfaces from which the resulting radiation may go directly to the detector. Such background is produced mostly in the window at the entrance to the spectrometer or from the diffractor itself if there is no window.

For a given diffracting crystal, not much can be done readily about the first three backgrounds. The effect of X-ray fluorescence in the crystal is usually negligible in crystals of low atomic number; one should avoid crystals containing elements whose characteristic lines are near those of the assayed elements, and the effect may then be virtually eliminated by rejection of the fluorescence quanta by a pulse-height analyzer (Section III.B.6). Finally, with respect to the background caused by electrons reaching the spectrometer, it should be noted that the size of this effect varies greatly among instruments, and that dramatic improvements in performance have been achieved in some instruments by the insertion of magnets to deflect stray electrons away from the spectrometer.

Another measure of performance, especially important at low X-ray intensities, is the efficiency or collecting power of the spectrometer. The efficiency for a particular characteristic radiation may be defined as follows: Suppose that the spectrometer is peaked on the X-ray line in question and that N_e quanta of this characteristic radiation emerge isotropically from the position of the specimen, resulting in a number of detected and counted quanta N_d. We then define the efficiency ϵ as the detected fraction

$$\epsilon = N_d/N_e \tag{2}$$

(Of course the radiation does not actually emerge isotropically from a real specimen because of absorption effects, but the definition of efficiency in terms of the hypothetical isotropic source is best suited to the discussion of instrumental performance.)

From the geometry of Fig. 11, we note that the number of quanta impinging on the diffracting crystal will be $N_e A \sin \theta \ (4 \pi s^2)^{-1}$ where $A = L_1 L_2$ is the area of the crystal face. If the crystal diffracts or reflects a fraction r of this radiation onto the detector, and the detector responds with its own efficiency ϵ_d, the spectrometer efficiency ϵ must then be

$$\epsilon = r\epsilon_d(A \sin \theta/4\pi s^2) \tag{3}$$

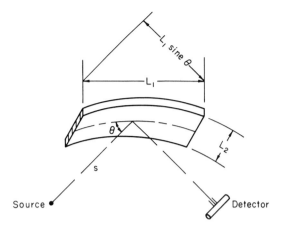

FIG. 11. Showing the diffractor projected area "seen" by the source, $L_1 L_2 \sin \theta$.

From Fig. 10 we may see that $s = 2R \sin \theta$, so that ϵ may finally be written in the form

$$\epsilon = r\epsilon_d \frac{1}{16\pi \sin \theta} \frac{A}{R^2} \tag{4}$$

For a given characteristic line and given crystal spacing d, the angle θ is determined by Eq. (1). Also, roughly speaking, the efficiency ϵ_d and reflecting power r may be regarded as properties of the individual detector and diffractor, respectively. Thus as a measure of the efficiency inherent in the spectrometer design itself, we may use the remaining variable factor in Eq. (4), A/R^2, A being the crystal area and R being the radius of the Rowland circle. For comparisons among commercial instruments, A/R^2 is a convenient figure of merit since A and R are usually both specified in manufacturers' literature.

Unfortunately, there are stringent limits on the values A/R^2 achievable in practice. In the first place, there is a formidable mechanical problem in bringing a large crystal close to the specimen without interfering with other components. But more fundamentally, if a large, cylindrically curved crystal were brought *too* close, the angle of incidence θ would no longer be sufficiently constant over the whole surface. Much of the area would fail to diffract (and would become worse than useless, since it would still contribute background). Thus the heightened value of A/R^2 would no longer signify a genuine improvement; the futility of a further increase in A/R^2 would be manifested in Eq. (4) in a corresponding decrease in the crystal reflection factor r. (This is why r can only roughly be regarded as a property solely of the diffractor itself.)

In view of the foregoing, it is clearly desirable to have some index of spectrometer efficiency based on actual measurements. Generally it is not convenient to measure the efficiency itself because there is no ready way to measure N_e, the number of emitted characteristic quanta in Eq. (2). A useful index is provided by the counting rates obtained under certain standard conditions, for example the counts per second per nA of probe current when a flat, thick specimen of pure copper is exposed to a beam of 30 kV electrons, and a lithium-fluoride diffractor is peaked for Cu K_α radiation (1 nA = 10^{-9} A). Most manufacturers quote performance for this or a closely similar set of conditions. Since the detector efficiency ϵ_d should be close to 100% for Cu K_α radiation (and indeed for all radiation up to much longer wavelengths), and lithium fluoride is among the more standard and reproducible diffractors, this set of conditions has the further advantage of indicating not only overall efficiency but also, fairly well, the efficiency inherent in the spectrometer design itself.

Sensitivities in counts per second per nA are best reported in terms of probe current, which can be measured definitively by replacing the specimen with a Faraday cup. However, specimen current is sometimes reported instead to avoid the bother of introducing the Faraday cup. Because of backscattering and secondary-electron emission, the specimen current is less than the probe current by approx. 30% when 30-kV electrons strike copper, but this figure is not always the same because the secondary emission is sensitive to surface finish, surface layers such as oxide, and local electrical fields.

With 30-kV electrons incident on flat, thick specimens of pure copper, high-performance probe spectrometers can produce of the order of 1000 counts per second per nA of probe current. Since the X-ray counting rates are often low in the study of biological specimens, any standard microprobe intended for general biological research should have at least one highly efficient diffracting spectrometer. A minimum of 500 cps/nA under the specified conditions is perhaps a reasonable criterion of suitability for broad applications to biology.

In fact, when a spectrometer yields 1000 cps/nA for the cited conditions of operation, it is working with an overall efficiency ϵ of the order of 10^{-4}. For the reasons already mentioned, there is not much hope of any future drastic improvement in the efficiency of diffracting spectrometers.

2. Semifocusing Crystals

Sometimes the diffracting crystal is cylindrically bent but not ground ("Johann geometry," Fig. 12. To continue with the notation of the preceding section, the radius of curvature of the bend is designated as $2R$). This arrangement is called semifocusing because the radiation orig-

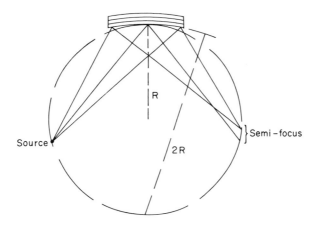

FIG. 12. Geometry of a semifocusing crystal diffractor.

inating from a line (seen as a point in Fig. 12) is not ideally focused into a conjugate line but converges less perfectly.

A semifocusing spectrometer can be constructed in an especially simple design in which the distance s from specimen to diffractor is kept constant and the crystal is simply rotated to obtain different angles θ. Also, since the reflected radiation does not converge completely, there is not much to be gained by keeping a detector slit on the dotted circle of Fig. 12 and the detector may be brought closer instead, with no entrance slit. This design has the following serious limitations:

a. In fully focusing spectrometers, the angle θ between incident radiation and diffracting planes is ideally perfectly constant over the full length of crystal shown in Fig. 10 (although there is slight variation along the direction normal to the figure). In the Johann geometry, the angle is only approximately constant and would be insufficiently constant over a large crystal. Hence the index of collecting power A/R^2 and the efficiency ϵ cannot be made as large as with fully focusing systems. In practice, the efficiency is commonly poorer by a factor of 10 or so.

b. The peak:background ratio is poorer than with fully focusing systems because a narrow entrance slit cannot be profitably employed at the detector, and also because the greater variation of θ over the crystal face implies the diffraction of a wider band of wavelengths. In practice the peak:background ratio is commonly poorer by a factor of the order of 10.

c. In order to have an approximately constant θ the geometrical condition is the same as with fully focusing crystals, namely $s = 2R \sin \theta$ But if s is kept constant, it is clear that this condition can be adequately satisfied for a crystal of given $2R$ only over a quite limited range of angles θ. Thus a given crystal cannot be used over the full range of mechanically accessible angles θ. A given crystal *spacing d* can be exploited over the full angular and wavelength range implied by Eq. (1) only by preparing three or more crystals with different radii $2R$.

To counterbalance these disadvantages the semifocusing design with constant s has the advantages of simplicity, compactness, and much less precision required in construction. Semifocusing spectrometers are therefore often used where X-ray spectrometers are fitted to pre-existing instruments not intended primarily for X-ray microanalysis. The semifocusing geometry is also used when one chooses to employ a diffracting crystal which cannot be ground.

3. Pseudo-Crystal Diffractors

According to Eq. (1), large spacings d are required for the diffraction of long wavelengths λ. Genuine crystals with sufficiently large d are not available for wavelengths longer than approx. 25 Å (i.e., characteristic radiation from elements of atomic number $Z < 8$). X radiation from elements $5 < Z < 9$ can be analyzed by means of "pseudo-crystals" of stearate, i.e., diffractors made by picking up and superposing successive layers of stearate film from a liquid surface onto a suitably curved base. The technique has been extended to the analysis of beryllium and boron ($Z = 4$ and 5) as well, by the construction of multilayered pseudo-crystals with spacings even larger than the stearates', for example, lead lignocerate, for which $d = 65$Å (Kimoto *et al.*, 1969).

It is not possible to grind the surface of these multilayered pseudo-crystals and they are generally used in the semifocusing mode.

4. Stepped Pseudo-Crystal Diffractors

Okano *et al.*, (1968) have recently produced a "stepped" diffractor consisting of stearate films built up on a specially designed base with several curved steps on its surface. Such diffractors conform approximately to the fully focusing geometry and seem to perform much better than the common semifocusing stearates, but the stepped type has not yet come into wide use.

5. Diffraction Gratings

Diffraction gratings can be used for the radiations from elements $2 < Z < 7$. They perform best at the long-wavelength end of this range. The production of efficient gratings and the operation of grating spectrometers in microprobes are highly specialized and difficult tasks; although the gratings seem to perform better than the pseudo-crystals over the entire quoted atomic-number range (Nicholson *et al.*, 1965; Nicholson and Hasler, 1966), gratings are not routinely used in any commercial microprobe instrument.

6. Nondiffracting Spectrometers (Pulse-Height Analysis)

Monochromatic electromagnetic radiation consists of monoenergetic quanta, the quantum energy E and the wavelength λ being related by the familiar Einstein relation

$$E = h\nu = hc/\lambda \qquad (5)$$

where h is Planck's constant, ν is the frequency of the radiation, and c is the speed of propagation ("the speed of light"). In units convenient for X-ray spectroscopy, this equation reduces to

$$\lambda E = 12.4 \qquad (6)$$

with λ in Ångstrom units (1 Å = 10^{-10} m) and E in kilo electron-volts (1 keV = 1.6 10^{-16} J). As Eq. (6) indicates, an analysis of an X-ray output in terms of its distribution of quantum energies is fully equivalent to the conventional spectrum provided by diffractors in terms of wavelength.

A distribution of quantum energies can be determined by means of a gas-filled proportional counter or a solid-state germanium or silicon counter. These detectors produce one electrical pulse for each detected quantum, with the property that the height of each pulse is proportional on the average to the quantum energy. Hence a spectrum of radiation can be analyzed by passing the radiation directly into the detector without using any diffractor and recording the resulting distribution of pulse heights by means of an electronic pulse-height analyzer. Figure 13 illustrates the process for the hypothetical case of a spectrum containing three characteristic lines with widely different quantum energies.

In the microprobe field spectroscopy based on diffracting crystals is often called "dispersive," and nondiffractive methods are therefore usually referred to as "nondispersive." We shall not use this terminology because it seems erroneous. Whereas prisms and gratings actually disperse polychromatic radiation, simultaneously refracting or diffracting radiation of different wavelengths along different directions, a diffracting crystal does not disperse; at a given angular setting, it selects and diffracts one wavelength and suppresses (absorbs or scatters) others.

In comparison with diffractive methods, a major attraction of non-diffractive spectroscopy is the possibility of much greater collecting power. We have already noted that the efficiency of diffracting systems

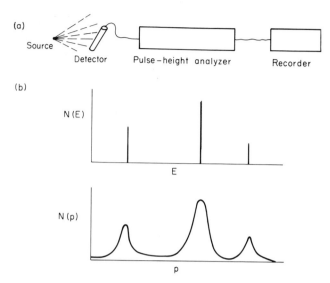

FIG. 13. The principle of nondiffractive spectrometry (pulse-height analysis). (a) Disposition of components. (b) A hypothetical spectrum consisting of three spectral lines incident onto the detector. $N(E)$ is the number of quanta incident with energy E. (c) The resulting spectrum of electrical pulses produced by the detector. $N(p)\ dp$ is the number of pulses recorded in the pulse-height range between p and $p + dp$.

is approx. 10^{-4} at best and that the detection of a higher fraction of the chosen radiation is prevented by the requirement of a very nearly constant angle of incidence at the surface of the diffractor. In nondiffractive spectroscopy, there is no such requirement and hence, in principle, a much larger fraction of the radiation can be intercepted by the detector. With gas counters, efficiencies of approx. 10^{-2} have been achieved. In fact, since it is easy to construct gas counters with large windows, an efficiency of 10^{-2} can be achieved without bringing the counter very close to the specimen. With the solid-state counters, unfortunately, high-quality performance has been achieved only with small detectors, so that high efficiency can be obtained only through a close approach to the specimen. To gain an efficiency ϵ of 10^{-2} with a solid-state detector presenting a typical area A of 0.5 cm² to the specimen, the distance S from specimen to detector must be no more than 2 cm (as calculated from $\epsilon \leqq A\ (4\pi S^2)^{-1}$). Certain technical difficulties must be overcome before such close spacings can be practical. To date, solid-state detectors have been fitted to microprobes only much farther from the specimen, indeed yielding efficiencies again of the order of 10^{-4}. However the application of the solid-state detectors is very recent and it is almost certain

that much closer spacings and higher efficiencies will soon be achieved
with them.

The chief limitation of nondiffractive pulse-height spectroscopy is implicit in Fig. 13. Monochromatic radiation produces not pulses of a single
height, but a set of pulses with a considerable spread about the average
height. In gas-filled counters, the spread is so large that radiations from
elements of adjacent atomic number are not visibly resolved; the pulse-
height distribution generated by the combination of two adjacent radiations shows only a single broadened peak. Although it has been shown
that even such distributions can be quantitatively analyzed, each com-
posite pulse-height spectrum being resolvable as a linear superposition
of the distributions produced by pure standards (Dolby, 1963; Cooke
and Duncumb, 1966; Hall *et al.*, in press; Trombka *et al.*, 1966), the
nondiffractive method has become popular only recently with the advent
of solid-state detectors which provide much better quantum-energy
resolution.

The resolution of a detector for a given characteristic radiation is
usually expressed in terms of the "full width at half height ('FWHH')."
This is defined in terms of the pulse-height distribution produced when
the detector is exposed solely to the one characteristic radiation and
is the quantum-energy band width corresponding to the width of the
pulse-height distribution at half the height of the peak (cf. Fig. 14).
The currently available lithium-drifted silicon or germanium detectors
function with a FWHH of approximately 250 eV. (This performance
figure has been rapidly improved in the past two years, the width being
due in fact mainly to electronic noise in the first stage of pulse amplification, and there are some grounds to expect that 100 eV FWHH will

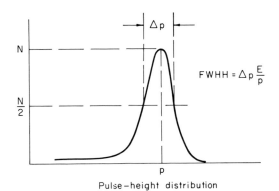

Fig. 14. Definition of the full width at half height *FWHH* of the pulse-height
distribution produced in response to radiation of quantum energy *E*.

be routinely available fairly soon.) Two hundred fifty electron volts is the separation between the main X-ray lines of aluminum and silicon, $Z = 16$ and 17, and the separation between adjacent atomic numbers increases with Z. Consequently a FWHH of 250 eV implies that in the pulse-height distribution obtained from a probed microvolume, all of the *major* constituents of atomic number $Z > 14$, will be manifest as distinct peaks.

However the quantum-energy resolution of the solid-state detectors is vastly inferior to that of diffractors. From single peaks like those in Fig. 6, by converting from angle θ or wavelength λ to quantum energy E, one may note that the FWHH values for fully focusing crystals are typically in the range 1–5 eV. In other terms, where peak:background for fully focusing crystals is of the order of 1000:1, it may be 30:1 for a nondiffractive solid-state system. This relatively poor discrimination of nondiffractive spectroscopy has the following consequences:

a. Among elements of atomic number $Z < 14$, even major constituents of a specimen may not stand out as distinct peaks in the pulse-height distribution.

b. Over the entire range of atomic numbers, minor constituents may often be obscured by interference with the K_β or L radiations of other constituents.

c. Even in the absence of any interfering characteristic radiation, the background of X-ray continuum generated by an electron probe sets a lower limit of the order of 1% to the elemental weight-fraction which may be measured.

In more practical terms, the situation currently seems as follows:

a. At this date, the detectors are usually mounted approx. 20 cm from the specimen in the solid-state nondiffractive systems which are offered as attachments to commercial microprobes. At this distance, the non-diffractive system has no greater collecting power than a high-performance fully focusing diffractor, while the wavelength discrimination of the nondiffractive system is very much poorer. The one substantial advantage of such a nondiffractive unit is that the entire pulse-height distribution can be recorded at once when the detector is coupled to a multichannel pulse-height analyzer. Hence major constituents may be rapidly and simultaneously identified, without the need to search for them one by one as a diffractor is rotated.

b. While gas-filled proportional counters can be mounted to work non-diffractively with high efficiency, their usefulness is greatly limited by their poor quantum-energy resolution. Coupled to the usual electronics, they can be satisfactory only for the assay of constituents present in weight-fractions of several per cent at least, and then only in the absence of interfering elements. One can improve greatly on this performance by the use of elaborate methods for deconvoluting pulse-height distributions, as described in the references cited above. But the greater potential of the solid-state detector has pre-empted the role of the gas counter in nondiffractive analysis.

c. Solid-state nondiffractive attachments will certainly be built with substantially improved efficiency, gained either by bringing the detector closer to the specimen or by increasing the area presented to the specimen. As compared to diffractors, these solid-state units will then offer higher efficiency at the cost of much poorer wavelength resolution and peak:background ratios. Hence they will reduce the minimum measurable amount of an element but only with an *increase* in the minimum measurable weight-fraction; this means that they will make possible certain assays where elements are relatively highly concentrated in extremely small volumes.

We have discussed nondiffractive analysis in detail, indeed out of proportion to its proper role, because the situation is complex and because the publicity attending the recent development of solid-state detectors may leave the impression that diffractive methods are obsolete. It should be clear that the nondiffractive method extends the range of microprobe analysis but the bulk of biological analyses must continue to be based on diffractors.

C. Instrumental Configurations and Their Performance

A complete instrumental configuration can be conceived by hypothetically combining an electron probe with any one or more of the modes of viewing described above, plus any one or more of the modes of X-ray spectroscopy just described. Here we shall not consider every one of these combinations. We want rather to compare the capabilities of the configurations which are commercially available.

1. Description of Configurations

(a) *Static Probe*. The first microprobe analyzer (Castaing, 1951) did not have electron scanning facilities. Since all of its capabilities have been incorporated into the modern "standard" probe discussed above in Section II, it is merely mentioned here as a matter of historical interest.

(b) *The Modern Standard Microprobe*. The conventional analytical instrument has been described in detail in Section II but the main features are recapitulated here to serve as a standard against which we shall compare the capabilities of other configurations. Modern standard microprobes offer both optical and scanning electron images of the specimen and include at least one fully focusing diffracting spectrometer, with which elements may be assayed in amounts down to approx. 10^{-16} g and weight-fractions down to approx. 10^{-4}. Spatial resolution in the scanning electron image cannot be finer than the size of the probe, which can be reduced to approx. 0.2 μm (0.1 μm in some instruments).

(c) *Combined Transmission Electron Microscope-microanalyzer* ("EMMA"). In an instrument which is designed from the start with microscopy and X-ray microanalysis equally in mind, high-resolution transmission electron microscopy can be provided together with high-performance X-ray analysis based on fully focusing diffractors. Spatial resolution in the electron image may be 5–10 Å while the minimum measurable amount of an element remains of the order of 10^{-16} g. If transmission electron microscopy is the only way of viewing the specimen, one may be limited to the study of specimens 2000 Å thick or less, in which the weight-fraction must usually be at least of the order of 10^{-3} in order to have 10^{-16} g or more of element under the probe. However, if an accessory mode of viewing is available, such as a conventional scanning electron image, it may also be possible to study thicker specimens in which weight-fractions of 10^{-4} are measurable just as in the standard probe.

Nondiffractive X-ray spectroscopes may be fitted to EMMA instruments to achieve better spatial resolution in the X-ray analysis and to measure elemental amounts less than 10^{-16} g, but only in microvolumes where the weight-fraction is of the order of a few per cent or more. Later (Sections IV and V.C), we try to assess the potential of the nondiffractive method more closely.

(d) *Conventional Electron Microscope with X-ray Attachment.* Several electron-microscope manufacturers offer X-ray attachments as accessories to their standard instruments. In comparison with EMMA instruments, there are usually two chief limitations to such apparatus. The electron beam usually cannot be focused to a spot finer than 1 or 2 μm at the specimen, precluding finer spatial resolution of the X-ray analysis. Secondly the pre-existing column usually does not accommodate a highly efficient diffracting spectrometer, though the available diffracting attachments may well suffice for the analysis of many precipitates observed in EM images of tissue sections.

Nondiffractive X-ray systems may be fitted to pre-existing columns more readily than diffractors, but the main *forté* of such systems—the assay of extremely small volumes—cannot be exploited if the probe cannot be made finer than 1 μm.

(e) *Standard Microprobe with Transmission EM Attachment.* With some microprobe instruments, electron lenses may be attached on the exit side of the specimen to provide conventional transmission electron images of thin specimens. The image is not as good as in high quality electron microscopes, especially with regard to spatial resolution. However, in comparison with the designs which stress conventional electron microscopy, the standard probe facilities are favorable for the study of thicker specimens. In particular, the specimen stage accommodates large specimens, and optical microscopy is available.

(f) *Scanning Electron Microscope (SEM) with X-ray Attachment.* The microprobe analyzer was designed for the purpose of chemical analysis and the SEM for microscopy, but the boundary is now disappearing between the standard microprobe and the SEM with X-ray attachments, since the most recent designs suggest the compatibility of optimum performance in both scanning microscopy and X-ray spectroscopy. But we shall continue to designate as "standard probes" the majority of the modern instruments, which possess two probe-forming lenses and cannot produce probes less than 0.1 μm in diameter; the term "SEM with X-ray attachment" will be reserved for instruments which can produce probes with diameters of the order of 100 Å, usually by means of three electron lenses.

X-ray analysis in the scanning electron microscope is a recent development so that actual experience with this configuration is still quite limited. But a few points seem clear and noteworthy:

1. All three modes of X-ray spectroscopy, i.e., fully and semifocusing diffractors and nondiffracting systems, have been fitted to scanning electron microscopes, with performance as described above. There is usually more free space around the specimen in the SEM than in the standard probe, so that it is easier to mount solid-state detectors close to the specimen.

2. As is the case with the EMMA instruments, the spatial resolution of the X-ray analysis cannot be nearly as good as the spatial resolution available in the electron image (see Section IV.C).

3. Since the SEM is itself basically a special mode of viewing a specimen, the incorporated optical microscope is usually not very versatile. With some specimens, depending on the type of contrast needed, a more satisfactory view may be obtained with the optical microscope in a standard probe in spite of the finer resolution in the scanning image of the SEM.

(g) *Scanning Electron Microscope with Attachments for X-Ray Analysis and Transmission Scanning Images.* An attachment for transmission scanning images can be fitted readily to most SEMs. When an

X-ray attachment is also fitted, the X-ray microanalysis of quite thin specimens can be based on the transmission electron image, as in the EMMA instruments.

For ultrathin specimens ($\leq$ 1000 Å), the SEM offers no advantage over an EMMA design, and the spatial resolution in the transmission SEM image is generally not nearly as good. However, since there is no need to focus the transmitted electrons, there is no problem of chromatic aberration (i.e., defocusing of electrons which have lost energy) so that the scanning mode can provide transmission images up to greater specimen thicknesses, of the order of 1 μm (Swift and Brown, 1970; also Swift *et al.*, 1969). The greater thickness implies a higher available X-ray intensity, which means that the minimum detectable weight-fractions in the transmission SEM mode should be better than in an EMMA (if the X-ray spectrometer is equally good).

The SEM with both transmission and X-ray attachments is extremely versatile: Thick specimens may also be studied, by means of the usual scanning modes; spatial resolution in the image is better than in the standard probe; and for very thin specimens, the contrast in the transmission image is extremely good—as good as in EMMA instruments. (For additional information, see Russ and McNatt, 1969; Sutfin and Ogilvie, 1970.)

2. Summary and Comparison of Configurations

Most of the preceding information is summarized in Table I. The numerical entries are obviously very approximate generalizations but the table may nevertheless enable the biologist to judge if any existing microprobe configuration can cope with his problem of the moment.

In the following sections (IV and V), the bases of the numerical entries will be explained, so that particular analytical problems may be evaluated more accurately.

IV. Spatial Resolution of the X-Ray Analysis

A. Thick Specimens

It is essential to make a clear distinction between thick and thin specimens. Here "thick" denotes specimens in which the impinging probe electrons cannot reach the far surface; "thin" implies that most of the electrons emerge from the far surface with the loss of only a small fraction of their initial energy. (We may note that in this terminology, thickness is not a property of the specimen alone; one and the same specimen may be "thick" with respect to a low-voltage probe and "thin" with respect to a high-voltage probe.) Not only the spatial resolution

TABLE I

COMPARISON OF PERFORMANCE OF SOME CONFIGURATIONS

Instrument	Viewing mode	X-ray mode	Specimen thickness (μm)	Minimum amount of element (g)	Minimum weight-fraction	Spatial resolution of X-ray analysis (μm)	Spatial resolution of the electron image (μm)
Standard	Optical *or* electron scan	Diffractive	any	10^{-16}	10^{-4}	~ 1	$\geqq 0.1$
		Nondiffractive		10^{-17}?	10^{-2}	?	
EMMA	Transmission EM	Diffractive	$\leqq 0.2\ \mu m$	10^{-16}	10^{-3}	$0.1-0.2$	0.001
		Nondiffractive		10^{-17}?	$>10^{-2}$	~ 0.04	
Conventional EM + X-ray attachment	Transmission EM	Diffractive	$\leqq 0.2\ \mu m$	$>10^{-16}$	$>10^{-3}$	$1-2$	0.001
		Nondiffractive		?	$>10^{-2}$	$1-2$	
Probe + attachment for conventional EM	Transmission EM	Diffractive	$\leqq 0.2\ \mu m$	10^{-16}	$\geqq 10^{-3}$	$0.1-0.2$	0.005
		Nondiffractive		10^{-17}?	$>10^{-2}$	~ 0.04	
SEM + X-ray attachment	Electron scan	Diffractive, ff[a]	any	10^{-16}	10^{-4}	~ 1	0.01
		Diffractive, sf[b]		10^{-15}	10^{-3}	~ 1	
		Nondiffractive		10^{-17}?	$>10^{-2}$	?	
SEM + transmission SEM and X-ray attachments	Electron scan by transmission	Diffractive, ff[a]	$\leqq 1\ \mu m$	10^{-16}	$\leqq 10^{-3}$	$0.1-0.2$	0.005
		Diffractive, sf[b]		10^{-15}	$\leqq 10^{-2}$	>0.2	
		Nondiffractive		10^{-17}?	$>10^{-2}$	~ 0.04	

[a] Fully focusing X-ray spectrometer.
[b] Semifocusing X-ray spectrometer.

but also the sensitivity and the theory of quantitation depend on radically different considerations in thick and in thin specimens, so we shall adhere to the above definitions throughout this chapter.

In *thick* specimens, a typical probe electron gradually loses energy as it is scattered, and it finally becomes unable to excite a given characteristic radiation when its energy drops below the amount needed to ionize the associated atomic shell. Although scattering and energy loss are random processes, the statistical result of the many small steps in the electron trajectories is a fairly well defined "active volume" of excitation. Thus, when a specimen is exposed to a static probe, each characteristic radiation is produced predominantly within a certain active volume which depends on the initial energy of the electrons, the stopping power and therefore the density of the specimen, and the ionization energy required to excite the radiation in question. The active volumes are different for different chemical elements and even for the different main spectral lines of one element because the ionization thresholds differ (see Fig. 15). The spatial resolution of an X-ray analysis obviously cannot be finer than the associated active volume.

In dense specimens like metals, the electron range may be quite short and resolution may depend on the initial diameter of the probe as much as on electron scatter. But in biological specimens, where the density is relatively low, the main problem in achieving good resolution is to avoid excessive scatter.

The X-ray spatial resolution in thick biological specimens can be roughly estimated by means of Fig. 16, adapted from Hall *et al.* (in press).

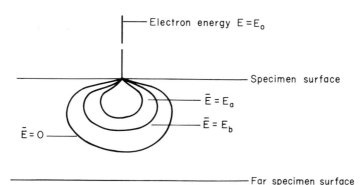

FIG. 15. Active volumes of excitation within a specimen. A spectral line with a high ionization threshold E_a is generally excited within the inner contour, where, on the average, the energy E of the probe electrons still exceeds E_a. A line of lower threshold E_b is generally excited within a larger volume. The probe electrons are brought to rest, on the average, within the outermost contour.

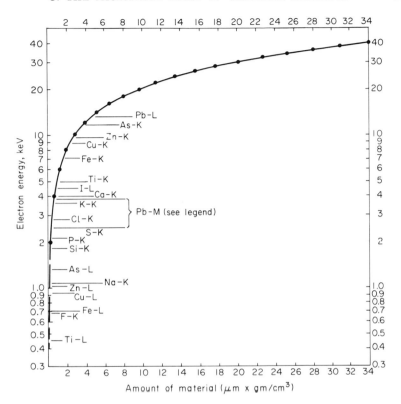

Fɪɢ. 16. Graph for the rough estimation of spatial resolution. The positions of the ionization thresholds of some common K and L X radiations are shown. (For M radiations, there are five thresholds covering a wide band of energies; for example, the highest and lowest thresholds for the Pb-M lines are shown. When M radiations are detected, estimates of resolution should be based on the threshold associated with the particular observed spectral line.) [From Hall *et al.* (in press), courtesy of Charles C Thomas, Publisher, Fort Lauderdale, Florida.]

The graph gives the average amount of material traversed as an electron loses its energy. An example will show how the figure is used: Suppose that 20-keV probe electrons are incident on a specimen of density 1.2 g cm^{-3}, and zinc ($Z = 30$) is assayed by means of its K$_\alpha$ radiation. The value corresponding to 20 keV is 9.7 μm g cm^{-3}; the value corresponding to the K-shell ionization threshold of element 30 is 2.7 μm g cm^{-3}. The difference, 7.0 μm g cm^{-3}, represents the material traversed as the electrons go from 20 keV to the ionization threshold, and the spatial resolution is approx. $7.0/1.2 \cong 6$ μm. On the other hand, if the probe is operated at 6 kV and zinc is assayed in the same specimen by means of its L radiation, the two corresponding values are 1.3 and 0.2 μm g cm^{-3} and the resolution is approx. $1.1/1.2 \cong 1$ μm.

The specimen is opaque to the probe electrons—"thick" in our terminology—when the product of thickness and density, μm $\times$ g cm^{-3}, is greater than the value corresponding to the probe voltage E in the graph. The thickness then exceeds the electron range.

Figure 16 was calculated from a well-known formula for the rate of electron energy loss (Bethe, 1930; Nelms, 1956). For the mean atomic number which appears in the formula, the composition of protein was assumed. However, the dependence on atomic number is weak and the same curve may be used for hard tissues with little error.

In soft biological tissues, the density ranges from far less than 1 g cm^{-3} in certain dried, unembedded tissues to approx. 1.2 g cm^{-3} in embedded tissues. If one must have an X-ray resolution of a few μm or better in such specimens, Fig. 16 shows that the probe voltage E_0 must be no more than a few kilovolts above the ionization threshold E_x. On the other hand, unfortunately, X rays are efficiently excited only when the ratio E_0/E_x is approx. 2 or more. The only way to keep $(E_0 - E_x)$ small and simultaneously have E_0/E_x large is to work with low values of the threshold energy E_x, which implies equally low energies for the detected quanta. In thick biological specimens, the spatial resolution of the X-ray analysis can usually be held to one or a few microns if one uses only radiations with quantum energies $E < 4$ keV, and wavelengths $\lambda > 3$ Å.

X-ray spectra are classified into groups of lines designated K, L, M, ... For a given element, the K group has the highest quantum energies, while all the lines increase in quantum energy with increasing atomic number. The restriction to quantum energies $E < 4$ keV implies the detection of K radiations from atomic numbers $Z \leq 20$, the detection of L radiations in the range $20 < Z \leq 53$, and the detection of M radiations for $Z > 53$.

Five points should be noted:

1. The preceding discussion applies to the spatial resolution of the X-ray analysis, but *not* to the spatial resolution of a scanning electron image. For reasons which will not be discussed here, the resolution in a scanning electron image is generally not affected much by the scattering of the probe electrons beneath the surface of the specimen.

2. We have so far considered only the X radiation produced directly by the probe. But this radiation itself can generate additional characteristic X rays, which are called "secondary" or "fluorescence" radiation. Since the range of the primary X rays in the specimen is much greater than the range of the electrons, fluorescence radiation may be produced

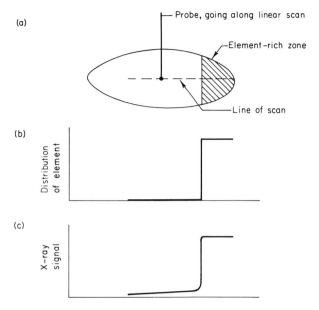

FIG. 17. Distortion of apparent elemental distribution due to X-ray fluorescence. (a) Relationship of specimen and line of scan. (b) Elemental distribution along line of scan (a hypothetical case). (c) Observed X-ray signal, with tail due to fluorescence.

far away from the probe, as far as 1 mm in some cases. The effect is negligible in most biological studies but can be very misleading in one situation, i.e., when one wants to determine the distribution of an element adjacent to a region where it is highly concentrated. As depicted in Fig. 17, a small but significant amount of the element's characteristic radiation may then be produced by fluorescence even when the probe itself is confined to a nearby region where the element is absent. The fluorescence naturally becomes more intense as the probe is brought closer to the rich region, so that the data take the form of Fig. 17c; this pattern is therefore to be interpreted very warily.

3. There is another way in which X rays may be generated within the specimen far from the position of the probe itself: generation by scattered or stray electrons. While the electron *brightness* (current:area) within the probe is immensely higher than outside, the area within the probe is tiny, so that a considerable part of the total flux and hence of the excitation may be outside of the probe.

This danger can be assessed with suitable test objects, and the effect can be suppressed by scatter-traps.

Although the quoted values for spatial resolution are not affected by excitation by fluorescence or by stray electrons, these two effects clearly degrade the resolution in the basic sense of reducing the precision of localization. The assessment and control of both effects have been discussed in detail by Dils *et al.* (1963); cf. also Reed (1966).

With reduction in specimen thickness, fluorescence becomes negligible because most of the potentially inducing radiation escapes the specimen. However, the problem of stray electrons is not reduced—indeed, it is at its worst in the study of ultrathin specimens, where the background may come chiefly from stray electrons striking thick objects like the bars of supporting grids.

4. Superior X-ray spatial resolution may sometimes be possible with a very fine probe—perhaps 0.1 μm in diameter or less, as is available in the SEM—and a very small overvoltage difference $(E_0 - E_x)$. The X-ray intensity would be quite low because of the limited current which can be focused into a fine probe and because the X-ray excitation is inefficient when E_0/E_x is small, but adequate signals might be obtained in some cases with a closely placed nondiffractive detector. At present this approach has not been explored enough to establish its potential. However, it cannot achieve better resolution than the application of nondiffractive analysis to ultrathin specimens (Section IV.C), where high probe voltages can be used, yielding higher probe currents and better excitation.

5. Spatial resolution refers to the ability to recognize slightly separated objects as distinct. It is, of course, possible to detect and analyze the X rays from individual objects which are smaller than the analytical spatial resolution, just as it is possible in microscope images to see individual objects which are smaller than the limit of resolution. The size of the smallest analyzable "hot spot" depends mainly on the minimum measurable amount of element, which will be considered later.

B. Thin Specimens (Thickness $\frac{1}{4}$–10 μm)

When a thin specimen is mounted on a thin support, the probe electrons spread out within the specimen as illustrated in Fig. 18. (A thin support is important to avoid widely spread, backscattered electrons.) For specimens more than 1 μm thick, the lateral resolution still depends primarily on the spread of the electrons due to scattering, but this spread is now very much less than the electron range.

Many biological specimens are "thin" with respect to ordinary useful probe voltages. From Fig. 16, we may note that a 30-keV electron loses only about 3 keV (actually slightly more due to scattering) in going through 4 μm g cm^{-3} of tissue. In a soft tissue with a dry-weight fraction of one-half, this corresponds to a dried section of unembedded fresh-frozen tissue cut at 8 μm; in an embedded soft tissue with density 1.2

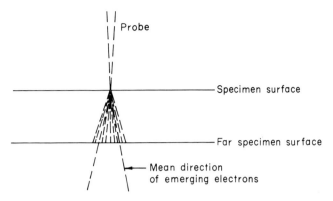

Fig. 18. The nature of the excited volume, and the spread of the probe, within a thin specimen.

g cm^{-3}, it corresponds to a 3-μm section. At higher probe voltages, even thicker specimens qualify as thin.

Since the electrons pass right through the specimen, the spatial resolution of the X-ray analysis in the direction of the probe is equal to the specimen thickness. The lateral resolution is more important than the axial and may be estimated from Fig. 19, which is based on the

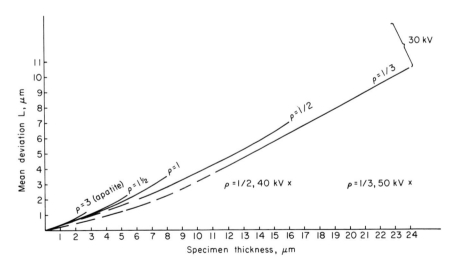

Fig. 19. The spread of the probe in thin specimens, calculated for various specimen thicknesses, specimen densities ρ(g/cm^3), and probe kilovoltages. $L \cong$ root-mean-square deviation of the probe from its axis at emergence from the specimen. [From Hall *et al.* (in press), courtesy of Charles C Thomas, Publisher, Fort Lauderdale, Florida, to be published.]

calculations of Bishop (1965) and adapted from Hall *et al.* (in press). Since the mean diameter at emergence is approx. twice the mean deviation L, but the mean diameter in the middle of the specimen is approx. half the diameter at emergence, the quantity L is a fair measure of the lateral resolution.

The lateral resolution is actually somewhat better than indicated in Fig. 19, for two reasons: First, the graph actually gives Bishop's calculation of the spread as a function of depth in thick specimens, including the effect of backscattering from deep layers which are not there in thin specimens. Second, the dashed lines are linear interpolations from Bishop's "thinnest" points back to the origin, whereas the correct curve must really be concave upwards.

In the case of sections cut frozen and then dried, a considerable uncertainty is introduced by the unknown degree of collapse after sectioning. Thus if the dry-weight fraction of a soft tissue is $\frac{1}{3}$, for example, and a 12-μm section is cut from a fresh-frozen piece and dried, if the section does not collapse the density will be approx. $\frac{1}{3}$ g cm^{-3} and the lateral resolution with 30-keV electrons is seen to be approx. 4 μm. It is more likely that the section will collapse to some extent. The product μm $\times$ g cm^{-3}, must then remain constant, but if the section collapses to a density of $\frac{1}{2}$ g cm^{-3} and a thickness of 8 μm, the lateral resolution becomes better than 3 μm; and extreme collapse to 1 g cm^{-3} and 4 μm would give a lateral resolution between 1 and 2 μm. Thus collapse, if it occurs, improves the resolution.

Figure 19 shows that lateral resolution can be held to the neighborhood of 1 μm in the study of thin specimens by restricting thickness to 5 μm or less while using probe voltages above 30 kV.

C. Ultrathin Specimens

For thicknesses $t \leqq 2000$ Å and probe voltages above 30 kV, electron scattering within the specimen has a negligible effect and the spatial resolution of the X-ray analysis is approximately equal to the diameter of the probe. The available probe current decreases rapidly as the diameter is reduced, and the minimal diameter is reached when the X-ray intensity becomes insufficient. Hence the spatial resolution in ultrathin specimens depends in practice on the minimum measurable amount of element. This minimum is considered in detail in Section IV.C. It is concluded there that one cannot expect to use probe diameters less than approx. 1000 Å when diffractive X-ray spectrometry is employed, and it may sometimes be possible to go down to approx. 300 Å with non-diffractive X-ray spectrometry.

V. Sensitivity

A. THICK SPECIMENS

1. The Minimum Measurable Weight-Fraction of an Element

The measured weight-fraction is the mass of assayed element divided by the mass of specimen within the analysed microvolume. The minimum measurable weight-fraction depends primarily on the peak/background performance of the X-ray spectrometer. As already mentioned, the peak: background ratio is usually in the range 500:1–1000:1 when an electron probe of optimum voltage impinges on a pure element and the X rays are analyzed with a fully focusing spectrometer. However the background is due mainly to X-ray continuum, the intensity of which is proportional to the mean atomic number. Hence in a specimen (as compared to a pure standard) the background is affected by a factor $\bar{Z}/Z$, while the signal $(P - B)$ is attenuated by a factor C_x, so that the ratio of signal $(P - B)$ to background B is approx.

$$\frac{P - B}{B} = C_x \frac{Z}{\bar{Z}} \left(\frac{P - B}{B} \right)_e, \tag{7}$$

where C_x is the weight-fraction of the assayed element, $(P/B)_e$ is the signal/background ratio for the pure element, and $\bar{Z}$ is the mean atomic number of the specimen. Since $Z/\bar{Z}$ is well above unity in most biological studies, and measurement becomes impractical when $(P - B)/B$ falls much below unity, Eq. (7) usually suggests a minimum measurable weight-fraction in the range 10^{-3}–10^{-4}.

Optimum operating conditions have been established and minimum weight-fractions have been determined experimentally for a number of elements by Andersen (1967a,b). His values confirm the quoted range.

2. The Minimum Measurable Amount of an Element

For a given element and specimen, one can determine the minimum measurable weight-fraction as a function of probe voltage. The minimum measurable amount of the element at a given voltage must then be the product weight-fraction (mass of element:mass of specimen) × density (mass of specimen:volume) × analyzed volume (the active volume defined above). Thus there is a different minimum at each probe voltage, and one can achieve the overall minimum by selecting the right voltage. The resulting minimum measurable amounts of different elements have been calculated and tabulated by Andersen in the papers cited above. In "thick" biological specimens, for most elements the overall minimum is in the range 10^{-14}–10^{-16} g.

This minimum does not depend simply on the inherent sensitivity of the technique. If the probe voltage is raised high enough for the most efficient excitation, the "active volume" in a thick specimen necessarily becomes rather large, and the analyzed amount of element must increase in proportion with the active volume. The overall minimum is therefore shifted towards lower probe voltages, where excitation may not be so efficient and the peak:background ratio may suffer but the active volume is smaller.

In a "thin" specimen, the situation is radically different since the active volume is almost independent of the probe voltage, being approx. equal to *probe area* $\times$ *specimen thickness*. The consequences of this are worked out in the next section.

B. Thin Specimens (Thickness $\frac{1}{4}$–10 μm)

1. The Minimum Measurable Weight-Fraction of an Element

The peak:background ratio is not affected much by specimen thickness. Hence the line of reasoning developed above for thick specimens may be applied as well to thin specimens, suggesting the same range of minimum measurable weight-fractions: 10^{-3}–10^{-4}.

2. The Minimum Measurable Amount of an Element

Minimum measurable amounts may be established for thin specimens in a relatively simple and general way. When the probe is focused onto an area containing S_x g/cm^2 of the assayed element, the spectrometer counting rate R, in counts per second, is

$$R = \text{(electrons incident/sec)} \times \text{(quanta generated/electron)}$$
$$\times \text{(detection efficiency)}$$
$$= (0.6 \ 10^{13}i)(wQ_xS_x)\epsilon \tag{8}$$

Here i is the probe current in μA; the subscript x denotes not only the element but also the spectral line for which the spectrometer is set; Q_x is the cross section for ionization of the relevant atomic orbit expressed in cm^2/g of element; w, the "fluorescence yield," is the probability that an ionization results in the emission of a quantum of the selected characteristic radiation; and ϵ is the spectrometer efficiency, which has been discussed earlier. The effect of X-ray absorption within the specimen is ignored in Eq. (8) since it is usually unimportant in thin specimens.

The maximum probe current which can be focused into a spot of diameter d μm is known to be proportional to $d^{8/3}$ (Duncumb, 1959), so we can write

$$i = i_0 d^{8/3} \tag{9}$$

where i_0 is the maximum current which can be focused into a 1-μm spot and we make the assumption, justified below, that the current i is set as large as possible in order to maximize the signal. The total amount of element under the probe, M_x, is

$$M_x \quad (g) = S_x(g/cm^2) \; (\pi/4)d^2 \times 10^{-8} \; (cm^2) \tag{10}$$

The substitution of Eqs. (9) and (10) into Eq. (8) gives the relation between elemental amount and the counting rate R_m at maximum current:

$$M_x \quad (g) = 1.3 \times 10^{-21} \frac{R_m}{\epsilon w Q_x i_0 d^{2/3}} \tag{11}$$

We shall apply Eq. (11) first to a specific set of conditions. Suppose that copper K_α radiation is detected with a 30-kV probe putting a maxmum current of 1 μA into a 1 μm spot and that the spectrometer efficiency is 10^{-4}. At 30 kV, the K ionization cross section is approx. 3.6 cm^2/g and the fluorescence yield is approx. 0.31. If we now set a practical minimum of 5 counts per second for the counting rate, Eq. (11) implies a minimum measurable amount M_x (with a 1-μm probe) of 6×10^{-17} g.

Direct observation has shown that this estimate is realistic. In a particular commercial instrument with a fully focusing spectrometer, a 1-μA 30-kV probe impinging on a thin film of copper of known mass per unit area yielded 5×10^8 counts/sec/g cm^{-2} of copper. Probably the most direct way to determine the spectrometer efficiency ϵ is to substitute this datum into Eq. (8) along with the quoted values of w and Q_x; the result in this case is $\epsilon = 7 \times 10^{-5}$, seven-tenths of the efficiency postulated above. However in the present context, the datum combined with Eq. (8) is best regarded as a direct measurement of the entire product $\epsilon w Q_x$, which can then be used in Eq. (11) without relying on the accepted values for w and Q_x. Thus the datum really signifies that $\epsilon w Q_x$ was seven-tenths of the product postulated above (no matter how reliably w and Q_x are known). The minimal mass M_x in this case, consequently ten-sevenths as large as the hypothetical value above, is still slightly below 10^{-16} g.

We may now generalize our considerations, first to other probe instruments, and then to other elements, probe voltages, and probe diameters.

Equation (11) includes only two factors which depend on the instrument itself, namely i_0 and ϵ.

a. i_0, the Maximum Current Which Can Be Focused into a 1-μm Spot.
In some instruments 1 μA at 30 kV can be focused into a spot slightly
smaller than 1 μm, and in others only into a slightly larger spot. However, because of the very steep dependence of available current on diameter, it is true for all standard probes that at 30 kV a maximum of
1 μA can be focused into a spot approx. 1 μm in diameter, just as
we have already assumed.

i_0 may be substantially different in nonstandard configurations where
the electron optical components are quite different. At 30 kV, i_0 is unfortunately likely to be much less than 1 μA in conventional electron
microscopes to which X-ray attachments have been added. In scanning
electron microscopes, many factors are involved and a wide range of
values of i_0 may be encountered.

b. ϵ, the Collecting Power of the X-ray Spectrometer. According to
the crude formulation of Eq. (11), the minimum measurable amount
M_x is inversely proportional to the efficiency ϵ of the X-ray spectrometer.
The most convenient way to compare the efficiencies of different instruments is to observe the counting rate obtained from a thick specimen
under a fixed set of conditions. In the case above, the rate was 500
counts per sec per nA of 30-kV electrons incident on a thick piece of
pure copper. By making the same measurement in another instrument,
one can readily take differing efficiencies into account.

Such a comparison of instrumental efficiencies is valid, strictly speaking, only for the spectral line which is observed. With semifocusing
diffractors, the efficiency varies in a complicated way as the diffracting
crystal is rotated. With fully focusing diffractors, a comparison at one
wavelength is generally usable through the crystals' angular range [see
Eq. (4)], but in any case the efficiency should be determined separately
for each diffracting crystal.

One may sometimes be able to use a nondiffractive spectrometer with
an efficiency of the order of 10^{-2}. According to Eq. (11), the minimum
elemental mass M_x may then be reduced to the order of 10^{-18} g. However,
in biological work, this performance is realizable only in a very few
cases, because of complications which will be discussed soon.

Besides the instrumental factors, the sensitivity depends on which
element is assayed and on factors open to the discretion of the operator:
the choice of spectral line, the probe kilovoltage, and the diameter of
the probe.

(a) The Choice of Spectral Line. We have seen that for thick specimens it is usually necessary to work with low probe voltages and long-
wavelength radiations. The situation is entirely different for thin speci-

mens. A specimen can be regarded as "thin" only when the probe voltage is high (see Section IV.B), and there is then no reason to concentrate on the longer wavelengths, while many technical and fundamental reasons favor the exploitation of shorter wavelengths (Hall, 1968). Hence with thin specimens one should normally detect K radiations, at least for the analysis of elements with atomic numbers up to approx. $Z = 40$ (quantum energies $E \leq 16$ keV; wavelengths $\lambda \geq 0.8$ Å). Most instruments do not provide voltages high enough for efficient excitation of the K radiations from elements with atomic number much more than 40, or crystals for the analysis of wavelengths much shorter than 0.8 Å, but elements with $Z > 40$ can be assayed by means of their L or M radiations.

(b) *Dependence of Sensitivity on Element and Probe Voltage.* In Eq. (11), the factors w, Q_x, and i_0 are affected by the atomic number of the assayed element and by the probe voltage. The optimum sensitivities have been worked out by Hall *et al.* (in press). Their analysis is summarized here:

For efficient excitation one should satisfy the condition

$$E_0 \geqq 2.7 E_x \qquad (12)$$

where E_0 is the energy of the probe electrons and E_x is the ionization threshold for excitation of the selected spectral line. Provided that $E_0 \geqq 2.7 E_x$, the quantity $(Q_x i_0)$ happens to vary with the atomic number Z of the assayed element in proportion to Z^{-3}, so that the signal in counts per second, according to Eq. (11), is proportional to (wZ^{-3}).

The factor (wZ^{-3}) happens to be remarkably unchanging for K radiations over the wide range of atomic numbers $8 \leq Z \leq 38$ (see Table II). Moreover, it can be shown on theoretical grounds and has been confirmed experimentally by Reed (1968) that the sensitivity is approximately the same for the assay of one element by virtue of its most intense K radiation, and the assay of another element by virtue of its most intense L radiation, when the two wavelengths are similar. Hence the sensitivity remains approx. the same for the assay of elements with $Z > 38$ based on L radiations. It follows that the limit deduced above for the assay of copper, namely 10^{-16} g, can be taken as valid over the entire range for which efficient diffractors are available, i.e., $Z \geqq 8$.

(c) *The Diameter of the Probe.* The relationship between probe diameter d and elemental amount M_x is given in Eq. (11). But some care is required for correct interpretation.

We must consider the situation, encountered fairly often, where the assayed element is not uniformly distributed but is concentrated within a small part of the probed volume. Equation (11) remains valid in

TABLE II

THE SENSITIVITY FACTOR wZ^{-3} FOR K RADIATIONS FROM
ELEMENTS OF ATOMIC NUMBER $6 \leqq Z \leqq 44$

Z	$10^6 wZ^{-3}$
6	6.5
8	8.5
10	11.0
12	8.7
14	10.1
16	12.2
18	12.7
20	13.3
24	13.7
28	13.1
32	11.9
38	9.7
44	7.1

this case in spite of the fact that a homogeneous distribution was postulated in the course of its derivation [see Eq. (10)]. In a thin specimen, a given total elemental mass produces the same signal no matter how it is distributed within the probed volume, because the increased X-ray yield per electron from the richer zones is balanced by the decreased yield from the poorer ones.

An interesting consequence, implicit in Eq. (11), is that a tiny "hot spot" may be undetectable when the probe is reduced to its size, but may be analyzed by means of a larger probe. The reason stems from the very steep dependence of maximum probe current on probe diameter: the larger probe can not only provide a larger total current, but even a larger maximum current onto the hot spot itself.

In specimens which are thin but not ultrathin, probe diameters much less than 1 μm are not often useful. Smaller diameters involve some loss in sensitivity as we have just seen, and any significant gain in spatial resolution is likely to be precluded by electron scatter within the specimen (see Fig. 18). On the other hand, an increase in probe diameter above 1 μm is rarely useful as the increased sensitivity implicit in Eq. (11) cannot be realized in thin biological specimens. This is because the maximum current tolerated by the specimen, even after optimum preparation, happens to be approx. equal to the maximum that can be focused into a 1-μm spot. [This is true at all probe voltages even though the maximum available current varies with probe voltage, since the maximum tolerated current varies in the same way, as shown by Hall (1968).] Hence for probe diameters above 1 μm, the specimen may well burn up if one applies the maximum current as was assumed in the derivation of Eq. (11).

In sum, in spite of the interaction of many complicated factors, a simple general result emerges in the diffractive analysis of thin biological specimens (thicker than approx. $\frac{1}{4}$ μm): The minimum measurable amount is in the neighborhood of 10^{-16} g for all elements of atomic number $Z > 7$, and the best sensitivity is obtained with probe diameters in the range $\frac{1}{2}$–1 μm.

We can now return to the question which was deferred earlier: the potential of *nondiffractive* analysis. Several possibilities should be considered:

a. *Ordinary tissue, 1-μm probe:* The minimum elemental amount of 10^{-18} g, suggested by Eq. (11) when the efficiency $\epsilon = 10^{-2}$, is clearly not realizable. Even if the tissue were only $\frac{1}{4}$-μm thick, the excited volume would contain approx. 2×10^{-13} g of tissue. With an elemental mass of 10^{-18} g, the weight-fraction would be less than 10^{-5}, hopelessly below the minimum fraction detectable nondiffractively. In fact, with a 1-μm probe, even 10^{-16} g of element corresponds to a weight-fraction below the nondiffractive limit.

b. *Isolated particle, 1-μm probe:* Since the minimum weight-fraction for nondiffractive probe analysis is of the order of 10^{-2}, 10^{-18} g of element should be measureable in an excited volume containing only 10^{-16} g of organic material. While a specimen consisting essentially of an isolated particle might itself present no greater mass, an analysis would not be possible because there is no specimen support with a low enough mass per unit area.

c. *Ordinary tissue, ordinary probe voltage, reduced spot size:* By reducing the probe diameter, even though M_a is increased in accord with Eq. (11), one might hope to reduce the analyzed total mass sufficiently to arrive at an analyzable weight fraction. The weight-fraction would indeed be approx. 10^{-2} *if* the probe diameter were reduced to approx. 0.05 μm; the analyzed volume were a cylinder with diameter 0.05 μm and height $\frac{1}{4}$ μm; and the elemental amount in that volume, in accord with Eq. (11), were 7×10^{-18} g. But in fact a 0.05-μm probe, even if instrumentally available, would be unsuccessful because electron scattering would make the excited volume much larger than the postulated cylinder.

d. *Ordinary tissue, reduced probe voltage, reduced spot size:* The excited volume can be made very small by reducing the probe diameter and reducing the probe voltage until $(E_o - E_x)$ is small. (The specimen, even if physically thin, would then be either intermediate between "thick" and "thin" or simply "thick" in our terminology.) There may be some situations where elemental amounts well under 10^{-16} g can be measured in this way nondiffractively in spite of the poor weight-fraction limit. The potential of this approach cannot yet be judged closely; the necessary experimental or theoretical work has not yet been done.

It is apparent that nondiffractive analysis has at best a very limited role in the study of the specimens considered so far. Later we shall see that with respect to ultrathin specimens the situation is more promising.

3. *The Relationship between Minimum Measurable Amount and Weight-Fraction*

In the consideration of thick specimens, we have followed Andersen (1967a,b) in using the simple relation *mass of element = weight-fraction × density × excited volume.* The minimum measurable elemental amounts and the minimum measurable weight-fractions deduced in this way are consistent in the sense that the presence of the minimum quantity of either type, calculated for a given specimen, guarantees that the other minimum condition is satisfied in the specimen as well.

For "thin" specimens in the thickness range $\frac{1}{4}$–10 μm, a similar approach is not convenient because there is no simple general way to take account of the excited volume, which varies with probe voltage, probe diameter, specimen thickness, and specimen density. We have accordingly deduced simple independent minimal masses and weight-fractions. In judging whether a given assay may be feasible, one must therefore realize that *both* minimum conditions must be satisfied.

In fact, when an elemental weight-fraction is 10^{-4}, 10^{-16} g of the element are contained in a cylindrical volume with diameter 1 μm and density 1.2 g cm^{-3}. In specimens which are thicker or denser, analysis may well be precluded because the weight-fraction is too low even when the amount of element in the analyzed volume is well above the minimum. When the specimen is thinner or less dense or the probe is smaller, the amount of element may well be insufficient even when the weight-fraction is above the minimum. When both conditions are satisfied simultaneously, analysis should be feasible.

4. Choice of High or Low Probe Voltage

Over a considerable range of specimen thicknesses, the analyst can choose to use a low probe voltage and apply the thick-specimen theory, or to use a high voltage and apply the thin-specimen theory. With an 8-kV probe, specimens are "thick" down to approx. 200 μg/cm^2, i.e., a thickness of approx. 2 μm at density 1.2 g cm^{-3}; and with a 50-kV probe, they are "thin" up to about 700 μg/cm^2, i.e., a thickness of approx. 14 μm at a density (or dry weight per unit volume in unembedded tissue) of $\frac{1}{2}$ g cm^{-3}. Whichever choice is made, the analytical performance may be anticipated in some detail from the preceding discussion. In sum, however, high probe voltages usually provide better minimum measurable elemental amounts, while spatial resolution is more readily controlled at low probe voltages.

C. Ultrathin Specimens

In ultrathin specimens (thickness $\leq$ 2000 Å), so long as the probe diameter is not much less than the thickness of the specimen, electron scatter is unimportant and the excited volume is well approximated by a cylinder with the diameter of the probe and the height (thickness) of the specimen. Under this assumption, the lateral spatial resolution of the X-ray analysis is equal to the probe diameter, and the relationship between spatial resolution, minimum analyzable elemental amount, and minimum analyzable weight-fraction can be formulated in a simple general way We shall summarize here the treatment by Hall (1968).

An instrumental sensitivity Y may be defined for an assayed element as the spectrometer counting rate per $\mu g/cm^2$ when the probe is set on a thin film of the element in pure state. Similarly a quantity B, which is intended to account for background, is defined as the counting rate per $\mu g/cm^2$ when the probe is set on a thin film of *organic* material containing none of the assayed element. Within the probed region of the specimen, it is supposed that there are S $\mu g/cm^2$ of organic material, and that all of the assayed element is present within an enclosed "hot spot" where there are fS $\mu g/cm^2$ of element. The local elemental weight-fraction C within the hot spot is then

$$C = \frac{fS}{S + fS} = \frac{f}{1 + f} \tag{13}$$

The hot spot may be smaller than the probe, so a geometrical factor g is defined:

$$g = \text{area presented by hot spot/area of probe} \tag{14}$$

From statistics, one obtains the following formula (Hall, 1968) for the elemental weight-fraction C which can be measured in a counting-time t sec with a probable fractional error p:

$$C = \frac{f}{1 + f} \quad \text{with} \quad f = \frac{1}{2tp^2SYg}[1 + (1 + 8tp^2SB)^{1/2}] \tag{15}$$

If the probe is no larger than the hot-spot, the factor g is simply unity. But the terminology has been established to deal as well with the situation, quite common in ultrathin specimens, where the assayed element is localized to a tiny region (a particle perhaps), and an adequate signal can be obtained only when the probe is considerably larger. In this case, in our notation, C refers to the elemental weight-fraction within the hot-spot, *not* the mean weight-fraction in the entire excited volume.

Table III has been calculated from Eq. (15) and gives estimated elemental amounts and elemental weight-fractions which can be measured in organic materials with 20% probable error in a running time of 20 sec, as functions of probe diameter and specimen thickness.

The sensitivity Y, according to Eqs. (8) and (9), is

$$Y = 0.6 \times 10^7 i_0 w Q_x d^{8/3} \epsilon \tag{16}$$

TABLE III

ESTIMATED MINIMUM MEASURABLE ELEMENTAL AMOUNTS M_x AND
WEIGHT-FRACTIONS C[a,b]

Entry No.	Detector	Hot-spot diameter (μm)	Probe diameter (μm)	Mass per unit area $S(\mu g/cm^2)$	Weight-fraction C	Elemental amount M_x (10^{-16} g)
1	↓diff[c]	1.	1.	↓5	0.0008	0.3
2		0.32	1.		0.0078	0.3
3		0.32	0.32		0.0074	0.3
4		0.1	1.		0.075	0.3
5		0.1	0.32		0.073	0.3
6		0.1	0.1		0.142	0.7
7		0.032	0.1		0.64	0.7
8		0.032	0.032		0.76	1.2
9	solid[d]	0.1	0.32		0.012	0.05
10	solid	0.1	0.1		0.0067	0.03
11	gas[e]	0.1	0.1		0.0089	0.04
12	solid	0.032	0.032		0.043	0.02
13	gas	0.032	0.032		0.051	0.02
14	solid	0.01	0.01		0.44	0.03
15	gas	0.01	0.01		0.45	0.03
16	↓diff	1.	1.	↓12	0.0005	0.5
17		0.32	1.		0.0046	0.5
18		0.32	0.32		0.0036	0.4
19		0.1	1.		0.045	0.5
20		0.1	0.32		0.036	0.4
21		0.1	0.1		0.065	0.7
22		0.032	0.1		0.41	0.7
23		0.032	0.032		0.56	1.3
24	solid	0.1	0.32		0.0079	0.08
25	solid	0.1	0.1		0.0041	0.04
26	gas	0.1	0.1		0.0056	0.06
27	solid	0.032	0.032		0.024	0.02
28	gas	0.032	0.032		0.029	0.03

[a] The amounts and weight-fractions have been calculated for a statistical probable error of 20% in a running time of 20 sec. The values are functions of hot-spot diameter, probe diameter, and specimen thickness.

[b] This table is similar to the one published by Hall (1968), but is based on more modern performance figures.

[c] Fully focusing diffractive spectrometer.

[d] Solid-state detector used nondiffractively.

[e] Gas counter used nondiffractively.

(A factor of 10^6 has been introduced because it is convenient here to change the definition of S from gm/cm^2 to $\mu g/cm^2$.) We have seen in Section IV.B.2 that $i_0 w Q_x$ does not vary much with elemental atomic number. According to the values quoted there,

$$Y \cong 7 \times 10^6 d^{8/3} \epsilon \qquad (17)$$

The estimations in the table are based on Eq. (17) with an efficiency $\epsilon = 10^{-4}$ assumed for diffractive spectrometry and $\epsilon = 10^{-2}$ assumed for nondiffractive. The probe diameter d, which is also the lateral spatial resolution, enters the basic Eq. (15) through Eq. (17).

B in Eq. (15) is obtained from typical measured peak:background ratios in thin films after due allowance for the mean atomic number of an organic matrix [see Eq. (7)]. The values assumed for Y/B in the calculation of the table were 2000:1 for diffractive spectrometers, 50:1 for nondiffractive systems with gas counters, and 100:1 for non-diffractive systems with solid-state counters.

In the interpretation of Table III, several points should be noted:

1. The limit of operation is manifested by the approach of the analyzable elemental weight-fraction C to unity. In any case, an actual weight-fraction, of course, cannot exceed unity, and in biological specimens the actual fraction C is usually well under unity no matter how localized the assay is.

2. In ultrathin specimens, one is limited primarily by counting rate, i.e., by the very small elemental amounts within the very small assayed volumes. The local elemental weight-fraction must be higher than in thicker specimens in order to have a measurable amount of element present.

3. As the probe diameter is reduced, diffractive spectrometry becomes impractical in the neighborhood of $d = 0.1 \mu m$.

4. In the assay of very small hot-spots by diffractive spectrometry, the minimum measurable amount may be improved, at some cost to spatial resolution, by making the probe somewhat larger than the hot-spot (see entries 5 and 6). The excellent wavelength discrimination of a diffractor permits this exploitation of the higher current density available in the larger probe, in spite of the extra background generated outside the hot-spot.

5. The table suggests that nondiffractive analysis may be applicable to hot-spots with mean dimensions as small as approx. 0.03 μm, for the measurement of elemental amounts as small as $\approx 10^{-18}$ g. Such assays have not yet been carried out and the estimation is speculative. One serious problem still must be overcome. Equation (15) accounts for the

background from the specimen itself, but there is always an additional background of continuum generated by stray electrons in surfaces near the specimen (lenses, apertures, specimen holder, and other hardware); it remains to be seen whether this extraneous background can be reduced sufficiently for the measurement of the very low X-ray intensity from $\approx 10^{-18}$ g of element.

6. The table indicates little difference between gas counters and solid-state counters as nondiffractive spectrometers. This is true so long as the limits in the assay of extremely small volumes are set primarily by the low counting rate rather than by the peak:background ratio. On the other hand, if the extraneous background cannot be thoroughly suppressed, or when there are problems of interference between spectral lines, the solid-state counters should be much superior.

7. With nondiffractive spectrometry, it is generally futile to make the probe larger than the hot-spot (see entries 9 and 10); the spectrometer cannot cope well with the background generated outside the hot spot.

8. Table III gives estimates for two values of specimen mass per unit area, 5 and 12 $\mu g/cm^2$. For conventional ultrathin sections, which are cut from embedded preparations weighing approx. 1.2 g cm^{-3}, the tabulations correspond to section thicknesses of approx. 400 and 1000 Å. For cryostat sections of freshly frozen organic material with a dry-weight fraction of $\frac{1}{2}$, the tabulated values of mass per unit area would result from the drying of sections cut at 1000 and 2400 Å.

9. To compare the sensitivity for embedded tissues versus nonembedded cryostat preparations, one should consider sections cut at the same *thickness*. One may then imagine *completely* identical tissues prepared in the two ways, with the same amount of the assayed element in an analyzed microvolume; the comparison then depends solely on the effect of the embedding medium which is present additionally in one of the two preparations. To the extent that embedding medium *is* actually present in the assayed microvolume, it can only add to the background and thereby increase the probable statistical error of the assay. Accordingly, in comparison with cryostat sections, a given statistical accuracy can be achieved in embedded material only where there is a larger amount of the assayed element. This effect is shown, for example, in entries 1 and 16 of the table, entry 16 representing 1000 Å of embedded tissue and entry 1 representing a cryostat section which is cut at the same thickness and has approx. one-third as much mass per unit area after drying.

Of the pair of entries, the lower elemental weight-fraction is listed for the embedded one. This does not signify inherently better sensitivity in terms of weight-fraction. For the same probable error, the amount of element per unit mass of *tissue* must actually be slightly higher in the embedded preparation, but the weight-fraction presented to the probe is reduced by the additional mass of embedding medium.

Equation (15) is itself fairly precise but it cannot be applied precisely. In cryostat sections, one generally does not know the local dry mass per unit area S accurately; S varies greatly within the specimen. In sections of embedded tissue, the mass per unit area may be uniform and the minimum measurable amount of element M_x can consequently be determined well, but the minimum measurable weight-fraction *with respect to the tissue alone* may be considerably larger than the tabulated value if the assayed microvolume contains much embedding medium. Hence Eq. (15) and Table III can serve only as rough guides.

VI. Quantitation

A. THICK SPECIMENS

The quantity which can be measured in thick specimens in the microprobe is local elemental weight-fraction, i.e., the mass of an element divided by the total mass within an analyzed microvolume. The theory of this measurement has been enthusiastically developed over the course of many years and has now reached a stage where, in favorable specimens, the probable error of measurement is only 1 or 2% of the measured weight-fraction. We shall not give a complete exposition of the theory here, first because it is so elaborate that too much space would be needed, and second because it cannot often be applied soundly to biological specimens. Our intentions in this section are to explain the physical foundations of the theory of quantitation, to describe the conventional approach, and to examine the underlying assumptions. The reader should be enabled to understand the technical literature and converse with microprobe technologists, to judge for himself the feasibility of any particular measurement (the technologists are not to be trusted on this topic!), and to carry out quantitation in favorable cases. Fuller expositions of the theory are available in many texts, including Birks (1963) and Philibert (1969).

1. The Physical Procedure of Measurement

For the quantitative assay of an element, only a single homogeneous standard is needed, of known composition, containing the element in question. One compares the intensity of the selected characteristic radiation from the chosen spot in the specimen with the intensity of the same radiation from the standard. The probe voltage must be the same for the two observations. The X-ray intensities are directly proportional to probe current, so one can either use the same probe current or readily correct for any difference.

For approximate quantitation [Eqs. (18) and (19) below], this extraordinarily simple procedure comprises all of the requisite observations. Even when the most complete and accurate theory of quantitation is used, most often no further observations are needed, but it is occasionally necessary to assay more than one element in order to make certain corrections.

2. The First Approximation to Quantitation

In first approximation,

$$C_x \cong (I_x/I_{x0})C_{x0} \tag{18}$$

where C_x and C_{x0} are the weight-fractions of the assayed chemical element x in the analyzed microvolume of the specimen and in the standard, and I_x and I_{x0} are the respective observed characteristic X-ray intensities obtained at the same probe voltage and current, with suitable subtraction of noncharacteristic background. It is often convenient to use a standard consisting of the assayed element in pure state, in which case $C_{x0} = 1$ and Eq. (18) reduces to

$$C_x \cong I_x/I_{x0} \tag{19}$$

Equations (18) and (19) represent far more than wishful thinking. We shall see that they have a sound physical basis and are often fairly accurate.

3. Physical Basis of the Approximate Eqs. (18) and (19)

In order to deduce elemental weight-fractions from observed X-ray intensities, it is necessary to establish the functional relationship between these quantities. This has been done on theoretical grounds.

In thick specimens most of the probe electrons are brought to rest, dissipating their energy chiefly in collisions with the electrons bound to the atoms of the specimen. To calculate the yield of any characteristic radiation, one may ask, "What is the average number of characteristic quanta generated by a probe electron before it is brought to rest?" From the standpoint of this question, the yield of any radiation is seen to depend on the competition between the excitation of that radiation and nonproductive modes of energy dissipation. Hence the law of energy loss of the probe electrons must be central to the theory of quantitation in thick specimens.

Let us consider the implications of a very simple energy-loss law,

$$dE = F(E) \, \rho \, ds \qquad (20)$$

(Here dE is the average energy lost by a probe electron of energy E as it traverses a distance ds in material of density ρ g cm^{-3} and $F(E)$ is an arbitrary function of E.) Equation (20) alleges, roughly speaking, that the average energy lost in a path length ds is proportional to the amount of matter traversed but is independent of the composition of the medium; a dependence $F(E)$ on electron-energy E is recognized but not specified.

The average number of quanta of a particular radiation x, generated as a probe electron travels a distance ds and loses average energy dE in a specimen, can be written

$$\begin{aligned} n_x \, dE &= wq_x N_x \, ds \\ &= wq_x N_x \, ds/dE \, dE \end{aligned} \qquad (21)$$

(Here the subscript x denotes both the assayed element and the particular selected spectral line; q_x is the cross section for ionization of the associated atomic level in cm^2 per atom; w is the fluorescence yield as defined earlier; and N_x is the number of atoms of element x per cm^3.)

Substitution of the loss-law (20) into Eq. (21) gives

$$n_x \, dE = \frac{wq_x N_x}{\rho} \frac{dE}{F(E)} \qquad (22)$$

For a pure standard (denoted by subscript 0), Eq. (22) reduces to

$$n_{x0} \, dE = \frac{wq_x N_{x0}}{\rho_0} \frac{dE}{F(E)} \qquad (23)$$

and the division of Eq. (22) by Eq. (23) gives

$$\frac{n_x}{n_{x0}} = \frac{N_x}{N_{x0}} \frac{\rho_0}{\rho} \qquad (24)$$

Now ρ_0 and ρ can be written

$$\rho_0 = N_{x0} A_x m \qquad (25)$$

and

$$\rho = \Sigma(NAm) \qquad (26)$$

where A_x is the atomic weight of element x, m is the mass of a hydrogen atom (the inverse of Avogadro's number), and the sum in Eq. (26) is extended over all of the constituent elements in the specimen. Substitution of (25) and (26) into Eq. (24) gives

$$\frac{n_x}{n_{x0}} = \frac{N_x A_x}{\Sigma(NA)} = C_x \tag{27}$$

We have shown that the energy-loss law (20) implies the validity of Eq. (27) for each part of the trajectory of a probe electron (each path interval ds or energy interval dE); a simple integration then shows that Eq. (27) should apply to each trajectory as a whole. The physical basis of the "first approximation," Eq. (19), can now be recognized from a comparison of Eqs. (19) and (27). The conclusion is:

The quantitation formula (19) depends on two conditions: the validity of the energy-loss law (20), and the equality of the ratios of observed intensities I_x/I_{x0} and directly generated intensities n_x/n_{x0}.

However, neither of these two conditions is exactly satisfied. We must now consider the extent of the departures from them, as well as other effects which reduce the accuracy of Eq. (19).

4. Corrections Required for Accurate Quantitation

(a). *Penetration Effects.* In the microprobe literature, the term "penetration effects" is used to encompass all consequences of deviations from the electron energy-loss law (20), which is the microprobe analyst's ideal because it leads to the nicest equations one can imagine, i.e., Eqs. (18) and (19). The law is imperfect for the following reasons:

(1). *The "Z/A correction."* Equation (20) postulates a direct relation between specimen mass and electron energy-loss along a trajectory. It is only to be expected that this idea is inaccurate, since half or more of the mass of a specimen consists of neutrons which have nothing to do with the energy loss of the probe electrons. Since the loss occurs primarily in collisions with bound electrons, a more realistic (though still approximate) loss law is

$$dE = F(E)\Sigma(NZ)\, ds \tag{28}$$

Here the specimen density ρ in Eq. (20) has been replaced by the number of electrons per cm^3 in the specimen, which is expressed in the form $\Sigma(NZ)$, N being atoms/cm^3, Z being atomic number, and the sum being taken over all constituent elements.

Equation (28) may be incorporated into the fundamental equation (21) in place of the naive law (20). The resulting more correct version of Eq. (22) is

$$n_x\, dE = wq_x \frac{N_x}{\Sigma(NZ)} \frac{dE}{F(E)} \tag{29}$$

In place of Eq. (23) for the pure standard one obtains

$$n_{x0}\, dE = wq_x \frac{N_{x0}}{N_{x0}Z_x}\frac{dE}{F(E)} = \frac{wq_x}{Z_x}\frac{dE}{F(E)} \tag{30}$$

and the division of Eq. (29) by Eq. (30) leads to the result which replaces Eq. (27),

$$\frac{n_x}{n_{x0}} = \frac{N_x Z_x}{\Sigma(NZ)} \tag{31}$$

Equation (31) has a remarkably simple meaning. The right-hand side is the number of electrons per cm^3 bound to the assayed element in the analyzed microvolume of the specimen, divided by the total number of electrons per cm^3 in the same microvolume. We shall refer to this quantity as the "electron-fraction" of the assayed element. Thus the quantity which is "read out" by microprobe analysis is fundamentally not the weight-fraction of an element, but rather its electron-fraction.

While an electron-fraction is basically just as significant as a weight-fraction, the common language of analytical chemistry refers to weight-fractions. Equation (31) is readily converted to weight-fractions by means of the relation between weight-fraction C and atomic weight A:

$$C_i = N_i A_i / \Sigma(NA) \tag{32}$$

After Eq. (32) is substituted into Eq. (31), a touch of algebra (involving no approximations) gives

$$\frac{n_x}{n_{x0}} = C_x \frac{Z_x/A_x}{\Sigma(CZ/A)} = C_x \frac{Z_x/A_x}{\overline{Z/A}} \tag{33}$$

(the bar in the denominator denoting an average over all constituent elements).

A comparison of Eq. (33) with Eq. (27) shows that the more accurate loss law (28) leads to a correction factor $[(Z_x/A_x)/(\overline{Z/A})]$. After this and other corrections have been lumped into a grand expression [Eq. (63) below], we shall consider the magnitude of this effect and how to reckon with it.

(2). *The effect of electron binding-energy (conventional treatment).* The energy-loss law (28) is itself only an approximation. The rate of energy loss of the probe electrons is known to depend not only on the concentration (number per cm^3) of bound electrons in the specimen, but also on their binding (or ionization) energies. If a specimen consisted of N atoms per cm^3 of a *single* chemical element of atomic number Z, and the bound electrons all had the same binding energy, then the average rate of energy loss dE/ds would be well represented by the equation of Bethe (1930), which can be written in the form

$$\frac{dE}{ds} = \frac{2\pi e^4}{E} NZ \ln \frac{E}{J} \tag{34}$$

(Here e is the electrical charge of the electron, J is a quantity proportional to the binding energy, and ln denotes the natural logarithm.) Of course, in real atoms the electrons do not all have the same binding energy; this fact is readily accommodated in Eq. (34) by inserting an appropriately averaged energy J. Finally, the energy-loss law for *multi-element* specimens may be written

$$\frac{dE}{ds} = \frac{2\pi e^4}{E} \sum_r \left(N_r Z_r \ln \frac{E}{J_r} \right) \tag{35}$$

Here the sum is taken over the constituent chemical elements of the specimen, with one term for each element, and for J_r one conventionally inserts the values calculated for independent atoms (i.e., neglecting the binding forces which hold a solid body together).

Equation (35) is the most accurate loss law in general use. When this law is inserted into the fundamental equation (21) and the ratio of excitations in specimen and standard is again formulated, in place of Eq. (27) or Eq. (33) one obtains

$$\begin{aligned}
\frac{n_x}{n_{x0}} &= C_x \frac{Z_x/A_x \ln (E/J_x)}{\Sigma_r[C_r Z_r / A_r \ln (E/J_r)]} \\
&= C_x \frac{Z_x/A_x \ln (E/J_x)}{Z/A \ln (E/J)}
\end{aligned} \tag{36}$$

(the average in the final expression being defined by the sum in the preceding one, which is again composed of one term for each constituent chemical element).

Equation (36) shows that the excitation ratio n_x/n_{x0} is really a complicated function of the energy E of the probe electron. Therefore, one cannot readily integrate to obtain the overall average ratio n_x/n_{x0} for complete electron trajectories. In those laboratories where quantitative microprobe analysis is routinely done by computer, the integration is part of the computer program. Otherwise, one may use an approximation which usually introduces an error of a few per cent or less. Although the factor $\ln E/J_r$ varies a great deal with atomic number and with energy E, the ratio $(\ln E/J_1)/(\ln E/J_2)$ for any two elements is only slightly dependent on probe energy. It follows that one may write as a good approximation

$$\frac{Z_x/A_x \ln (E/J_x)}{Z/A \ln (E/J)} \cong \frac{Z_x/A_x \ln (\bar{E}/J_x)}{Z/A \ln (\bar{E}/J)} \tag{37}$$

where $\bar{E}$ is the average energy of the probe electrons during excitation,

$$\bar{E} = \tfrac{1}{2}(E_0 + E_x) \tag{38}$$

E_0 being the energy at entry into the specimen and E_x being the excitation threshold energy.

For the factor $\ln E/J$ several expressions are in current use. We shall insert the widely employed form (Nelms, 1956)

$$\ln \frac{E}{J_r} = \ln \frac{1166E}{11.5Z_r} \cong \ln \frac{100E}{Z_r} \tag{39}$$

with the energy E expressed in keV. The combination of Eqs. (39), (37), and (36), plus a bit of algebraic rearrangement, gives the expression

$$\frac{n_x}{n_{x0}} \cong C_x \frac{Z_x/A_x}{Z/A} \frac{2 + \log \bar{E} - \log Z_x}{2 + \log \bar{E} - \overline{\log Z}} \tag{40}$$

Equation (40) is the "penetration correction" which is incorporated into the overall formula (63) below.

(3) *The effect of electron binding-energy (further considerations).* Even if the fundamental loss law (34) were perfectly correct, Eq. (35) as commonly used would be imperfect for three reasons. First, in a single-element solid specimen, the electron binding energies are not exactly the same as in isolated atoms because of the interactions between neighboring atoms. Second, in multielement specimens the binding energies are further affected by the altered interactions between different types of atoms. Finally, in many solids, the pattern of electron binding is so affected by interactions between atoms that many electrons are not bound to individual atoms but exist rather in collective "volume plasma" or "surface plasma" states associated with the structure of the solid as a whole.

We need not be concerned here with the difference between isolated atoms (i.e., a gaseous specimen) and a solid single-element specimen, since the biological specimen and the standard are *both* in the solid state. Also it can be shown that the change in atomic binding energies caused by the presence of other chemical elements is generally unimportant. But the situation with respect to "plasma losses" is much less clear. The plasma population and energy levels are affected not only by chemical composition but by *physical* condition—dispersed or porous specimens have more surface-plasma states than do compact ones. Quite recently Ichinokawa (1969) has reported a strong dependence of X-ray yield on the degree of compaction of the specimen, a dependence which he has attributed to variation in the amount of energy lost by the probe electrons to surface-plasma states.

The effect is remarkably large. For example, the carbon K X-ray yield from pure carbon at a density of 1.4 g cm³ was found to be only 70% of the yield at density 2.25 g cm^{-3}. The conventional thick-specimen theory predicts no change in X-ray yield with compression, and such effects have not been reported earlier, presumably because highly quantitative work has generally been limited to compact speci-mens. Ichinokawa's findings still must be repeated by others and his interpretation is still controversial, but if his view is confirmed, accurate quantitative analysis by the conventional thick-specimen theory will be especially difficult in biological specimens where the degree of compac-tion is ill-defined.

(b) *The Backscattering Correction.* In applying Eq. (27) to entire electron trajectories, we have implicitly assumed that each probe electron dissipates all of its energy within the specimen. Actually some incident electrons are scattered back out of the specimen while they still have enough energy to excite the selected X radiation. One usually accounts for this waste in terms of a backscatter coefficient R, which is defined as the number of the selected characteristic quanta actually generated within a specimen or standard, divided by the number which *would* be generated in the absence of backscattering. (In this definition, we refer only to the radiation generated directly by the probe electrons; the indirectly generated fluorescence radiation is to be considered sep-arately.) The correction factor R is to be applied to both the specimen and the standard, as indicated below in Eq. (63).

R is a well-established function of specimen mean atomic number $\bar{Z}$ and of the ratio of probe voltage to excitation threshold E_0/E_x and may be read out from published tables (Duncumb and Reed, 1968). The best way to weight the average $\bar{Z}$ has not been determined, but it is probably soundest to use the weighting

$$\tilde{Z} = \frac{\Sigma(NZ)}{\Sigma N} = \frac{\Sigma(CZ/A)}{\Sigma(C/A)} \tag{41}$$

Because backscattering is weak in specimens of low atomic number, the correction factor in biological tissues is close to unity. (This statement does not apply, of course, to inorganic inclusions!) In soft tissue, $\bar{Z}$ is approx. 3.9 and R, according to the cited tabulation, is generally in the range 0.98–1. In mineralized tissue $\bar{Z}$ is less than or equal to the value for apatite, which is 11.4, and R is generally within the range 0.96–1. However it may be quite important to correct for backscatter in the *standard*, especially when the atomic number of the assayed element is large and a pure standard is used. Values of R can be interpolated from Table IV, excerpted from Duncumb and Reed's paper.

TABLE IV

VALUES[a] OF THE BACKSCATTER COEFFICIENT R

Z E_0/E_x: 100	10	5	$2\frac{1}{2}$	$1\frac{2}{3}$	$1\frac{1}{4}$
0	1.00	1.00	1.00	1.00	1.00	1.00
10	0.93	0.94	0.95	0.97	0.98	0.99
20	0.86	0.87	0.89	0.92	0.95	0.98
30	0.79	0.81	0.83	0.87	0.91	0.96
40	0.74	0.76	0.78	0.83	0.88	0.94
50	0.69	0.72	0.74	0.79	0.85	0.92
60	0.66	0.69	0.71	0.76	0.83	0.90
70	0.64	0.66	0.69	0.74	0.81	0.89
80	0.61	0.64	0.67	0.72	0.79	0.87
90	0.59	0.61	0.64	0.70	0.77	0.86

[a] Excerpted from Duncumb and Reed (1968).

For given operating conditions (given E_0 and E_x), the coefficient R depends only on the mean atomic number. Hence in the literature one finds the backscatter correction lumped together with the "penetration" correction factor Z/A which was derived above, under the common heading of "the atomic number effect."

(*c*) *The Absorption Correction.* A large part of the generated X radiation may be absorbed within a specimen. Hence the ratio of observed intensities I_x/I_{x0} may be quite different from the ratio of quanta directly generated in specimen and standard, n_x/n_{x0}. Absorption is conventionally taken into account by means of a factor $f(\chi,h,\xi)$, which may be defined as the observed intensity of the selected characteristic radiation, divided by the intensity which *would* be observed if there were no X-ray absorption in the specimen. (Again we refer only to the directly generated radiation; the fluorescence radiation remains to be considered.) Like the backscatter coefficient, the absorption factor f is to be applied to both specimen and standard, as written below in Eq. (63), and the absorption effect vanishes if f is the same in specimen and standard.

The function $f(\chi,h,\xi)$ has been well determined by extensive theoretical and experimental work. We shall not review the theoretical derivation but merely quote the result, cookbook style, to enable the reader to do absorption corrections. Philibert's formula (1963) as modified by

Duncumb and Shields (1966) is

$$f(\chi, h, \xi) = \left[\left(1 + \frac{\chi}{\xi}\right)\left(1 + \frac{h}{1 + h}\frac{\chi}{\xi}\right)\right]^{-1} \tag{42}$$

with the definitions

$$h = 1.2\overline{A/Z^2} \tag{43}$$

$$\xi = 2.39\ 10^5/(E_0^{1.5} - E_x^{1.5}) \tag{44}$$

(E still expressed in keV), and

$$\chi = \sigma\ \csc\ \theta \tag{45}$$

θ is the so-called "takeoff angle" (see Fig. 20a), and σ is the X-ray absorption cross section of the specimen for the selected spectral line, in cm^2/g.

In the case of instruments where the direction of the incident electrons is not normal to the specimen surface, Eq. (45) must be replaced by the more general formula

$$\chi = \sigma\ \cos\ \beta\ \sec\ \gamma = \sigma\ \cos\ \beta\ \csc\ \theta \tag{46}$$

where β and γ are angles measured with respect to the normal to the specimen surface (see Fig. 20b). Quantitative analysis with an oblique probe can be as accurate as with the more usual normal incidence, as shown by Colby et al. (1969).

The absorption cross section σ of a multielement specimen is

$$\sigma = \sum_r (C_r\sigma_r) \tag{47}$$

with one term in the sum for each constituent element. The cross sections σ_r for the individual elements may be taken from a modern tabulation [for example, Heinrich (1966); and Henke et al. (1967) for wavelengths $\lambda > 12$ Å].

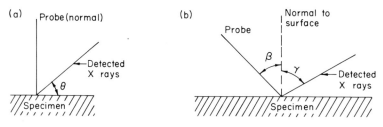

Fig. 20. The definition of the angles involved in the absorption correction. (a) Definition of θ [Eq. (45)] when the probe is normal to the specimen surface. (b) The definitions of β and γ [Eq. (46)] when the probe is not normal to the specimen surface. N.B. In Fig. 20b, the two rays and the normal are not necessarily all in one plane.

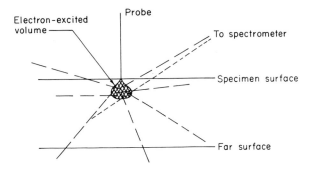

FIG. 21. The generation of fluorescence radiation. The primary X radiation, represented by dashed lines, is generated directly by the probe electrons within the shaded region. Only one ray of fluorescence radiation is shown (short dashes), but fluorescence radiation may be generated in any material reached by the primary X rays.

The foregoing Eqs. (42)–(47) provide a complete recipe for the calculation of the absorption function $f(\chi,h,\xi)$.

(d) *The Fluorescence Correction.* A particular X-ray spectral line can be excited by any radiation whose quantum energy exceeds the relevant ionization threshold. So far we have considered only direct excitation by the probe electrons, but additionally secondary "fluorescence" characteristic X-rays are excited by sufficiently energetic primary characteristic and continuum X-ray quanta, themselves generated within the specimen by the probe electrons (see Fig. 21).

The standard equations for the magnitude of the fluorescence effect are much too complicated to be presented fully here, especially in view of the fact that they do not apply accurately to any interesting specimens. In their derivation, it is assumed that the specimen is homogeneous, not only throughout the microvolume reached by the probe electrons but throughout the much larger volume reached by the primary X rays. If a specimen is that homogeneous, there is no interest in microanalyzing it! It is further assumed that the specimens are thick enough to absorb fully the down-going primary X rays. Biological specimens rarely satisfy this assumption. Consequently some of the potentially inducing radiation traverses and escapes the specimen, and the fluorescence effect is therefore usually smaller than indicated by the standard formulas. However, on the basis of the standard expressions, we shall be able to establish that X-ray fluorescence is negligible in most biological specimens, and we shall identify the situations where the effect may be significant. We shall also arrive at modified formulas applicable to some inhomogeneous specimens.

The excitation of fluorescence by characteristic and by continuum X rays must be considered separately. We can write

$$I = I_p + I_c + I_b \tag{48}$$

where I is the observed intensity of the selected spectral line, and I_p, I_c, and I_b are the contributions due respectively to primary excitation, characteristically induced fluorescence, and continuum-induced fluorescence. (The subscript b, stands for "bremsstrahlung," which is another word for the X-ray continuum.) Following the conventional notation, we define

$$\gamma_c = I_c/I_p \quad \text{and} \quad \gamma_b = I_b/I_p \tag{49}$$

The fluorescence correction factor I_p/I can then be written

$$\frac{I_p}{I} = \frac{1}{1 + \gamma_c + \gamma_b} \tag{50}$$

This factor is often written as

$$\frac{I_p}{I} = \frac{1}{1 + \gamma} \quad \text{with} \quad \gamma = \gamma_c + \gamma_b \tag{51}$$

but γ_c and γ_b must be evaluated separately.

(1) *Fluorescence Induced by Characteristic X rays in Normal Tissues.* The standard correction has been put into convenient form by Reed (1965). His tabulations show that for characteristically induced fluorescence to be significant, the inducing radiation must come from a major constituent in the specimen; the excitation threshold of the assayed radiation E_x must not be very low (the requirement is approximately that $E_x \geq 2$ keV, corresponding to a limit at the phosphorus K radiation); and the quantum energy of the inducing radiation must not be much greater than E_x. In normal soft tissue, the only major constituents are the elements C, N, O, and H, and the quantum energies of their characteristic radiations are too small to induce fluorescence radiation with $E_x \geq 2$ keV. Hence *characteristically induced fluorescence is always negligible in normal soft tissue.*

In normal hard tissue, the sole elemental constituent whose characteristic radiation may induce appreciable fluorescence is calcium, and the only normal constituents which may be at all appreciably excited by this radiation are the elements K, Cl, S, and P. The size of the effect depends strongly on operating conditions (take-off angle and probe voltage), but Reed's tabulations show that in homogeneous hard tissue, the potassium fluorescence intensity could well be as much as 7% of the primary potassium signal; the value is less than 2% for chlorine, and progressively less still for S and P. In specimens of hard tissue, the thickness is often enough to absorb almost all of the downwards primary characteristic radiation, so that these figures may be regarded as fairly accurate rather than as gross overestimations. Of course the constituents are *not* distributed homogeneously in real specimens, but we shall soon consider the implications of the estimation in more realistic terms.

For quantitative analysis, one should correct for characteristically induced fluorescence in the standard as well as in the specimen. However, for single-element standards this correction is *nil*, because the excitation of fluorescence by an element's own characteristic radiation is negligible. Ionization of the K shell by the K radiation is precluded because the quantum energy of the radiation is too small. The excitation of the L by the K, or of the M by the L radiation is insignificant because of the poor match of the quantum energies. As to L-L fluorescence, the commonly assayed L_α radiation *can* in fact be excited by the more energetic spectral lines in the L group, but for elements of high atomic number the intensity of the sufficiently energetic lines is extremely weak, while for the lower atomic numbers the effect is rendered negligible by the low values assumed by the fluorescence yield factor w_L.

(2) *Fluorescence Induced by Continuum X Rays in Normal Tissues.* We shall discuss continuum-induced fluorescence in terms of γ_b, the ratio of continuum fluorescence to primary radiation, defined precisely above by Eqs. (49) and (48).

An accurate expression for γ_b in thick homogeneous specimens has been obtained by Hénoc (1968). As this expression is too complicated for convenient use, we shall base our estimations instead on a modification of a simple formula derived by Green and Cosslett (1961). For a *single-element specimen*, their equations show that the ratio γ_b is approx.

$$\gamma_b \cong 10^{-11}(kE_x/R)ZA \tag{52}$$

In this formula R is the backscatter correction factor described above (Section VI.A.4.b), k can be interpolated from Table V (taken from Green and Cosslett), E_x is the ionization threshold for the assayed radia-

TABLE V

THE NUMBER k IN EQ. (52) AS A FUNCTION OF MEAN
ELECTRON ENERGY[a] $\bar{E}$

$\bar{E}$ (keV)[b]	1	2	5	10	20	50
$10^{-5}k$	1.0	1.3	1.8	2.3	2.9	4.4

[a] From Green and Cosslett (1961).
[b] $\bar{E} = \frac{1}{2}(E_0 + E_x)$.

tion in keV, and Z and A are the atomic number and the atomic weight of the element.

Equation (52) can be generalized for *multielement* specimens. The derivation will not be given here, but the result is

$$\gamma_b \cong 10^{-11} \frac{E_x k}{R} \frac{\overline{Z^2/A}}{\overline{Z/A}} A_x \frac{\sigma_{bx}}{\sigma_b} \tag{53}$$

Here the averages are weighted with respect to the weight-fraction C of each constituent (i.e., $\bar{F} = \Sigma CF$), σ_{bx} is the mass absorption coefficient of the assayed element x for the inducing continuum radiation, and $\sigma_b = \Sigma C_r \sigma_{br}$ [Eq. (47)] is the absorption coefficient of the specimen for the same radiation. Although σ_{bx} and σ_b depend strongly on wavelength, the ratio σ_{bx}/σ_b is almost independent of wavelength, so that one may use the ratio calculated for any quantum energy slightly above E_x (this energy region being responsible for most of the fluorescence).

In biological tissues, both soft and hard, Z/A is close to one-half for all of the major constituent elements and Eq. (53) reduces accurately to

$$\gamma_b \cong 10^{-11} \frac{E_x k}{R} \bar{Z} A_x \frac{\sigma_{bx}}{\sigma_b} \tag{54}$$

We shall use Eqs. (52) and (54) as guides to the importance of continuum-fluorescence, even though they produce overestimates of γ_b for two reasons. As already considered, they involve the assumption that the down-going inducing radiation is fully absorbed. Second, these equations neglect the absorption of the observed spectral line within the specimen; since on the average the fluorescence radiation is generated deeper down, it is more absorbed than the primary radiation and this neglect adds to the overestimation of γ_b. (Absorption of the selected spectral line *is* taken into account in Hénoc's more accurate continuum-theory, as well as in Reed's formula which was used above for characteristically induced fluorescence.) However the absorption of the fluorescence radiation usually does not change the order of magnitude of the effect because most of the fluorescence is induced by primary X radiation of similar penetrating power, so that the fluorescence is usually not generated too far down to get out.

In instruments with a very low take-off angle θ, the mean absorption path may be made so large by the geometrical factor $\csc \theta$ that the emergent fluorescence may be mostly repressed, but no modern commercial instrument has such a low take-off angle.

Overestimates of the ratio γ_b are listed in Table VI, for a variety

TABLE VI

THE RATIO γ_b OF CONTINUUM FLUORESCENCE TO PRIMARY RADIATION[a]

Assayed element x	Z_x	Spectral line	E_x	γ_b for single element	$\gamma_b(0)$ for soft tissue	$\gamma_b(0)$ for apatite	$C_x(\tfrac{1}{2})$ soft	$C_x(\tfrac{1}{2})$ apatite[b]
S	16	K	2.5	0.0025	0.010	0.0073	0.11	0.41
Cl	17	K	2.8	0.0035	0.016	0.012	0.089	0.33
Ca	20	K	4.0	0.0076	0.042	0.0096	0.057	1.0
Fe	26	K	7.1	0.031	0.28	0.059	0.025	0.35
Fe		L	0.71	0.0019	0.0020	0.0026	0.27	0.58
Zn	30	K	9.7	0.065	0.71	0.15	0.018	0.22
Zn		L	1.0	0.004	0.0053	0.0067	0.16	0.31
Mo	42	K	20.	0.36	—[c]	—	—	—
Mo		L	2.5	0.023	0.051	0.038	0.060	0.20
Ag	47	K	26.	0.57	—	—	—	—
Ag		L	3.4	0.044	0.12	0.028	0.042	0.63
Ag		M	0.40	0.0042	0.0018	0.0023	0.34	0.79
Sn	50	K	29.	0.80	—	—	—	—
Sn		L	3.9	0.064	0.19	0.044	0.035	0.49
Sn		M	0.51	0.0064	0.0031	0.0040	0.26	0.57
I	53	K	33.	1.05	—	—	—	—
I		L	4.6	0.092	0.26	0.059	0.035	0.49
I		M	0.63	0.0090	0.0049	0.0063	0.21	0.43
Os	76	K	74.	7.65	—	—	—	—
Os		L	11.	0.70	4.6	0.98	0.009	0.11
Os		M	2.0	0.070	0.081	0.060	0.056	0.19
Au	79	L	12.	0.89	6.1	1.3	0.009	0.097
Au		M	2.3	0.092	0.11	0.084	0.049	0.16
Pb	82	L	13.	1.09	6.8	1.5	0.009	0.10
Pb		M	2.6	0.11	0.13	0.098	0.050	0.17
U	92	L	17.	2.1	12.	2.5	0.008	0.092
U		M	3.6	0.25	0.34	0.077	0.036	0.50

[a] The values are according to Eqs. (52) and (54). For the determination of k in Eqs. (52) and (54), it is assumed that $E_0 = 3E_x$, so that $E = \bar{E} = \tfrac{1}{2}(E_0 + E_x) = 2E_x$.
[b] The definitions of $\gamma_b(0)$ and $C_x(\tfrac{1}{2})$ are explained in note h of the text.
[c] The blank spaces are due to a lack of data on absorption coefficients.

of assayed elements in the pure state (standards), in soft tissues, and in the mineral apatite $(3Ca_3(PO_4)_2 \cdot Ca(OH)_2)$. For the pure state, the ratio has been calculated from Eq. (52) and for soft tissue and apatite from Eq. (54).

To interpret Table VI properly, the following points should be noted:

a. In the derivation of Eqs. (52) and (54), a formula is needed for the cross section q_x for ionization by the probe electrons. For K, L, and M-shell excitation, we have used the formula quoted by Green and Cosslett for K-shell ionization,

$$q_x \quad (\text{cm}^2/\text{atom}) \cong 8 \; 10^{-20} \frac{1}{E_0 E_x} \ln \frac{E_0}{E_x} \tag{55}$$

The use of this expression for L and M as well as K-shell ionization introduces some error, which can be shown to be not excessive for the purpose of rough estimation.

b. For any given atomic shell (K, L, or M), and any given column of the table (single-element, soft tissue, or apatite matrix), the ratio γ_b rises smoothly and steeply with the atomic number Z_x of the assayed element. The only exceptions to this rule occur for spectral lines of low excitation threshold E_x, in soft tissue or apatite, where an increase in Z_x may increase the factor σ_b in Eq. (54) as the absorption edge of a major matrix constituent is crossed.

c. Since the X-ray continuum intensity is proportional to the mean atomic number of the specimen, there is a tendency to suppose that continuum-fluorescence must be negligible in specimens of low mean atomic number, including biological tissues. The table shows that this notion is false.

In *single-element specimens*, since E_x is proportional to Z^2 and A is proportional to Z, Eq. (52) shows that γ_b varies approx. as Z^4, and the continuum-fluorescence is indeed negligible at low Z.

But the situation can be much worse for the assay of heavy elements in a light matrix. Dividing Eq. (54) by Eq. (52), we compare the continuum-fluorescence from element x in a matrix to the fluorescence from a specimen of pure x:

$$\frac{\gamma_b}{\gamma_b(\text{pure})} = \frac{R_{\text{pure}}}{R_{\text{mix}}} \frac{\bar{Z}}{Z_x} \frac{\sigma_{bx}}{\sigma_b} \tag{56}$$

In this equation $R_{\text{pure}}/R_{\text{mix}}$ does not differ greatly from unity while the factor $\bar{Z}/Z_x$, which corresponds to the relative continuum intensity in the two specimens, favors a weaker fluorescence effect in the light matrix. But the dominant factor is σ_{bx}/σ_b, which can be extremely large for a heavy element. As a result, the fluorescence effect in a biological matrix may be much greater than in the pure element. Physically, the reason is that the heavy element can absorb a very disproportionate share of the inducing continuum radiation because of the low absorption cross section σ_b that goes with the low atomic number of the matrix.

d. Equation (54) overestimates the fluorescence in most soft-tissue specimens because they are usually not thick enough to absorb all of the downwards continuum X rays. But where Eq. (54) indicates a large correction, the effect is usually at least substantial in the specimens which we have termed "electron-thick." Furthermore it is not possible to make large fluorescence corrections accurately in real specimens; even if one worked out a formula for specimens which are less than "X-ray thick," it could not take account of the inhomogeneities in biological specimens. Hence the only hope for accurate quantitation in "electron-thick" biological specimens is to choose operating conditions for which Table VI indicates that continuum-fluorescence is slight.

e. The parameter which most influences γ_b is the excitation threshold E_x. When E_x is reduced, the mean quantum energy and the mean penetrating power of the inducing continuum radiation are reduced, and the ratio σ_{bx}/σ_b in Eq. (54) falls very steeply.

One should refer to Table VI in choosing the spectral line to be used for an analysis, but a rough rule is to limit E_x to $E_x \leq 4$ keV. This means switching to L radiations for $Z_x > 20$, and from L to M in the vicinity of $Z_x = 50$. (The situation near $Z = 50$ is complicated. One may have to observe L radiation and accept a large fluorescence error when the M lines of the assayed element are close to the characteristic radiations from carbon and nitrogen, because the wavelength resolution of diffractors is not good at these long wavelengths.)

The condition $E_x \leq 4$ keV has already been suggested above (Section IV.A) as necessary for the preservation of good spatial resolution in thick specimens of soft tissue. It is now seen to be essential for accurate quantitative analysis in such specimens as well.

f. Continuum fluorescence in highly mineralized tissue is weaker than in soft tissue. Let us use Eq. (54) to compare the fluorescence of a given spectral line in soft tissue and in apatite. The factors E_x, k, and A_x are the same in both cases and R is only slightly different. In apatite, $\bar{Z}$ is 14.1 as against $6\frac{1}{2}$ in soft tissue, and σ_{bx}/σ_b in apatite happens to be one-tenth of its value in soft tissue over the range where fluorescence may be important, i.e., $E_x \geq 4$ keV. The net result is that for a given spectral line, γ_b in an apatite matrix generally has approx. one-fifth of its value in a matrix of soft tissue. Thus one can afford to detect K radiations up to somewhat higher values of Z_x before the fluorescence is intolerable. But the effect is still not to be ignored: γ_b for zinc K radiation in an apatite matrix, for example, is approx. 15%.

g. In single-element standards, the thickness is usually enough to absorb all of the downwards continuum radiation, while the absorption of the observed spectral line may be considerable. Consequently Eq. (52) is inaccurate, but fluorescence in pure standards may be accurately calculated by Hénoc's formula. We shall not discuss this complicated formula because a biologist is rarely driven so far as to use it. Table VI shows that when there is substantial fluorescence in a single-element standard, there will usually be a still larger fluorescence effect in the biological specimen. Since accurate correction for the specimen is impossible, there is generally little point to making the accurate correction for the standard.

h. In computing Table VI from Eq. (54), we have neglected the contribution of the assayed element to the total absorption cross section σ_b. In fact, at the usual low concentrations of assayed elements this is not too bad an approximation in spite of the commonly high value of σ_{bx}. But the values tabulated for γ_b in specimens are therefore really the limits asymptotically approached as $C_x \to 0$; this is the meaning of the notation $\gamma_b(0)$ as a column heading in Table VI.

For the general case we may write

$$\sigma_b = C_x \sigma_{bx} + (1 - C_x)\sigma_b' \tag{57}$$

where σ_b' would be the absorption cross section in the absence of the assayed element. As C_x is imagined to increase, σ_b increases (assuming $\sigma_{bx} > \sigma_b'$); hence according to Eq. (54), γ_b becomes progressively less than $\gamma_b(0)$. It is convenient to define a weight-fraction $C_x(\frac{1}{2})$, the value of C_x at which γ_b falls to half of the tabulated value $\gamma_b(0)$. From (57) it is readily deduced that

$$C_x(\tfrac{1}{2}) = \sigma_b'/(\sigma_{bx} - \sigma_b') \tag{58}$$

Values of $C_x(\frac{1}{2})$ are listed in Table VI. As long as the weight-fraction C_x is well under $C_x(\frac{1}{2})$, the tabulated quantity $\gamma_b(0)$ is a reasonable approximation to γ_b as defined in Eq. (54). But when C_x is too large for this approximation to be acceptable, the legitimate γ_b can easily be estimated from the relation

$$\gamma_b = \gamma_b(0) \frac{1}{1 + C_x/C_x(\tfrac{1}{2})} \tag{59}$$

The approximation $\gamma_b \approx \gamma_b(0)$ is poor especially in one important category of studies—assays of the heavy-metal stains Os, Pb, and U. In tissues stained with these elements, their weight-fractions C_x may easily exceed the low values of $C_x(\frac{1}{2})$ listed in the table. In this case, $\gamma_b(0)$ may be much larger than γ_b, and estimations of γ_b should be based on Eq. (59).

To sum up: When the continuum-fluorescence effect in biological specimens is large, there is no hope of correcting for it accurately, but the effect can usually be kept small or entirely negligible by restricting X-ray analysis to spectral lines with low excitation thresholds E_x.

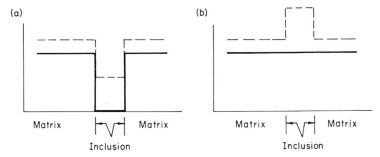

FIG. 22. Possible specious fluorescence effects at inclusions. Solid curves: Elemental distribution (hypothetical). Dashed curves: X-ray signal vs. position of probe. (a) When the probe is confined to the inclusion and the inclusion contains *none* of the assayed element, a signal is still obtained due to fluorescence in the matrix. (b) The elemental concentration is the same in matrix and inclusion, but the signal rises as the probe comes on to the inclusion because of the increase in continuum intensity and hence in continuum-induced fluorescence.

(3) *Fluorescence in certain inhomogeneous specimens.* We have already talked about fluorescence in the neighborhood of a large plane boundary (Section IV.A, Fig. 17). For qualitative discussions of this case, the reader may refer to Dils *et al.* (1963) and Reed (1966).

There is another situation where seriously misleading effects may occur: when isolated inclusions are present in a tissue and the assayed element is not predominantly localized to the inclusions. As the probe is moved from the matrix onto an inclusion, the intensity of the continuum changes, and the resulting change in continuum-induced fluorescence may give a completely false impression. The assayed element may speciously appear to be present in the inclusions (Fig. 22a) or may even speciously seem to be especially concentrated there (Fig. 22b).

The equations for homogeneous specimens can be modified readily to apply to isolated inclusions, provided that the inclusions are larger than the range of the probe electrons and much smaller than the range of the primary X rays (for example, inclusions of high mean atomic number, $\frac{1}{2}$–1 μm in diameter). Then when the probe is confined to the inclusion, the primary signal and the continuum intensity will depend only on the composition of the inclusion, while the fluorescence will be generated almost entirely in the surrounding matrix (cf. Fig. 23). Hence in place of Eq. (54) one gets

$$\gamma_b' \cong 10^{-11} \frac{kE_x}{R'} \bar{Z}' A_x \frac{\sigma_{bx}}{\sigma_b} \frac{C_x}{C_x'} \tag{60}$$

where γ_b' is the ratio of continuum-fluorescence to primary signal with the probe on the inclusion and all of the primed quantities refer to the

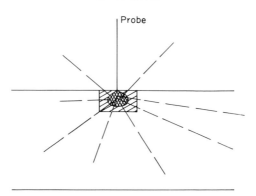

FIG. 23. Conditions for the use of Eq. (60) when the probe strikes an inclusion. Singly shaded region: the inclusion. Doubly shaded region: the electron-excited volume. Dashed rays: primary X rays.

inclusion, while σ_{bx}/σ_b and C_x refer to the surrounding matrix. (The assumption $Z/A \sim \frac{1}{2}$, inherited from Eq. (54), may not be accurate in the inclusion, but the associated error is unimportant in this context.)

Equation (60) shows that the continuum fluorescence is grossly amplified when the probe is put onto a heavy inclusion in a light matrix. The factor Z' represents the intensified continuum produced in such an inclusion, while the large factor σ_{bx}/σ_b indicates the efficient conversion of that continuum into fluorescence in the light matrix. Comparing Eqs. (54) and (60), we see that as the probe moves from matrix to inclusion, the continuum-fluorescence effect, loosely speaking, is amplified by the factor $\bar{Z}'/\bar{Z} = \bar{Z}(\text{inclusion})/\bar{Z}(\text{matrix})$.

It is more useful to consider the total effect on observed intensity as the probe moves from matrix to inclusion. We now write $(I_p + I_b)$ for the sum of the observed primary signal plus continuum-fluorescence when the probe is on the matrix, and $(I_p' + I_b')$ for the corresponding quantities with the probe on the inclusion. In a straightforward way, one obtains

$$\frac{I_p' + I_b'}{I_p + I_b} = \frac{1}{1 + \gamma_b}\left(\frac{R'C_x'}{RC_x} + \gamma_b\frac{\bar{Z}'}{\bar{Z}}\right) \tag{61}$$

(primed quantities referring to the inclusion and unprimed, including γ_b, to the matrix).

If the assayed element is completely absent from the inclusion (i.e., $C_x' = 0$), I_p' is zero but Eq. (61) shows that one may still observe a substantial signal I_b'. For instance, if calcium were assayed in a matrix of soft tissue and the probe were restricted to an inclusion of gold containing *no* calcium, γ_b would be 0.04, $\bar{Z}'/\bar{Z}$ would be $79/6.5 = 12$, and Eq. (61)

shows that the calcium signal "from the inclusion" might be approx. half of the intensity observed in the matrix! This situation has been sketched in Fig. 22a.

If the weight-fraction of the assayed element is actually the same in matrix and inclusion (i.e., $C_x' = C_x$), Eq. (61) shows that the observed intensity may deceptively rise considerably as the probe comes onto the inclusion. This is the situation sketched in Fig. 22b.

Similar effects may occur in mineralizing tissues where small mineral plaques are surrounded by organic tissue. Here however the effect can only be slight since the value of $\bar{Z}$ in calcium phosphate is only approx. twice the value in soft tissue.

In the discussion of isolated inclusions, we have not yet considered fluorescence induced by characteristic X rays. Earlier we noted that such fluorescence is generally unimportant in normal biological tissues, because of the lack of elements with efficiently inducing spectral lines. However if such elements are present in inclusions, the ratio γ_c of characteristically induced fluorescence to primary signal may be quite high in biological specimens. Indeed, when the inclusion is larger than the electron range and much smaller than the range of the inducing X rays, the effect may be "amplified" in the manner already noted with continuum fluorescence: The inducing radiation may be generated at maximum intensity within the inclusion, while the conversion to fluorescence may be exceptionally efficient because of the low absorption cross section of the biological matrix, which competes with the assayed element for the absorption of the inducing radiation.

As in the case of continuum-fluorescence, we should consider the total effect on observed intensity as the probe moves from matrix to inclusion. An equation similar to (61) can readily be derived:

$$\frac{I_p' + I_c'}{I_p + I_c} = \frac{1}{1 + \gamma_c} \frac{R'}{R} \left(\frac{C_x'}{C_x} + \gamma_c' \right) \tag{62}$$

The quantities I_c and γ_c can be omitted from Eq. (62) since the characteristically induced fluorescence is generally negligible when the probe is on the matrix. The main job in the use of Eq. (62) is the calculation of the quantity γ_c', which is to be obtained by the following rule: Evaluate the formula given by Reed (1965) for γ_c in homogeneous specimens, substituting for each symbol its value in the matrix, except for the weight-fraction of the fluorescence-inducing element—for this parameter substitute the value in the inclusion.

The characteristically induced fluorescence of any spectral line is to be estimated separately for each possible inducing element, but the effect is significant only when the quantum energy of the inducing radiation is slightly above the excitation threshold of the observed spectral line. A close match is not encountered often in biological studies, even when inclusions are involved. However the possibility should not be forgotten, as a highly concentrated, efficiently inducing element can produce the artifacts of Fig. 22 on the scale there indicated. Equation (62) together with Reed's paper is sufficient for estimating the danger.

(e) *Overall Equations for Quantitation.* When the naive equation (18) is amended to take account of "penetration effects," backscattering, absorption within the specimen, and fluorescence, we get

$$C_x(\text{sp}) = C_x(\text{st}) \frac{I_x(\text{sp})}{I_x(\text{st})} \left[\frac{P(\text{st})}{P(\text{sp})} \frac{R(\tilde{Z}_{\text{st}})}{R(\tilde{Z}_{\text{sp}})} \frac{f(\chi_{\text{st}}, h_{\text{st}}, \xi)}{f(\chi_{\text{sp}}, h_{\text{sp}}, \xi)} \frac{1 + \gamma_b(\text{st}) + \gamma_c(\text{st})}{1 + \gamma_b(\text{sp}) + \gamma_c(\text{sp})} \right]$$

(63)

Here the notations sp and st refer respectively to the analyzed spot in the specimen and in the standard; C_x is the weight-fraction of the assayed element; and I_x is the observed intensity of the selected spectral line (obtained after suitable subtraction of spectrometer background). The penetration correction $P(\text{st})/P(\text{sp})$ comes from Eq. (40):

$$\frac{P(\text{st})}{P(\text{sp})} = \frac{\overline{Z/A}(\text{sp})}{\overline{Z/A}(\text{st})} \frac{2 + \log \bar{E} - (\overline{\log Z})_{\text{sp}}}{2 + \log \bar{E} - (\overline{\log Z})_{\text{st}}}$$

(64)

The backscatter correction $R(\text{st})/R(\text{sp})$ can be obtained from Table IV in Section VI.A.4.b above; the absorption correction $f(\text{st})/f(\text{sp})$ is given in Eqs. (42)–(47); and the fluorescence correction $[1 + \gamma_b(\text{st}) + \gamma_c(\text{st})]/[1 + \gamma_b(\text{sp}) + \gamma_c(\text{sp})]$ has just been discussed. The averages denoted by a straight bar are all weighted with respect to weight-fraction (i.e., $\bar{F} = \Sigma_r C_r F_r$), while the average $\tilde{Z}$ is weighted with respect to atom-fractions [i.e., $\tilde{Z} = \Sigma N_r Z_r / \Sigma N_r$, Eq. (41)].

Equation (63) differs from the approximate Eq. (18) only in the presence of the bracketed correction factors. The approximate equation (18) is extremely simple to use—only one spectral line has to be observed, and the computation is trivial. But in order to work out the correction factors in Eq. (63), the composition of the assayed spot must be postulated. In metallurgy and mineralogy, it is commonly necessary to observe two or more spectral lines to establish the composition adequately, and an iterative procedure must be used if the end-result of Eq. (63) is too inconsistent with the postulated composition. However in biology, one often can work out the correction factors well enough on the basis of the nominal composition of soft tissue or bone-mineral, with no need for iteration.

Equation (63) is somewhat simplified if a single-element standard is used. Then $C_x(\text{st}) = 1$ and $\gamma_c(\text{st}) = 0$; $\bar{Z}_{\text{st}} = \tilde{Z}_{\text{st}} = Z_x$; and the penetration correction reduces to

$$\frac{P(\text{st})}{P(\text{sp})} = \frac{\overline{Z/A}(\text{sp})}{Z_x/A_x} \frac{2 + \log \bar{E} - (\overline{\log Z})_{\text{sp}}}{2 + \log \bar{E} - \log Z_x} \tag{65}$$

For specimen spots consisting predominantly of soft tissue or apatite, the labor of making corrections can be reduced with the help of values worked out in advance for several of the quantities involved in Eqs. (63) and (64). In a *soft tissue* with a composition 7% H, 50% C, 16% N, 25% O, and 2% (S + P) by weight, containing no other elements in substantial amounts, among the quantities involved in Eq. (63) $\tilde{Z}_{\text{sp}} = 3.9$, $\bar{Z}_{\text{sp}} = 6.5$, and $h_{\text{sp}} = 0.42$, while the penetration correction (64) reduces to

$$\frac{P(\text{st})}{P(\text{sp})} = \frac{0.535}{Z/A(\text{st})} \frac{1.226 + \log \bar{E}}{2 + \log \bar{E} - (\overline{\log Z})_{\text{st}}} \tag{66}$$

In a specimen region containing *apatite* $(3\text{Ca}_3(\text{PO}_4)_2 \cdot \text{Ca}(\text{OH})_2)$ as the only significant constituent, $\tilde{Z}_{\text{sp}} = 11.4$, $\bar{Z}_{\text{sp}} = 14.1$, and $h_{\text{sp}} = 0.20$, while the penetration correction reduces to

$$\frac{P(\text{st})}{P(\text{sp})} = \frac{0.501}{Z/A(\text{st})} \frac{0.889 + \log \bar{E}}{2 + \log \bar{E} - (\overline{\log Z})_{\text{st}}} \tag{67}$$

The presence of other calcium phosphates instead of apatite makes virtually no difference in these values. Even when the hard-tissue constituent is not a phosphate at all but is calcium carbonate, the only significant change from the values quoted for apatite is that h (calcium carbonate) = 0.24. Similarly, the values quoted above for soft tissue are scarcely affected by normal variations in composition. The important variable in the composition of the soft tissues is the weight-fraction of hydrogen C_H, and in place of the value $\overline{Z/A} = 0.535$ in Eq. (66), one may want to use the form

$$\overline{Z/A} \quad (\text{soft tissue}) = 0.500 + C_H/2 \tag{68}$$

In tissues where inorganic and organic phases are finely intermingled (some regions of bone for example), the corrections are obviously somewhat uncertain.

(*f*) *The Magnitude of the Corrections.* The penetration correction [Eq. (64)] may differ significantly from unity, especially when Z_x is large and a single-element standard is used. However this correction is easily evaluated accurately.

The backscatter factor R (specimen) is generally close to unity in biological tissues. For single-element standards, R(st) may be considerably less than unity but is easily obtained from Table IV.

The absorption factors f(st) and f(sp) are usually in the range $\frac{1}{2}$–1 but are sometimes *very* different from unity—in extreme cases absorption reduces the observed intensity by 90% or more.

As to fluorescence, one can usually choose operating conditions which make it negligible or slight in biological specimens. Under these conditions, the fluorescence in a single-element standard is almost always negligible.

(*g*) *The Use of Multielement Standards.* If the standard matches the composition of the specimen, the correction factors for specimen and standard in Eq. (63) may virtually cancel. Then one may be able to use the uncorrected equation (18) as an accurate formula even when the individual correction factors are far from unity.

The main difficulty in this strategy is to find or make suitable standards, which must be stable under the probe and very homogeneous. Satisfactory standards are gradually being developed.

For assays of calcium and phosphorus in highly mineralized tissue like dental enamel, apatite is a good standard. Fluorapatite $(Ca_5(PO_4)_3F)$ has been used by Wörner and Wizgall (1969) for the assay of fluorine in enamel, and Johnson (1969) has used apatite loaded with strontium for studies of strontium in bone.

For assays in soft tissues, standards have been prepared by dissolving salts in protein solutions. Thus, as standards for K, Na, and Cl, Ingram and Hogben (1968) have dissolved KCl or NaCl in 20% albumin, ejected droplets of solution into liquid propane, and dried the frozen droplets in vacuum. Similarly, as a standard for iodine, Zeigler *et al.* (1969) have dissolved KI in hot aqueous gelatine, dried the cooled gel in vacuum, and finally used an evacuated die and press to obtain nonporous pellets.

Standards can also be prepared by loading ion-exchange beads with the element of interest (Freeman and Paulson, 1968). This promising technique, still under development, should be applicable to a wide range of elements.

(*h*) *Assumptions Underlying Quantitation in Thick Specimens.* The thick-specimen theory of quantitation rests on certain assumptions which have not yet been considered carefully here. If the equations are applied mechanically when the necessary conditions are not satisfied, one may get fallacious results (which cannot be checked by any other analytical technique). The assumptions have to do with electron opacity, homogeneity, surface finish, and "chemical shift."

The thick-specimen theory requires that the specimen be "electron-opaque," meaning roughly that the probe electrons must not be able to reach the far side of the specimen. More precisely, the probe electrons should not escape from the sides or bottom of the specimen while still capable of excitation, i.e., with energies above the threshold E_x. One must be especially careful about this requirement when analyzing porous specimens, or individual thin fibers or particulates. The sufficiency of the electron opacity in any assay can be judged on the basis of Section IV.A and Fig. 16.

The corrections for absorption and fluorescence rest on assumptions about specimen homogeneity. As we have already noted, the intensity of fluorescence can be calculated accurately only when the specimen is homogeneous throughout the range of the inducing X rays; since this requirement is generally not satisfied in real biological specimens, the only practical policy is to choose operating conditions which render fluorescence unimportant. As to the absorption correction, it has been implicitly assumed that the specimen is homogeneous throughout the region excited by the probe electrons—otherwise the X rays are not generated with the distribution in depth that is assumed in the derivation of Eq. (42). In fact, the equation depends on homogeneity throughout the somewhat larger region of the specimen traversed by the detected X rays, since inhomogeneity along the exit path obviously changes the degree of absorption. Such inhomogeneity can produce very misleading effects, as indicated in Fig. 24. For example, there may be an illusion of elevated concentrations in the enamel adjacent to the enamel-dentin junction in teeth, if the X rays from the enamel traverse the less absorbent dentin on their way to the spectroscope (Frazier, 1966).

The requisite homogeneity is not only chemical (uniformity of compo-

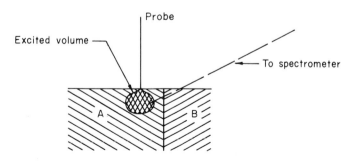

Fig. 24. Possible specious absorption effect at phase boundaries. If the absorption coefficient of the detected radiation in phase B is less than in phase A, the concentration of the assayed element near the boundary may spuriously seem to be higher than elsewhere in phase A.

sition) but physical as well: For the absorption formulas to be valid, the *density* (mass/volume) must be uniform. Specifically, porous specimens are *not* homogeneous in the present sense.

The requirements of electron-opacity and local homogeneity are naturally satisfied in certain tissues, notably dental enamel and fully mineralized compact bone. But the quantitative analysis of thick *porous* specimens is not reliable, because there is substantial X-ray absorption when the probe electrons penetrate deeply through the pores, and the inhomogeneity precludes an accurate absorption correction. Hence most tissues, which become porous if they are merely dehydrated, must be rendered realiably opaque and adequately homogeneous by embedding in a medium of similar composition and density, such as epoxy resin in the case of soft tissues.

When embedding medium is present during an analysis, two possibilities must be distinguished. If one knows that there is no embedding medium within the excited volume, then one knows that the analysis refers to the tissue itself. (This conclusion is not upset by the presence of embedding medium along the exit path of the X rays *outside* of the excited volume.) But more often, the excited volume contains or may contain a mixture of tissue and embedding medium. The measured weight-fraction is then the mass of element divided by the total mass within the excited volume, the total mass consisting of tissue plus medium. This weight-fraction may be much less than the value for the tissue proper.

In specimens with high X-ray absorption coefficients, quantitation requires that the top surface be flat and smooth. The criterion is that there must not be substantial absorption within path lengths corresponding to the deviations from flatness. In metallurgical work, where X-ray absorption coefficients are usually quite high, the specimens are commonly polished. However, sections of embedded soft tissue are generally sufficiently flat after being cut in the usual way without polishing. Among specimens of hard tissue, the adequacy of the surface finish must be evaluated separately for each case.

Finally, another assumption of the theory of quantitation involves chemical binding. We are referring here not to the effect of binding on the law of electron energy-loss (already discussed in Section VI.A.4), but to a direct effect on the excitation of the observed spectral line. In comparing the excitation in specimen and standard (Eqs. (29)–(31)], it was implicitly assumed that the fluorescence yield w and the ionization cross section q_x are independent of chemical state. This assumption is not entirely valid.

Characteristic X radiation is excited in a two-step process. First an "inner" electron is ejected from the atom; then an "outer" electron radiates the X ray as it "jumps" down to the vacated inner level. The first step, the ionization, depends mainly on the inner electron orbits which are scarcely affected by chemical state. But the second step, the emission, depends on electrons which start from outer levels which may be substantially affected.

One effect of an altered chemical state is to change slightly the energy levels of the bound electrons and hence to produce a "chemical shift" in the quantum energy and the wavelength of each spectral line. If the only effect is to shift the detected line by a fixed amount in the specimen, there is no serious problem; one simply has to peak the spectrometer separately for specimen and standard. But if the element is present in different chemical states in the specimen, the single spectral line may turn into two or more close lines, muddling the quantitation.

Aside from a shift in wavelength, an altered chemical state can also change the value of the fluorescence yield w for the selected spectral line. This may occur because the inner vacancy can be filled from different outer-electron states, each corresponding to a different spectral line, and the perturbation in the outer states may affect the relative transition probabilities. Thus the relative intensities of the different spectral lines may be changed.

In practice, chemical shift and chemically induced alterations in w and q_x are negligible except for radiations of long wavelength, approx. $\lambda > 5$ Å. In the range $5 < \lambda < 20$ Å, the effects are only slight. At longer wavelengths, one must hope that the assayed element is present in only one chemical environment (as is often the case). If the same environment prevails in the standard, quantitation should be feasible; otherwise w in specimen and standard may differ to a degree which is not yet well known.

(*i*) *A Critical Comment.* We have considered quantitation at length because it is a major part of the microprobe literature, and localized quantitative analyses are extremely desirable, but one must be aware of the many pitfalls and limitations. The contribution of microprobe analysis to biology is mainly the identification and localization of constituents, and approximate rather than accurate quantitative analysis.

B. THIN SPECIMENS

In thin specimens, the intensity of a characteristic radiation cannot suffice as a measure of weight-fraction, since the intensity obviously depends not only on weight-fraction but also on the local thickness or mass per unit area. Microprobe data may be completely misinterpreted if this point is not kept in mind.

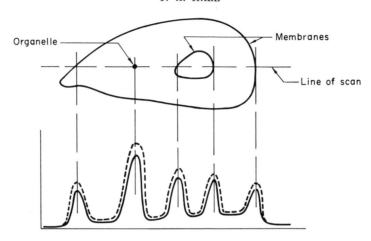

FIG. 25. Relationship between weight-fraction and X-ray intensity in thin specimens. Solid curve: Local mass per unit area along a line of scan (hypothetical).
Dotted curve: Characteristic X-ray signal along the line of scan (hypothetical).
The elemental weight-fraction is constant along the line of scan in this hypothetical
case in spite of the great variation in intensity of the X-ray spectral line.

Consider for example an ordinary thin section of soft tissue. Since
analysis is done in vacuum, the tissue must be dehydrated, leaving an
uneven distribution of tissue dry mass concentrated at regions of insoluble protein or lipoprotein (membranes and certain cell organelles, promi-
nently). Suppose now that the dry mass of tissue per unit area and
the observed intensity of a spectral line varied over a region of the
specimen as sketched in Fig. 25. Judging solely from the intensity of
the spectral line, one might naively conclude that the weight-fraction
of the assayed element was especially high at the positions of the peak
counting rates, but actually the elemental weight-fraction (elemental
mass/tissue dry mass) is constant throughout the linear scan in this
hypothetical case.

We shall recapitulate here the theory of quantitation developed for
thin biological specimens by Marshall and Hall [(1968), see also Hall
(1968), Hall and Höhling (1969), Hall et al. (in press)].

1. The Measurement of Elemental Weight-Fractions

The theory is conceived in terms of the quantities S_x (the local mass
of the assayed element per unit area), S (the local total mass of specimen
per unit area), and C_x (the local weight-fraction of the assayed element). (The precise definition of a local mass per unit area is as indicated in Fig. 26: the mass within a small cylinder coaxial with the probe

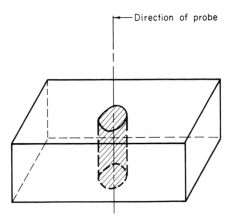

FIG. 26. Definition of a local mass per unit area S. S is mass within shaded cylinder/cross-sectional area of the cylinder.

divided by the cross sectional area.) The three quantities are related by

$$C_x = \frac{\text{elemental mass}}{\text{total mass}} = \frac{\text{elemental mass/area}}{\text{total mass/area}} = \frac{S_x}{S} \qquad (69)$$

The observed intensity of the spectral line is in essence proportional to S_x, the elemental mass/area. From Eq. (69) we see that in order to determine the weight-fraction C_x, one wants additionally a measure of S, the total mass/area. This measure is provided in the Marshall Hall theory by the intensity of the continuum radiation.

Instrumentally two new requirements appear. First, a solid-state or gas counter is fitted to the column to receive X rays nondiffractively from the specimen (Fig. 27); the count from this detector is restricted

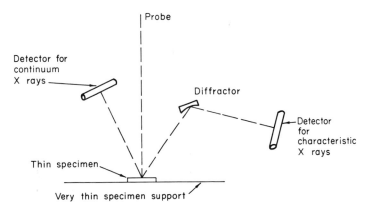

FIG. 27. Configuration for the detection of characteristic and continuum X rays from a thin specimen.

to a band of quantum energies in the X-ray continuum, excluding the characteristic X rays by means of pulse-height selection (Section III. B.6). Secondly, the specimen must be mounted on a very thin supporting film; the film then adds only a small constant background which can be separately measured and subtracted from the X-ray counts.

As a probe electron travels a distance ds in a specimen, it generates an average number $\eta_x \, ds$ of quanta of the selected spectral line, given by

$$\eta_x \, ds = w_x q_x N_x \, ds \tag{70}$$

(All notation will be defined directly below.) Simultaneously the probe electron generates an average number of continuum quanta $N_b \, dV \, ds$ in the quantum energy range between V and $(V + dV)$. According to the theory of Kramers (1923), which has recently been checked experimentally by Rao-Sahib and Wittry (1969), this number can be expressed in the form

$$\eta_b \, dV \, ds = \frac{k}{E} \frac{dV}{V} \sum_r (N_r Z_r{}^2) \, ds \tag{71}$$

When one compares the intensities from the specimen and from a standard, many factors cancel and the division of Eq. (70) by Eq. (71) gives the following equation for the ratio of characteristic and continuum intensities:

$$\frac{(\text{characteristic count/continuum count})_{\text{specimen}}}{(\text{characteristic count/continuum count})_{\text{standard}}} = \frac{(N_x/\Sigma_r N_r Z_r{}^2)_{\text{specimen}}}{(N_x/\Sigma_r N_r Z_r{}^2)_{\text{standard}}} \tag{72}$$

The symbols in these equations are defined as follows:

x subscript identifying the assayed element

N_x atoms of element x/cm³

q_x ionization cross section, cm²/atom of x, for ionization of the atomic shell associated with the observed spectral line

w_x fluorescence yield (fraction of the ionizations of the associated shell terminating in the emission of a quantum of the observed spectral line)

k a constant

E energy of the probe electron

Z atomic number

r index running over all constituent elements of specimen or standard

The only requirements on the standard are that it must contain the element x; its composition must be known; and it must be thin in the sense defined earlier. Beyond this specification, there is no need to know its thickness.

Equation (72) can be put into terms of weight-fractions C by means of the identity $C_x = N_x A_x / \Sigma(NA)$. It is convenient to introduce the following symbols:

R_x $\dfrac{\text{(characteristic count for element } x/\text{continuum count)}_{\text{specimen}}}{\text{(characteristic count for element } x/\text{continuum count)}_{\text{standard}}}$

("Characteristic count" refers to the genuine characteristic signal, i.e., with spectrometer background subtracted, and "continuum count" refers to the count from the specimen proper, with the count from the supporting film subtracted.)

G_x $N_x / \Sigma(NZ^2)$ in the standard for element x.

A_x atomic weight of element x.

Equation (72) then takes the form

$$C_x = R_x A_x G_x \sum_r (C_r Z_r^2 / A_r) \tag{73}$$

It follows that for any two elements a and b,

$$\frac{C_a}{C_b} = \frac{R_a A_a G_a}{R_b A_b G_b} \tag{74}$$

so that

$$C_x = R_x A_x G_x / \Sigma_r R_r A_r G_r \tag{75}$$

In order to use Eq. (75), one would have to measure R for every element that might contribute substantially to the denominator on the right-hand side. This would mean observing one spectral line and having a standard for each such element, but the measurements can be greatly simplified by a further development of the equations.

Typically, especially in soft biological tissue, the analyzed volume may be regarded as consisting of a standard matrix plus the element to be assayed, with a few other elements possibly present in important unknown amounts. If we use the subscript m for all of the matrix elements, and the subscript u for all other important elements (including the element x), Eq. (75) can be written

$$C_x = \frac{R_x A_x G_x}{\Sigma_m(RAG) + \Sigma_u(RAG)} \tag{76}$$

From Eq. (73) it can be shown, without approximation, that

$$\sum_m (RAG) = \frac{1 - \Sigma_u(RGZ^2)}{\Sigma_m(C'Z^2/A)} \tag{77}$$

where the quantity C_k' for a matrix-element k is defined by

$$C_k' = C_k/\Sigma_m(C) \tag{78}$$

Thus Eq. (76) becomes

$$C_x = \frac{R_x A_x G_x}{\dfrac{1 - \Sigma_u(RGZ^2)}{\Sigma_m(C'Z^2/A)} + \Sigma_u(RAG)} \tag{79}$$

Equation (79) may seem forbidding but it is really quite easy to use. There is no need to observe the spectral lines of the elements of the postulated matrix. The intensity of the characteristic radiation of the assayed element x must of course be recorded, but other spectral lines must be observed only if they belong to nonmatrix elements with high local weight-fractions. In the common case where there are no such elements (aside possibly from element x itself), Eq. (79) reduces to

$$C_x = \frac{R_x A_x G_x}{\dfrac{1 - R_x G_x Z_x^2}{\Sigma_m(C'Z^2/A)} + R_x A_x G_x} \tag{80}$$

When C_x itself is sufficiently low, Eq. (80) reduces further to

$$C_x = R_x A_x G_x \sum_m (C'Z^2/A) \tag{81}$$

Of the quantities in the formulas (80) and (81), only R_x has to be measured. The others are all known or readily calculable without any measurements. In the more general case [Eq. (79)], it may be necessary to measure one or two additional ratios R_u, but rarely more than two.

The postulated matrix enters the Eqs. (79)–(81) only through the factor $\Sigma_m(C'Z^2/A)$. For each matrix element k, C_k' is the weight-fraction of the element *in the matrix* (i.e., mass of element/mass of matrix). Thus C_k' and $\Sigma_m(C'Z^2/A)$ depend solely on the postulated composition of the matrix. For example, in a soft tissue with a postulated matrix composition 7% H, 50% C, 16% N, 25% O, and 2% (S + P), one has $C_H' = 0.07$, etc., and in fact $\Sigma_m(C'Z^2/A) = 3.28$. The value of $\Sigma_m(C'Z^2/A)$ does not depend on whether the matrix is actually a major or a minor part of the analyzed region of the specimen.

The results obtained from Eqs. (79)–(81) are rather insensitive to small variations in the postulated matrix or to the omission of minor elements from the monitored elements u. For example, it makes virtually no difference whether electrolytes like Na, K, or Cl are included in their normal amounts among the elements m or are omitted from both groups m and u. There is a good physical basis for this analytically favorable circumstance: All of the terms in Eqs. (79)–(81), except for terms with subscript x, have to do only with representing correctly the continuum intensity generated per unit of specimen mass, which depends in fact on the mean atomic number $\bar{Z}$ (or more precisely, on $\overline{Z^2/A}$). The equations incorporate an exact representation of Z^2/A for those elements whose spectral lines are actually observed. All of the other elements—all elements m and the remaining elements u—are represented by the sum $\Sigma_m(C'Z^2/A)$, which is simply $\overline{Z^2/A}$ for the postulated matrix. This procedure is inherently erroneous to the extent of the discrepancy between $\overline{Z^2/A}$ in the postulated matrix, and the genuine $\overline{Z^2/A}$ of all of the unmonitored elements in the analyzed region, but this discrepancy is usually slight in biological specimens. The chief unmonitored elements are generally the three main constituents of soft tissue (C, N, and O); since their atomic numbers are similar, the normal variations in the relative amounts of these elements do not strongly affect $\overline{Z^2/A}$. And the usual concentrations of other elements such as Na and K are too low to have much effect.

We should note the following points with respect to the method of quantitation embodied in Eqs. (79)–(81):

1. The formula (79) is valid over the entire range of weight-fractions, and it does *not* require that the analyzed region should consist mainly of the postulated matrix. Indeed the formula is valid even for regions where the postulated matrix is totally absent. [In this case $1 - \Sigma_u(RGZ^2) = 0$ and (79) reduces to (75).]

The use of the concept of the matrix may be clarified by consideration of a special case: binary specimens. We label the two elements x and y, x representing the assayed element as always. We now have the option of regarding both x and y as unknown elements u with no matrix elements m, or of regarding x as the sole element in group u while the "matrix" consists solely of element y. If we choose the second course, $Cy' = 1$, $\Sigma_m(C'Z^2/A) = Z_y^2/A_y$, and $\Sigma_u(RAG) = R_xA_xG_x$. Now from Eq. (73), one may readily derive the general result $\Sigma_r(RGZ^2) = (\Sigma_u + \Sigma_m)(RGZ^2) = 1$, so that here $1 - \Sigma_u(RGZ^2) = \Sigma_m(RGZ^2) = R_yG_yZ_y^2$, and Eq. (79) leads to

$$C_x = \frac{R_xA_xG_x}{\dfrac{R_yG_yZ_y^2}{Z_y^2/A_y} + R_xA_xG_x} = \frac{R_xA_xG_x}{R_yA_yG_y + R_xA_xG_x} \tag{82}$$

This is exactly what one gets if one regards both x and y as unknowns and applies the fundamental Eq. (75), so we have illustrated how the matrix concept is used and we have confirmed the identity of (79) with the simple formulation (75). Note however that the analysis of the binary specimen can be performed with the observation of only one spectral line when Eq. (79) is used with element y regarded as the matrix, whereas Eq. (82) calls for the observation of characteristic radiations from both elements! This magic results from the replacement of the term $R_yA_yG_y$, which contains the spectral intensity of element y, by the equal quantity $(1 - R_xG_xZ_x^2)A_y/Z_y^2$, which does not. It is just the same process [cf. Eq.

(77)] which makes it unnecessary in the general case to observe the spectral lines of any of the matrix elements when Eq. (79) is used.

2. The data enter the equations only as *ratios* of characteristic to continuum intensities, which are generated and recorded simultaneously as the probe electrons traverse the specimen. A minor advantage is that such ratios are independent of fluctuations in probe current. The major advantage of the continuum normalization is that in thin specimens, the ratios are independent of specimen thickness or mass per unit area. Hence there is no need to know the specimen thickness, and the measurements are unaffected by local variations in thickness or density.

The characteristic and the continuum intensities do not depend on the energy E of the probe electrons in the same way. The characteristic intensity per unit path length varies as q_x [Eq. (70) and (55)], i.e., as $E^{-1} \ln E/E_x$, while the continuum intensity varies as E^{-1} (Eq. 71). Therefore the ratio changes with E and is independent of thickness only for thin specimens, in which (according to our definition of "thin"!) the electron energy remains almost constant over the average trajectory. This condition is not difficult to satisfy in sections of soft tissue (see Section IV.B).

In the analysis of thin specimens by the continuum method, there is also no need for the backscatter and "penetration" corrections which are sometimes essential in the analysis of thick specimens. The backscatter correction-factor would be the same for characteristic and continuum radiations, leaving their ratio unaffected. The "penetration" effects depend on the law of electron energy-loss, which we saw to be fundamental to the thick-specimen theory but which is completely irrelevant here.

3. Fluorescence in thin specimens is negligible because almost all of the potentially inducing X rays escape the specimen.

4. Absorption corrections are also usually negligible or else slight. Small corrections can be made by means of the simple equations which we shall now present.

The flux of probe electrons in thin specimens is virtually independent of depth below the surface. To calculate the absorption correction, one assumes further that the composition is independent of depth. A simple integration then gives the relation between the observed X-ray count c_0, and the count which would be observed in the absence of absorption c_g:

$$c_g = c_0 \frac{\sigma \rho t \csc \theta}{1 - \exp(-\sigma \rho t \csc \theta)} \tag{83}$$

where σ is the X-ray absorption cross section ($\sigma = \Sigma_r \sigma_r C_r$, cm^2/g), ρ is the density (g/cm^3), t is the specimen thickness (cm), and csc θ is the cosecant of the take-off angle (Fig. 20a). The quantity to be inserted in the calculation of the ratios R in Eqs. (79)–(81) is c_g.

When $\sigma\rho t$ csc $\theta \ll 1$, Eq. (83) reduces to

$$c_g = c_0(1 + \tfrac{1}{2}(\sigma\rho t \text{ csc } \theta)) \tag{84}$$

In the worst case, if the composition were not homogeneous and all of the observed X rays came from the bottom of the specimen, the mean absorption path would be doubled and the relation between c_g and c_0 would be the same as Eq. (84) except that the number "$\tfrac{1}{2}$" would disappear. Thus, so long as $\sigma\rho t$ csc $\theta \ll 1$, the absorption correction is slight and it is reasonable to use Eq. (84). In practice this condition is usually satisfied in thin specimens except when the observed radiations are quite soft; indeed $\sigma\rho t$ csc θ is commonly small enough to make the absorption correction entirely negligible.

For larger absorption corrections, Eq. (83) may be used, but an accurate correction is not to be expected since the analyzed region may not be homogeneous and ρ and t are generally not well known; there may also be the annoyance of iteration if the composition cannot be guessed well enough to establish σ with sufficient accuracy initially.

The absorption correction applies to the standard as well as to the specimen, and to the continuum as well as to the characteristic radiation. Generally only a narrow band of continuum quantum energies is observed, so that the absorption cross section for the continuum may be reckoned adequately at the middle of this band. Absorption of the continuum is usually completely negligible when a high energy band is monitored. Alternatively, if a continuum band is chosen near the observed spectral line, the absorption corrections for characteristic and continuum radiations may be almost the same, tending to cancel in the ratio.

We concluded earlier that substantial absorption corrections are usually necessary in thick biological specimens; now it has been stated that absorption is usually slight in thin specimens. This may not seem to jibe with our chosen terminology, according to which one and the same specimen may be "thin" (with respect to high-energy probe electrons) and "thick" (with respect to low-energy electrons). But there is no real inconsistency. A low-energy probe can excite only low-energy characteristic quanta, which are readily absorbed, whereas one usually observes quanta of higher energy when the probe voltage is raised.

5. There is wide latitude in selecting the band of observed continuum radiation. The upper end of the quantum-energy band should be below the average energy of the probe electrons leaving the specimen, so that

throughout their trajectories the electrons remain capable of generating quanta over the entire monitored band. It is usually undesirable to set the low end of the band so far down that substantial absorption occurs. The only remaining restriction is that the band should exclude any characteristic radiations prominent in specimen or standard. Generally there is an ample energy band available between the probe voltage and the highest intense spectral line, and the continuum is so intense that only a narrow band has to be sampled (Hall and Werba, 1969).

6. Just as in the case of thick specimens, the measured weight-fraction can only be the local elemental mass divided by the local total mass, the latter including embedding medium if any is present. However, the reasons for embedding thick specimens (to gain electron opacity and render absorption calculable) do not apply to thin specimens. Moreover a section of soft tissue of given thickness has less mass per unit area and is therefore more satisfactorily "thin" when it is unembedded. Hence it is best to study thin specimens unembedded, thus removing the ambiguity in the measurement of weight-fractions.

The continuum method is not highly accurate even in the best circumstances. In addition to the error associated with the postulation of the composition of the matrix, the main uncertainty comes from the assumption that k in Eq. (71) is really constant; in fact k varies slightly with atomic number and the dependence is not well established at present. The overall probable error of a measurement of weight-fraction is generally in the neighborhood of $\pm 10\%$ of the measured value, but the continuum method avoids many of the difficulties of the thick-specimen theory of quantitation and is not affected by variations in specimen thickness. Furthermore, one can measure other significant quantities besides weight-fractions, as we shall now discuss.

2. The Measurement of Spatial Concentrations

Roughly speaking, the continuum intensity from a thin specimen is proportional to local total mass per unit area and the intensity of a spectral line is proportional to the local amount per unit area of the corresponding element. Relative amounts per unit area in different parts of a specimen or in different specimens can then be obtained directly from the X-ray counting rates. In specimens of uniform thickness, such data give directly the relative *spatial* concentrations (i.e., amounts per unit *volume*), since

$$\rho = S/t \qquad (85)$$

where ρ is mass/volume, S is mass/area, and t is specimen thickness.

The measurements of amounts per unit area can be made absolute by reference to a standard film: either a thin organic film with known total mass/area S or a thin film with a known amount S_x of the assayed element per unit area (usually prepared by evaporation). A film of *either* type can suffice to establish both S and S_x in the specimen through the relationship $C_x = S_x/S$ [Eq. (69)], provided that C_x also is measured by the method of the preceding section. However absolute measurements of S_x and S are not often desired—they can be converted to the intrinsically more interesting amounts per unit volume only through the factor t (thickness), which is usually not well known.

Since specimens in the vacuum of the microprobe are in a dry state, measurements of S over a uniformly thick specimen give the distribution of *dry* mass per unit volume, which is recognized to be one of the important parameters characterizing biological tissues. It is interesting to compare the continuum-emission method with the microradiographic method developed by Engström and his colleagues for the measurement of dry mass (Engström, 1966).

In the microradiographic technique, the specimen is exposed to X rays and the measure of specimen mass is the degree of X-ray absorption, determined from the degree of blackening of an underlying photographic emulsion.

In both methods, the quantity directly measured is not amount per unit volume but amount per unit area. In other words, these techniques are not sensitive to thickness as such but only to the mass of specimen in the path of the incident beam. However in neither case does the signal depend on the mass/area S alone. With the continuum-emission method, it may be seen from Eq. (71) that the signal is proportional to

$$\sum_r (N_r Z_r^2)\, ds \propto \sum_r (S_r Z_r^2/A_r) = S \sum_r (C_r Z_r^2/A_r) = S \overline{Z_r^2/A_r} \cong S\bar{Z} \quad (86)$$

With the microradiographic method, since the X-ray mass absorption cross section σ_r for each constituent element varies approx. as Z^3, the observed transmission is a measure of the quantity

$$\sum_r (S_r \sigma_r) \propto S \sum_r (C_r Z_r^3) = S\overline{Z^3} \quad (87)$$

Thus these methods give a good relative measure of S only when the mean atomic number is virtually constant. However it has been shown by Lindström (1955) that the microradiographic cofactor $\overline{Z^3}$ is virtually constant at least in ordinary soft tissues. The continuum-emission cofactor $\bar{Z}$ varies much less than $\overline{Z^3}$, so as a measure of mass/area, the continuum-emission method is much less sensitive than the microradiographic method to variations in composition.

Further advantages of the continuum-emission method are its convenience and speed (with no need to develop a photographic emulsion), and the fact that the relative measurement of mass/area is obtained along with the measurement of elemental weight-fraction. Indeed one set of microprobe data suffices for both measurements.

The microradiographic method provides a picture of the distribution of specimen dry mass in the form of the developed emulsion. Similar pictures can readily be obtained in the microprobe in the form of scanning images, the brightness of the display tube being modulated by the X-ray continuum intensity as the probe is scanned over the specimen.

The continuum-emission measurements of mass/area lose validity for several reasons if the specimen is too thick. The first difficulty to appear with increasing thickness results from the scatter of the probe electrons, which makes the length of the average trajectory and hence the continuum intensity more than directly proportional to mass/area. Scatter does not in itself affect the measurement of weight-fractions since characteristic and continuum intensities are augmented to the same extent. Consequently the thickness range of the continuum method is limited in measurements of mass/area more than in assays of weight-fractions. Satisfactory measurements of mass/area can usually be obtained in specimens up to about 500 $\mu g/cm^2$ so long as the probe voltage is kept above 30 kV.

The microradiographic method can be applied to much thicker specimens, it being a simple matter to use incident X rays with suitable penetrating power. In principle, the range can be similarly extended in the microprobe by the use of higher probe voltages. Although the maximum voltage available in most standard probes is 40 or 50 kV, some combination electron microscope-microanalyzers can provide 100 kV. At the moment, we cannot quote a maximum mass/area measureable at 100 kV because experience is lacking, but the value is certainly above 1 mg/cm^2.

C. Ultrathin Specimens

The theory of quantitation which we have outlined for thin specimens is entirely valid for ultrathin specimens. In fact the thinner the specimen, the more perfectly satisfied are the requirements of the theory (low energy-loss and little scatter of the probe electrons; low X-ray absorption). Nevertheless, quantitation in ultrathin specimens is more difficult than in specimens a few micrometers thick because of several practical difficulties:

1. Ultrathin sections are almost always studied with embedding medium present. This reduces the significance of measurements of weight-fraction and of total mass/area, leaving the determination of elemental ratios and of elemental amounts/area as the quantitative results of most significance to the analyst. (However with recent developments in microtomy (Appleton, 1968), it is becoming practical to study unembedded ultrathin cryostat sections.)

2. In the determination of specimen mass/area by the continuum-emission method, the background due to the supporting film must be subtracted from the total continuum intensity in order to determine the intensity from the specimen proper. In the case of ultrathin specimens, the supporting film and the specimen are usually similar in mass/area, so that nonuniformity in the supporting film can produce a considerable error—10% perhaps in practice.

3. Ultrathin specimens are generally studied in electron microscope-microanalyzers, where the probe-forming and objective lenses and other metal objects are close to the specimen. The collision of scattered electrons with these things produces a high extraneous background of continuum radiation, which cannot be determined simply by removing the specimen because the specimen does most of the scattering. This background can be minimized by setting the band of observed continuum quantum energies quite high, since in comparison with the spectrum from the thin specimen, the spectrum of continuum radiation from the thick surrounding objects is biased towards low quantum energies. There has not yet been enough practical experience to judge how successfully the extraneous background can be suppressed by this tactic and by careful collimation.

The determination of local mass per unit area in electron-microscope specimens is important not only to analysts; electron microscopists have been much concerned with the problem. The best established procedure is probably that of Zeitler and Bahr [(1962), cf. also Bahr (1968), Bahr and Zeitler (1965)], who measure the attenuation of the electron current passing through the objective aperture. With respect to linear dependence on specimen mass/area and relative independence of atomic number, the continuum generated in a specimen is a better measure of mass/area than is the attenuation of the electron beam, but the problem of extraneous continuum background has still to be overcome.

VII. Specimen Preparation

Microprobe studies have been carried out on thick blocks and on thin and ultrathin sections of both hard and soft tissues, and on isolated cells and microincinerated tissues. Since the preparative technique must vary not only with the type of specimen but also with the objectives of the particular study, a thorough discussion of specimen preparation is not possible in the space of this chapter, and we shall try simply to consider the main points. A more extensive treatment may be found elsewhere (Hall *et al.*, in press).

A. THE INTEGRITY OF THE SPECIMEN

In the preparation of material for microanalysis, the ideal is that the analyzed elements should be distributed within the prepared specimen in the same way as in the living tissue. It is important to recognize straight away that with respect to certain elements at least, one cannot expect to realize this ideal precisely in a tissue which has been dried at any stage in its preparation. The elements which were originally dissolved in an aqueous phase (in particular, electrolytes like Na, K, and Cl) cannot be expected to "stay put" as their solvent is removed. One is doing well if the electrolytes end up precipitated on the nearby surfaces of membranes, cytosomes, or macromolecules, so that they may be localized to the compartments which they occupied in the living state.

It is in fact possible to study specimens in the microprobe without first removing the water. The vacuum in the column is not too badly spoiled if the specimen does not release water vapor too rapidly and if there is only a narrow pumping path between the specimen chamber and the rest of the column. It is even possible to counteract evaporation from the specimen by sending gas or vapor jets across the surface, or a very cold stage may be used to hold the water as slowly subliming ice. These methods may yield interesting scanning images in which some of the artifacts of dried tissues may be avoided, but their value for microanalysis is dubious. With wet specimens, whether thick or thin, it is very difficult to be sure that the water remains in the thin superficial region to which the analysis is confined, and it must be remembered that chemical reactions and diffusion may continue in the wet state and alter the elemental distributions. On the other hand if one resorts to freezing, the solutes are inevitably dislocated as the water-ice boundary advances through the tissue; also it may not be possible to keep a superficial region frozen under a probe current high enough for microanalysis. Hence biological microprobe analysis will probably continue to be based on specimens from which the water has been removed, in spite of the limitations discussed above.

We may classify the prevalent preparative techniques, in which the water is removed, according to whether or not liquid solvents are used. Among the nonsolvent methods, three will be considered: freeze-drying of tissue blocks, drying of cryostat sections, and air-drying.

1. Methods Based on Liquid Solvents

In preparing tissues for study, liquid solvents are widely used as fixatives (alcohol, acetone, formalin, other aldehydes, etc.), as vehicles for stains, and to remove embedding medium (xylol, toluol, etc.). All of the standard liquid-bath techniques may be used to produce specimens which are quite amenable to microprobe analysis. But one must always judge whether the analyzed specimen is sufficiently similar to the original tissue. In other words, have the liquids removed or displaced the elements which are to be analyzed? An intolerable loss or displacement of elements has been documented in several cases (Andersen, 1967a,b; Hall and Höhling, 1969). When the action of the liquids is sufficiently known and acceptable, the standard "wet" preparative techniques may be best; otherwise it is preferable to use one of the "dry" techniques, to which we now turn.

2. Freeze-Drying of Tissue Blocks

In one well-established procedure, a block of tissue is first rapidly frozen by being plunged into very cold liquid, commonly propane or ethylene glycol. The immediate solidification of the outside of the block prevents the coolant from penetrating to the inside. The frozen block is then dried by sublimation in vacuum, after which it may be infiltrated with embedding medium and sectioned if desired. The embedding medium should be selected to be stable under the probe to avoid the necessity of de-embedding with a liquid solvent; in this respect, epoxy resins are toughest, and methacrylate may be used for ultrathin sections, but paraffin is unsatisfactory.

This procedure has been used quite successfully in microprobe work (Andersen, 1967a,b; Hall and Höhling, 1969; Ingram and Hogben, 1968), yielding high-quality sections which have not been exposed to liquid solvents. The main drawbacks are the long time needed to freeze-dry the tissue blocks (hours to days), and the limits placed on quantitation by the presence of embedding medium (cf. Section VI.A.4.h).

3. Drying of Frozen Sections ("Cryostat Sections")

A block of quickly frozen tissue may be sectioned directly while frozen, without embedding, and the frozen sections may then be readily dried by sublimation or evaporation. Histological detail is generally not preserved in such sections as finely as in sections of freeze-dried embedded blocks. However the quality is easily compatible with the spatial resolution of the X-ray microanalysis, and the procedure has important advantages: It is very simple and rapid; the tissue is exposed to no liquid, not even embedding medium; and weight-fractions and dry mass per unit area may be measured by the methods of Sections VI.B.1,2.

Cryostat sections have long been produced routinely in thicknesses down to approx. 2 μm. Recently, effective techniques seem to have been developed to produce ultrathin sections with no exposure of the tissue to liquids at any stage (Appleton, 1968; commercial literature of the LKB Corporation).

4. Air-Drying

Thin specimens which need not be sectioned, for example, smears of cells, could be inserted wet into the microprobe instrument where they would rapidly dry in the vacuum. Since such material deteriorates rapidly if left wet and unfixed, a prompt preliminary drying is usually desirable. This can be achieved in a minute or so by exposure to a stream of warm dry air.

In order to avoid the exposure of a specimen to liquid baths, one must foreswear most of the standard histological stains. One should make a determined effort to see the relevant features in optical or electron images of unstained specimens. If a stain is necessary, a vapor stain like vaporous osmium tetroxide may be sufficient [cf. Ingram and Hogben, (1968)] and is less threatening than liquid stains to the integrity of the specimen.

The procedures outlined above are described in detail in the standard texts, such as Pearse's (1968).

B. The Specimen Support

Specimens may be mounted on glass slides when it is important to examine them by transmission optical microscopy. Otherwise metallic slides or stubs may be the most convenient support.

When very small elemental amounts must be studied in specimens which are not absolutely opaque to the electron beam, the background of continuum and characteristic X rays from the support may be a limiting factor. The many recognized elements and the impurities in ordinary glass may then seriously interfere, and slides of Lucite or similar plastic may serve as a cheap transparent substitute for glass if high probe currents are not needed; pure quartz slides are a more expensive but much tougher alternative. When transparency is unnecessary, the continuum background can be reduced with supports of low atomic number, notably beryllium or carbon, and single-crystal silicon has been recommended for its relative freedom from impurities (Carroll and Tullis, 1968).

Thin specimens can also be mounted on very thin plastic films like the ones used in electron microscopy (Hall et al., in press). These films, 500–3000 Å in thickness, are made in the laboratory out of cellulose acetate, Formvar, or nylon. They can be supported by EM-type grids or can stand freely over areas up to many millimeters in diameter.

The continuum background from such films is very low and does not increase with probe voltage, so that they enable one to analyze thin specimens quantitatively at high voltages in accord with the theory of Section VI.B.

One can also drape sections across the bars of EM-type grids with no continuous support. However such preparations seem to be less stable under the probe than sections mounted on thin film, and the elimination of the background from the supporting film is generally not helpful. In the geometry of electron microscope-microanalyzers, there remains the high background generated by stray electrons, and in standard microprobes the background from the thin film is not large anyway compared to the signal from the usual specimens.

C. ELECTRICAL AND THERMAL CONDUCTION

The probe should not be allowed to build up concentrations of electrical charge in the specimen, as a strong local electrostatic field may disrupt or distort the tissue. A field which extends above the specimen surface can also deflect or defocus the probe itself.

There is of course no danger of a buildup of charge in electrically conducting specimens as long as the mounting does not insulate the specimen from ground potential. In nonconducting materials, including almost all biological specimens, the buildup of charge may be prevented in one of four ways:

1. There is usually a probe voltage at which the current of absorbed primary electrons just equals the flux of emitted secondary electrons so that there is no net accumulation of charge. This fact has been exploited in scanning electron microscopy, but the critical voltage, usually 2–3 kV, is too low for most X-ray analytical work.

2. With quite thin specimens, an electrically conducting support is sufficient to remove the charge, as the electrical resistance from any point in the specimen to the support is not too great.

3. Antistatic sprays have been used to make specimens conducting for scanning electron microscopy (Sikorski et al., 1968).

4. The prevalent method for conferring conductivity is to evaporate a layer of conducting material onto the specimen in vacuum. For X-ray microanalysis, a coating of low atomic number is desirable and a layer of carbon 100–200 Å thick or aluminum 50–100 Å thick is usually satisfactory. Such layers do not seriously interfere with inspection by transmission optical microscopy.

It is important to provide an unbroken electrical path from the coated surface to ground potential. With thick specimens, one should be sure that the sides are coated, either by evaporating obliquely while the specimen is rotated (Boyde, 1967), or by painting a streak of conducting material such as Aquadag (colloidal graphite) from the top to the specimen support. If the support is a glass slide or similar nonconductor, it may be well to paint another streak along this support back onto the metal holder.

When high probe currents are necessary, there is a danger of damaging the specimen by overheating. If the specimen is not very thick, it is then obviously advantageous to use a support with high thermal conductivity, for example beryllium or carbon. It may also be helpful to use heavier evaporated coatings. When thin sections are mounted on thin plastic films, if an aluminum coating of 500 Å is applied to *each* side of the preparation one may be able to use probe currents up to 1 μA (Hall *et al.*, 1966). It must be noted however that metallic supports and heavy coatings are unfortunately incompatible with transmission optical microscopy.

D. PROBLEMS PECULIAR TO SECTIONS OF HARD TISSUE

X-ray absorption is generally substantial in specimens of hard tissue, except in ultrathin sections, because the relatively high mean atomic number of the tissue implies a high absorption coefficient. The specimen surface must therefore be smooth and flat to avoid large, irregular absorption effects.

The most common preparative procedure is to embed according to standard histological practice, section with a heavy-duty microtome, and grind and polish according to standard mineralogical practice. Details have been given by several workers, including Söremark and Grøn (1966), Frazier (1966, 1967), Mellors (1964), and Boyde and Switsur (1963). But embedding entails the risk of elemental displacement unless a "dry" method is used; and grinding and polishing entail a risk of contaminating the surface with abrasive, distorting the surface, or redistributing the surface material in ground form. For microprobe analysis, the procedure of Andersen (1967a) may therefore be most suitable: simple cutting of the unembedded block followed by drying in vacuum or in a drying oven. He uses a diamond saw (diamond < 1 μm) for sections of approx. 100 μm, and for section thicknesses under 10 μm he freezes the tissue for the sake of rigidity and does the cutting in a cryostat.

Valuable morphological studies are being carried out in the scanning electron microscope (SEM) on *fractured* hard tissues with surfaces which are far from flat. Useful qualitative information should be available from the microanalysis of such specimens by means of the SEM X-ray attachments which are now available, although the surface irregularities must preclude quantitative analysis and raise problems in the interpretation of the X-ray data. Preparative procedures for such specimens have been outlined by Boyde and Jones (1968) and Boyde and Hobdell (1969).

E. PROBLEMS PECULIAR TO SECTIONS OF SOFT TISSUE

In soft tissues, the atomic number of the assayed element is usually well above the mean atomic number of the tissue, and the X-ray absorption coefficient for the monitored characteristic radiation is consequently usually low. Hence the surface of a section of soft tissue is usually sufficiently smooth without further treatment. It is only when a section is grossly irregular (curled up or with projecting filaments, for example) or when X rays of low penetrating power are to be recorded (roughly speaking, wavelengths longer than the 8-Å K radiation of aluminum, atomic number $Z = 13$), that the irregular X-ray absorption associated with an irregular surface becomes important.

The study of *embedded* sections of soft tissue is technically straightforward, although one should be concerned about the integrity of the specimen (Section VII.A) and the possibilities of quantitation are limited (Section VI.A.4.h). By means of the theory of Section VI.B, quantitative analysis can be carried out more fully on *unembedded* cryostat sections, but only when they are thin (preferably 5 μm or thinner) and mounted on very thin plastic supporting films. The same analytical theory can be applied to sections cut from paraffin-embedded blocks and then deparaffinized prior to probe study, although in this case one must seriously consider whether the deparaffinized section adequately represents the original tissue. Techniques for mounting cryostat or de-embedded sections onto thin film have been reviewed by Hall *et al.* (in press).

F. PROBLEMS PECULIAR TO ULTRATHIN SECTIONS

Sections prepared for conventional transmission electron microscopy are suitable for electron probe microanalysis. It is necessary to prevent the buildup of electrical charge within the specimen, but this is accomplished simply by mounting the section on a carbon film or coating with a thin layer of evaporated carbon according to conventional EM practice. The shape of ultrathin sections is favorable for the dissipation of heat so that it is usually unnecessary to take extra measures to increase thermal conductivity.

Very fine supporting grids are undesirable because excessive background may be generated in a grid bar when the probe is focused nearby, or the bar may block the path of the X rays to the spectrometer if the section is on the "wrong" side of the grid. A reasonable spacing for grids of rectangular pattern is 80 lines or less per centimeter. Grid materials of high atomic number produce a relatively high background of continuum, so copper is preferable to gold or palladium, and nickel may be used instead when copper is to be assayed in the specimen.

The search for good fields in combination electron microscope-microanalyzers generally consumes a great deal of time, as does the analysis of these fields once they are found. Hence it is extremely desirable to mount the specimens on "field-finder" grids—commercial grids with code markings on the bars or intersections—in order to be able to return readily to interesting areas.

We have already noted the risks in using liquid solvents for specimen preparation. The dangers are greatest in ultrathin sections where weakly bound elements can quickly be completely washed away, and it is therefore especially important to use one of the "dry" techniques—frozen-dried blocks or cryostat sections—if there is any question of liquid solvents dislocating the interesting elements. It may also be important to transfer specimens directly from the microtome knife to the grid, or else to recover them as rapidly as possible from the liquid in the microtome trough.

VIII. Damage to Specimens in the Electron-Probe Instrument

Certain types of specimen damage are inherent to the study of specimens in a vacuum. Besides the inescapable consequences of drying (Section VII.A), there is the possible shrinkage or distortion of the tissue, well known from electron microscopy. But the main concern must be with damage induced by the probe. Four types should be noted:

1. There may be gross damage such as curling or disintegration, due mainly to overheating of the specimen.

2. Volatile elements like sodium may be selectively removed at elevated temperatures.

3. In thick specimens, some elements may be driven away from the surface towards the interior. The effect is not well understood but is believed to be associated with electrostatic forces due to accumulated charge (Vassamillet and Caldwell, 1968). The effect can occur with high probe voltages in thick specimens even if the surface has a conducting coating, since charge may accumulate at some depth; in these circumstances, one might not be alerted to the danger by a defocused or deflected probe because the probe would be shielded from the electrostatic field by the coating.

4. The probe cracks organic vapors and deposits a layer of carbon on the specimen surface just as in electron microscopy. The rate of deposition of this "contamination" is quite variable, depending strongly

on the cleanliness of the vacuum system, on the probe current, and on the temperature of the specimen. Cold traps near the specimen can greatly reduce the deposition rate (Ranzetta and Scott, 1966) but in fact the contamination is rarely a hindrance and may conveniently mark the probed areas. It does interfere of course with the assay of carbon and can sometimes build up enough to significantly attenuate very soft radiations like the nitrogen and oxygen K lines. In ultrathin sections, the contamination deposited during a 100-sec static-probe analysis may have a mass per unit area of the same order as the specimen itself; the spectrometer background and the continuum intensity then increase during the analysis and the correct average values should be determined by two or more observations at recorded times.

Probe damage is generally negligible in biological specimens when the probe current is 0.01 μA or less. Damage is not very likely with currents under 0.05 μA but will probably occur at higher currents unless steps are taken to avoid it. The following hints may help to avoid damage at high currents:

1. For the effective removal of heat, there should be intimate contact between the specimen and its support. Therefore thick rigid supports should have a polished surface, and sections mounted on very thin flexible films must adhere well.

2. Quite high currents can be tolerated by thin sections mounted on very thin films, coated heavily on both faces with aluminum (Hall et al., 1966).

3. There is some evidence that very thin nylon films are more stable under the probe than similar films of cellulose acetate or Formvar.

4. Specimens are often mounted on thin films which rest in turn on metal grids. When such preparations are found to be unstable under the probe, one may omit the grid and work with specimens on free-standing film. A grid may actually impair the stability of the film, presumably because of stresses set up by thermal gradients.

5. Paraffin melts easily under the probe. For the study of embedded sections, epoxy embedding media are recommended.

Commonly one cannot be sure in advance that a preparation will not be damaged by the probe. It is therefore desirable to observe X-ray intensities immediately on a count-rate meter as the probe is brought onto fresh areas and to watch for any drift in the readings.

IX. Other Types of Microprobe Analysis

Besides the electron-probe X-ray method, there are several other probe techniques for performing elemental microanalyses. Here we shall outline the principles of these methods and compare their capabilities. I do not know enough to give an up-to-date, accurate description of the capability of each technique but the rough comparison should still help the reader to choose the right method for his problem, and more information can be found in the references which will be cited.

A. ELECTRON ENERGY-LOSS ANALYSIS AND OTHER ELECTRON-PROBE METHODS

The interaction of an electron probe with a specimen gives rise not only to X rays but also to several other kinds of radiation which can be exploited analytically. The probe generates visible light, termed "cathodoluminescence" (cf. Kyser et al., 1967; Kyser and Wittry, 1966), and there are some prospects for taking advantage of this light in biology (Pease and Hayes, 1966). Another kind of radiation consists of "Auger electrons," which are released with characteristic energies from individual atoms as a result of secondary ionization processes (Riviére, 1969; Harris, 1968). The Auger electrons have very low energies—tens or hundreds of electron volts—so that their range is extremely short and Auger spectroscopy is consequently suited to the analysis of surface monolayers. We shall not discuss cathodoluminescence or Auger-electron spectroscopy further as neither technique seems ready for extensive biological application. Indeed the Auger technique has not even been developed in a microprobe form, though there seems to be no obstacle to this. Here we shall consider another type of readout available with an electron probe: electron energy-loss analysis.

When an electron probe impinges on an ultrathin specimen, most of the electrons emerge from the far side having undergone either no energy loss, or else one or two interactions with energy losses characteristic of the material they have traversed. Thus one can learn about the composition of the specimen by placing an energy analyzer on the output side to determine the spectrum of electron energy losses (Fig. 28).

The method is not applicable to thicker specimens where the average number of loss-interactions is more than approx. 2 or 3 because the spectrum of losses becomes hopelessly smeared.

There are two main types of energy-loss interaction. Most losses are transfers of energy to collective states involving large numbers of atoms in the specimen—excitations of electron plasma or of a crystal lattice. These losses do not signify atoms as such but rather the presence of a molecule or crystal lattice of a particular structure and composition. Secondly, there are the simple ionizations of inner-electron shells, which do correspond to the energy levels of independent atoms. These inner-shell ionizations are just the interactions which provide the basis for

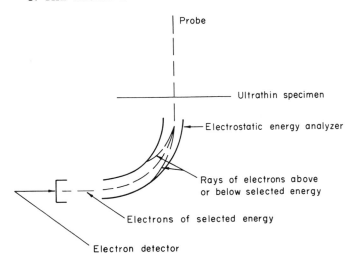

FIG. 28. The principle of electron energy-loss microanalysis (probe method). (The cylindrical electrostatic analyzer is the most common type of electron energy analyzer, but other types also may be used.)

X-ray microanalysis, only here they are recognized in terms of the consequences for the probe electron. However this mode of observation is feasible only when the ionization cross section is not too much less than the cross section for collective excitation—otherwise the characteristic ionization loss is obscured by overlaid collective losses. This requirement unfortunately is satisfied only at low atomic numbers where the ionization cross sections happen to be quite large [cf. Eq. (55)]. Hence most energy-loss analysis has been directed to the identification of molecular and crystalline aggregates. However elemental analysis based on inner-shell ionization has been carried out for oxygen, atomic number $Z = 8$ (Wittry et al., 1969) and there are hopes for extending the method up to calcium, $Z = 20$.

The energy-loss technique has been used only where the analyzed phase or element constitutes the major part of the analyzed microvolume. The collective states of excitation cannot be expected to exist unperturbed when the collective phase is finely mixed with other constituents, and the inner-shell ionization cross sections are not large enough to give intense signals from elements present in low weight-fractions in ultrathin specimens. Hence the method is naturally suited to the identification of the *major* constituents of extremely small volumes. The analyzed microvolumes may be of the order of 10^{-20} cm³ (100 Å thick $\times$ 10 Å diameter). The range of possible biological applications for such a technique is complementary to the range of X-ray microanalysis, where minor constituents can be analyzed but only in much larger microvolumes.

There are actually two approaches to energy-loss analysis in an electron microscope. With the probe method (Crewe *et al.*, 1968), one can do static-probe analysis by determining the energy spectrum at a fixed point in the specimen, or one can set the analyzer for a particular electron energy and scan a small area with the probe to get a selected-energy transmission scanning image. The other approach is not a probe method at all; electron-optical components are inserted between lenses in the image space of the electron microscope to disperse and remove from the beam all electrons outside of a narrow band of energies, again giving a selected-energy image (Castaing *et al.*, 1966). The sensitivities of the two approaches are basically similar. The probe method gives a more flexible control of contrast in the image and easier quantitation. In Castaing's approach the spatial resolution is the same as in ordinary electron microscopy; Crewe's probe method seems now to offer similar spatial resolution when one uses the very fine probe obtainable from a field-emission electron gun.

Castaing's arrangement for obtaining energy-selected electron images is commercially available (see Appendix I) but apparatus for the energy-analysis of the transmitted electron probe is not yet on the market.

B. The X-Ray Probe (X-Ray Fluorescence Microanalysis)

One arrangement for X-ray fluorescence microanalysis is shown in Fig. 29. A beam of electrons is focused into a fine spot on one side of a thin metal target, generating a "point source" of X rays which can traverse the target and emerge from the other side. The thin target serves also as part of the vacuum wall so that the electron beam is

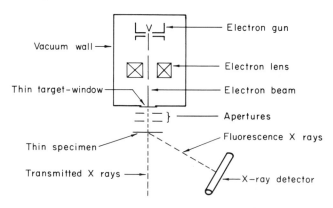

FIG. 29. An arrangement for X-ray probe (X-ray fluorescence) microanalysis.

in vacuum while the specimen does not have to be. Collimating apertures define the X-ray probe impinging on the thin specimen, the irradiated specimen area usually being 10–200 μm in diameter. The signal consists of the X-ray fluorescence excited within the specimen, and this is analyzed nondiffractively with the detector offset to avoid the transmitted rays of the probe.

The arrangement of Fig. 29 has been used to measure elemental amounts down to approx. 10^{-10} g and weight-fractions of the order of 1%, in specimen areas of the order of 100 μm in diameter (Long and Röckert, 1963; Röckert, 1966). Straightforward modifications are expected to lead to limits of sensitivity of approx. 10^{-12} g and measurable weight-fractions down to less than 0.1% in areas down to 10 μm in diameter (Hall et al., in press). Such sensitivities are suitable for the measurement of the total amounts of a number of elements in individual cells or in groups of a few cells, isolated or in tissue sections. While the performance figures are not nearly as impressive as those for the electron-probe method, the fact that the specimen does not have to be in vacuum should be important for many studies.

Zeitz (1961, 1969) has developed instrumentation which differs from that of Fig. 29 in just one important respect: the insertion of a diffracting crystal between specimen and detector to provide diffractive X-ray spectroscopy. This leads to much better values for minimum measurable weight-fractions at the expense of much poorer values for minimum measurable elemental amounts. Zeitz' system is therefore suited for the assay of a wider range of elements but only in larger volumes of specimen.

The instruments of Long and Röckert and of Zeitz are not commercially available at present. There are two commercial types of X-ray fluorescence analyzer: the well established systems with standard X-ray tubes and diffractive spectrometers, and the newer portable units in which a radioactive source is used instead of an X-ray tube and the fluorescence is analyzed by a nondiffractive spectrometer. Neither of these types is suitable for the analysis of microvolumes of biological tissue because they do not deliver a high enough exciting flux (X-ray quanta per unit area per second) to the specimen.

C. PROTON-PROBE X-RAY MICROANALYSIS

Schematically proton-probe X-ray microanalysis is just like the electron-probe method, except that a beam of protons is used in place of the beam of electrons. Protons being much heavier than electrons, they experience smaller accelerations in atomic collisions and consequently generate much less background continuum radiation. The favorable ratio of characteristic to continuum X radiation leads to minimum measurable weight-fractions of the order of one or a few parts per million.

Proton-probe analysis is at an early stage of development. Poole and Shaw (1969) judge that it may not be possible to produce useful probes with diameters less than 10 μm, and that the proton energies should be as high as 1 or 2 MeV for adequate excitation, implying a depth of penetration well above 10 μm in biological material. Thus in comparison with the electron probe, the proton probe may become useful for the measurement of lower elemental weight-fractions within larger microvolumes.

The evaluation by Poole and Shaw is based on the use of standard diffractive X-ray spectrography. The relative absence of continuum background and of backscattered probe particles should permit the use instead of closely placed solid-state nondiffractive spectrometers with very high efficiency of detection. One might then be able to afford the weaker excitation given by protons of lower energy for the sake of finer focusing and reduced penetration of the probe. A report of a test using this approach has been published by Johansson et al. (1970).

In order to generate as much characteristic X radiation as an electron probe, a proton probe must dissipate a much greater amount of energy in the specimen—greater by a factor of ten or more. Specimen damage from overheating may then be a serious obstacle to biological applications.

D. Ion-Probe Analysis

In an ion-probe analyzer, the specimen is exposed to a beam of ions, commonly argon or oxygen ions, at energies of the order of 10 keV. Those atoms which are driven off ("sputtered") from the surface of the specimen in ionic form are then analyzed by a mass spectrometer which is an integral part of the instrument (Fig. 30).

Ion-beam microanalysis is currently practiced in two forms which differ in just the same way as the two arrangements for electron energy-loss analysis in the electron microscope (Section IX.A, above). In the form originally introduced by Castaing and his co-workers (Rouberol et al., 1969), the incident ion beam is not focused to a probe but impinges onto a small area of the specimen, and the detecting apparatus not only selects a particular mass-number from the sputtered ions but also focuses the selected ions to give a magnified image of the distribution of the chosen constituent. The analysis of a *micro*area can then be carried out with the help of an aperture in the image plane. In the other form of ion-beam microanalysis (Fig. 30), the ion beam is focused to a probe; once the spectrometer is set for a particular constituent, the probe may

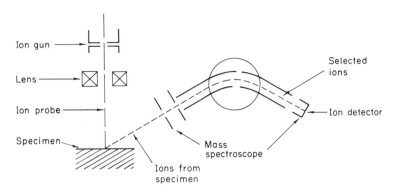

FIG. 30. The principle of ion-beam microanalysis (probe form).

be scanned to map the distribution of that constituent or may be fixed on a spot for the assay of a microarea (cf. Robinson *et al.*, 1968; Andersen, 1968). The two forms of ion-beam microanalysis are similar in their basic capabilities, and both forms are now commercially available.

The salient features of the technique are as follows:

1. The lateral spatial resolution seems at best to fall in the range 1–10 μm.

2. The penetrating power of the beam and of the emerging ions is much less than the range of the electrons in an electron probe. The analysis is consequently more confined to the surface of the specimen.

3. The analyzed layer is destroyed as it is studied, since the constituents are literally blasted away to be analyzed. However the specimen need not be damaged beyond the range of the ion beam. There is also the possibility of following a variation of composition with depth as the beam eats through the specimen.

4. The sensitivity of the technique is extraordinarily high. Minimum detectable elemental weight-fractions seem typically to be in the range of a few parts per million to a few parts per billion.

5. Since the analysis is based on mass spectroscopy, the technique can distinguish between the different isotopes of an element. Thus it may be possible to determine separately the amount of an administered radioisotope and the amount of the element in nonradioactive form in an analyzed microvolume.

6. The sensitivity varies widely among elements. While all constituents are ejected from the surface of the specimen, atoms or molecules which emerge un-ionized are not picked up by the mass spectrometer. The fraction emerging from the specimen as ionized atoms varies greatly from one element to another.

7. It is difficult to do quantitative analysis because the composition of the specimen strongly affects the distribution of an element with respect to its possible sputtered states: atomic or molecular, neutral or ionized, and singly or multiply ionized. This distribution is affected not only by the "bulk" composition of the specimen but also by the nature of any superficial layer, such as an oxide or deposited contaminant.

Biological studies based on ion-beam analysis are just beginning to appear in the literature (Galle *et al.*, 1969).

E. Laser-Probe Analysis

The principle of laser-probe analysis is shown in Fig. 31. The analyst looks at his specimen with an optical microscope and centers the micro-region which he chooses to analyze. A pulse of laser light is then sent down the microscope tube, striking the specimen as a focused probe. The energy in the pulse is enough to vaporize a small amount of tissue, and the optical spectra of the constituents of the vapor are excited by an arc discharge and analyzed by a conventional optical emission spectrograph.

The salient features of the technique are:

1. Among probe techniques, the laser method shares with X-ray fluo-rescence microanalysis the advantage that the specimen does not have to be put into a vacuum. Furthermore, the specimen can conveniently be viewed under ordinary conditions of high-quality optical microscopy.

2. The method destroys the analyzed microregion.

3. The analyzed volume is not precisely defined. The laser pulse produces a crater in the tissue, generally 50–100 μm in diameter.

4. The method is not well suited to quantitative analysis.

5. The limits of detectable elemental weight-fractions are those char-

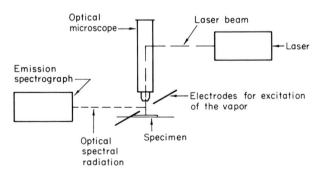

Fig. 31. The principle of laser-probe microanalysis.

acteristic of optical emission spectroscopy. This means that there is great variability among the elements, but the weight-fraction limits are often much better than attainable by electron-probe analysis. The spatial resolution however is not nearly so good, and the minimum measurable *amounts* of element are consequently less impressive.

Given this constellation of characteristics, the laser probe seems suited to the rapid identification of a large number of constituents in small volumes of tissue, but not to the quantitative analysis of very small elemental amounts in very small volumes. Apparatus for laser-probe analysis is commercially available. For further information see Brech (1969), and also Rosan *et al.* (1963).

ACKNOWLEDGMENTS

I want to thank my colleague J. Mulhern for clarification of many points in this article, and my colleague H. J. Höhling for close collaboration in many years of microprobe work.

REFERENCES

Andersen, C. A. (1967a). *Methods Biochem. Anal.* **15,** 147.
Andersen, C. A. (1967b). *Brit. J. Appl. Phys.* **18,** 1033.
Andersen, C. A. (1968). *Proc. Nat. Conf. Electron Microprobe Anal., 3rd, Chicago, 1968,* Paper 27.
Appleton, T. C. (1968). "Ultra-thin Frozen Sections for Electron Microscopy." LKB Instruments, Inc., Rockville, Maryland
Bahr, G. F. (1968). *Electron Microsc., 1968, Proc. Eur. Reg. Conf., 4th, 1968,* **I,** pp. 567–572. Tipografia Poliglotta Vaticana, Rome.
Bahr, G. F., and Zeitler, E. (1965). *In* "Quantitative Electron Microscopy" (G. F. Bahr and E. Zeitler, eds.). Williams & Wilkins, Baltimore, Maryland.
Bethe, H. A. (1930). *Ann. Phys. (Leipzig)* **5,** 325.
Birks, L. S. (1963). "Electron Probe Microanalysis." Wiley (Interscience), New York.
Bishop, H. (1965). Ph.D. Thesis, Univ. of Cambridge, Cambridge, England.
Boyde, A. (1967). *J. Roy. Microsc. Soc.* **86,** 359.
Boyde, A., and Hobdell, M. H. (1969). *Z. Zellforsch.* **93,** 213.
Boyde, A., and Jones, S. J. (1968). *Z. Zellforsch.* **92,** 536.
Boyde, A., and Switsur, V. R. (1963). *In* "X-ray Optics and X-ray Microanalysis" (H. Pattee, V. E. Cosslett and A. Engström, eds.), pp. 499–506. Academic Press, New York.
Brech, F. (1969). *Analysis Instrumentation* **6.** Plenum, New York.
Carroll, K. G., and Tullis, J. L. (1968). *Nature (London)* **217** (n5123), 1172.
Castaing, R. (1951). Ph.D. Thesis, Univer. of Paris (O.N.E.R.A. Publ. No. 55).
Castaing, R., El Hili, A., and Henry, L. (1966). *In* "X-ray Optics and Microanalysis" (R. Castaing, P. Deschamps, and J. Philibert, eds.), pp. 77–82. Hermann, Paris.
Colby, J. W., Wonsidler, D. R., and Conley, D. K. (1969). *Proc. Nat. Conf. Electron Microprobe Anal., 4th, Pasadena, 1969,* Paper 9.
Cooke, C. J., and Duncumb, P. (1966). *In* "X-ray Optics and Microanalysis" (R. Castaing, P. Deschamps, and J. Philibert, eds.), pp. 467–476. Hermann, Paris.

Crewe, A. V., Wall, J., and Welter, L. M. (1968). *J. Appl. Phys.* **39**, 5861.

Dils, R. R., Zeitz, L., and Huggins, R. A. (1963). *In* "X-ray Optics and X-ray Microanalysis" (H. Pattee, V. E. Cosslett, and A. Engström, eds.), pp. 341–360. Academic Press, New York.

Dolby, R. M. (1963). *J. Sci. Instrum.* **40**, 345.

Duncumb, P. (1959). *Brit. J. Appl. Phys.* **10**, 420.

Duncumb, P., and Reed, S. (1968). *In* "Quantitative Electron Probe Microanalysis" (K. F. J. Heinrich, ed.), pp. 133–154. *Nat. Bur. Stand. Spec. Publ.* **298.**

Duncumb, P., and Shields, P. K. (1966). *In* "The Electron Microprobe" (T. D. McKinley, K. F. J. Heinrich, and D. B. Wittry, eds.), pp. 284–295. Wiley, New York.

Engström, A. (1966). *In* "Physical Techniques in Biological Research" (A. W. Pollister, ed.), Vol. III A, pp. 87–171. Academic Press, New York.

Frazier, P. D. (1966). *Norelco Rep.* **13**, 25.

Frazier, P. D. (1967). *Arch. Oral Biol.* **12**, 25.

Freeman, D. H., and Paulson, R. A. (1968). *Nature* (*London*) **218**, 563.

Galle, P., Blaise, G., and Slodzian, G. (1969). *Proc. Nat. Conf. Electron Microprobe Anal., 4th, Pasadena, 1969,* Paper 36.

Green, M., and Cosslett, V. E. (1961). *Proc. Phys. Soc. London* **78**, 1206.

Hall, T. A. (1968). *In* "Quantitative Electron Probe Microanalysis" (K. F. J. Heinrich, ed.), pp. 269–299. *Nat. Bur. Stand. Spec. Publ.* **298.**

Hall, T. A., and Höhling, H. J. (1969). *In* "X-Ray Optics and Microanalysis" (G. Möllenstedt and K. H. Gaukler, eds.), pp. 582–591. Springer, Berlin.

Hall, T. A., and Werba, P. (1969). *In* "X-Ray Optics and Microanalysis" (G. Möllenstedt and K. H. Gaukler, eds.), pp. 93–98. Springer, Berlin.

Hall, T. A., Hale, A. J., and Switsur, V. R. (1966). *In* "The Electron Microprobe" (T. D. McKinley, K. F. J. Heinrich, and D. B. Wittry, eds.), pp. 805–833. Wiley, New York.

Hall, T. A., Röckert, H. O., and Saunders, R. L. deC. H. (in press). "X-Ray Microscopy in Clinical and Experimental Medicine." Thomas, Fort Lauderdale, Florida, to be published.

Harris, L. A. (1968). *J. Appl. Phys.* **39**, 1419.

Heinrich, K. F. J. (1966). *In* "The Electron Microprobe" (T. D. McKinley, K. F. J. Heinrich, and D. B. Wittry, eds.), pp. 351–377. Wiley, New York.

Henke, B. L., Elgin, R. L., Lent, R. E., and Ledingham, R. B. (1967). U.S. Doc. No. AFOSR 67-1254 (obtainable from Clearinghouse for Federal Sci. and Tech. Inform., Springfield, Virginia).

Hénoc, J. (1968). *In* "Quantitative Electron Probe Microanalysis" (K. F. J. Heinrich, ed.), pp. 197–214. *Nat. Bur. Stand. Spec. Publ.* **298.**

Ichinokawa, T., Kobayashi, H., and Nakajima, M. (1969). *Japan. J. Appl. Phys.* **8**, 1563.

Ingram, M. J., and Hogben, C. A. M. (1968). *Develop. Appl. Spectrosc.* **6**, 43.

Johansson, T. B., Akselsson, R., and Johansson, S. A. E. (1970). X-ray Analysis: Elemental trace analysis at the 10^{-12}-g level. *Nucl. Instrum. Methods* **84**, 141.

Johnson, A. R. (1969). *Proc. Nat. Conf. Electron Microprobe Anal., 4th, Pasadena, 1969,* Paper 38.

Kimoto, S., Hashimoto, H., and Uchiyama, H. (1969). *In* "X-Ray Optics and Microanalysis" (G. Möllenstedt and K. H. Gaukler, eds.), pp. 369–372. Springer, Berlin.

Kramers, H. A. (1923). *Phil. Mag.* **46**, 836.

Kyser, D. F., and Wittry, D. B. (1966). *In* "The Electron Microprobe" (T. D. McKinley, K. F. J. Heinrich, and D. B. Wittry, eds.), pp. 691–714. Wiley, New York.

Kyser, D. F., McCoy, J., and Wittry, D. B. (1967). *Proc. Nat. Conf. Electron Microprobe Anal., 2nd, Boston, 1967,* Paper 50.

Lindström, B. (1955). *Acta Radiol. Suppl.* **n125.**

Long, J. V. P., and Röckert, H. O. E. (1963). *In* "X-ray Optics and X-ray Microanalysis" (H. Pattee, V. E. Cosslett, and A. Engström, eds.), pp. 513–521. Academic Press, New York.

Marshall, D. J. (1967). Ph.D. Thesis, Univ. of Cambridge, Cambridge, England.

Marshall, D. J., and Hall, T. A. (1968). *Brit. J. Appl. Phys.* **1,** 1651.

Mellors, R. C. (1964). *Lab. Invest.* **13,** 183.

Nelms, A. T. (1956). *Nat. Bur. Stand. (U.S.) Circ.* No. 577.

Nicholson, J. B., and Hasler, M. F. (1966). *Advan. X-Ray Anal.* **9,** 420.

Nicholson, J. B., Mooney, C. F., and Griffin, G. L. (1965). *Advan. X-Ray Anal.* **8,** 301.

Okano, H., Tomura, T., Hara, K. (1968). *Proc. Nat. Conf. Electron Microprobe Anal., 3rd, Chicago, 1968,* Paper 31.

Pearse, A. G. E. (1968). "Histochemistry," 3rd ed., Vol. I. Churchill, London.

Pease, R. F. W., and Hayes, T. L. (1966). *Nature (London)* **210,** 1049.

Philibert, J. (1969). *In* "Vth International Congress on X-Ray Optics and Microanalysis" (G. Möllenstedt and K. H. Gaukler, eds.), pp. 114–131. Springer, Berlin.

Philibert, J. (1963). *In* "X-ray Optics and X-ray Microanalysis" (H. Pattee, V. E. Cosslett, and A. Engström, eds.), pp. 379–392. Academic Press, New York.

Poole, D. M., and Shaw, J. L. (1969). *In* "X-Ray Optics and Microanalysis" (G. Möllenstedt and K. H. Gaukler, eds.), pp. 310–321. Springer, Berlin.

Ranzetta, G. V. T., and Scott, V. D. (1966). *J. Sci. Instrum.* **43,** 816.

Rao-Sahib, T. S., and Wittry, D. B. (1969). *Proc. Nat. Conf. Electron Microprobe Anal., 4th, Pasadena, 1969,* Paper 2.

Reed, S. J. B. (1965). *Brit. J. Appl. Phys.* **16,** 913.

Reed, S. J. B. (1966). *In* "X-ray Optics and Microanalysis" (R. Castaing, P. Deschamps, and J. Philibert, eds.), pp. 339–349. Hermann, Paris.

Reed, S. J. B. (1968). *Brit. J. Appl. Phys.* **1,** 1090.

Rivière, J. C. (1969). *Phys. Bull. (Gt. Brit.)* **20,** 85.

Robinson, C. F., Liebl, H. J., and Andersen, C. A. (1968). *Proc. Nat. Conf. Electron Microprobe Anal., 3rd, Chicago, 1968,* Paper 26.

Röckert, H. O. E. (1966). *Methods and Achievements Exp. Pathol.* **1,** 143. S. Karger, Basel/New York.

Rosan, R. C., Healy, M. K., and McNary, W. F., Jr. (1963). *Science* **142,** 236.

Rouberol, J.-M., Guernet, J., Deschamps, P., Dagnot, J.-P., and Guyon de la Berge, J.-M. (1969). *In* "X-Ray Optics and Microanalysis" (G. Möllenstedt and K. H. Gaukler, eds.), pp. 311–318. Springer, Berlin.

Russ, J. C., and McNatt, E. (1969). *Proc. Electron Microscopy Soc. Amer.* **27,** 38.

Sikorksi, J., Moss, J. S., Newman, P. H., and Buckley, T. (1968). *J. Sci. Instrum.* **1,** 29.

Sims, R. T., and Hall, T. A. (1968). *J. Cell Sci.* **3,** 563.

Söremark, R., and Grøn, P. (1966). *Arch. Oral Biol.* **11,** 861.

Sutfin, L. V., and Ogilvie, R. E. (1970). *Proc. Ann. Symp. Scanning Electron Microscope, 3rd, Chicago, 1970.* To be published.

Swift, J. A., and Brown, A. C. (1970). *Proc. Ann. Symp. Scanning Electron Microscope, 3rd, Chicago, 1970,* Illinois Inst. of Technol., Chicago. To be published.

Swift, J. A., Brown, A. C., and Saxton, C. A. (1969). *J. Sci. Instrum.* **2,** 744.

Trombka, J. I., Adler, I., Schmadebeck, R., and Lamother, R. (1966). Non-dispersive X-Ray emission analysis for lunar surface geochemical exploration. NASA Rep. X-641-66-344. Goddard Space Flight Center.

Vassamillet, L. F., and Caldwell, V. E. (1968). *Proc. Nat. Conf. Electron Microprobe Anal., 3rd, Chicago, 1968,* Paper 40.

Wittry, D. B., Ferrier, R. P., and Cosslett, V. E. (1969). *Brit. J. Appl. Phys.* **2,** 1767.

Wörner, H., and Wizgall, H. (1969). *In* "X-Ray Optics and Microanalysis" (G. Möllenstedt and K. H. Gaukler, eds.), pp. 601–607. Springer, Berlin.

Zeigler, D. C., Zeigler, W. H., Harclerode, J. E., and White, E. W. (1969). *Proc. Nat. Conf. Electron Microprobe Anal., 4th, Pasadena, 1969,* Paper 37.

Zeitler, E., and Bahr, G. F. (1962). *J. Appl. Phys.* **33,** 847.

Zeitz, L. (1961). *Rev. Sci. Instrum.* **32,** 1423.

Zeitz, L. (1969). *Progr. Anal. Chem.* **3,** 35–72.

Appendix I. Manufacturers and Availability of Probe Instruments

The price of standard electron-probe instruments is upwards of $50,000, usually in the range $70,000–100,000. While many probes have been bought for metallurgical and mineralogical laboratories, there is not yet much time available on standard probes for biology, and the probes of special types are generally even more expensive and less accessible. This raises a practical problem for the biologist who wants to use a microprobe but is not ready to buy one. Three courses of action seem open: to request time on an instrument which has been purchased for nonbiological work, to consult with people who have done biological microprobe work (the appended bibliography may be helpful in this respect), and to consult with probe manufacturers, who can often make some time available in their applications laboratories. A list of manufacturers follows.

A. ELECTRON PROBES

Standard electron-probe instruments are made by the following companies, listed alphabetically:

1. Applied Research Laboratories, P. O. Box 129, Sunland, California 91040, U.S.A.

2. Associated Electrical Industries: GEC-AEI (Electronics) Ltd., Scientific Apparatus Division, P. O. Box 1, Harlow, Essex, England.

3. Cambridge Scientific Instruments, Chesterton Road, Cambridge, England.

4. Cameca, 103 Bd. Saint-Denis, Courbevoie (Seine), France.

5. Hitachi Ltd., Tokyo, Japan. (European agency: Marubeni-Iida Co., Machinery Division, 164 Clapham Park Road, London S.W. 4, England.)

6. Japan Electron Optics Laboratory Co., New Tokyo Building, 3-2 Marunouchi, Chiyoda-ku, Tokyo, Japan.

7. Materials Analysis Company, 1060 East Meadow Circle, Palo Alto, California 94303, U.S.A.

8. Philips Electronics Instruments, 750 South Fulton Avenue, Mount Vernon, New York 10550, U.S.A.

9. Siemens Aktiengesellschaft, Wernerwerk für Messtechnik, Karlsruhe, Federal Republic of Germany.

10. Technisch Physische Dienst TNO-TH, Delft, The Netherlands.

11 Some of the microprobes made in the Soviet Union may be available for purchase. Inquiries may be directed to: Techsnabexport, Moskva G-200, Moskva, U.S.S.R.

Most of these manufacturers have agencies and demonstration centers outside of their own countries, in the U.S.A., Great Britain, continental Europe, and Japan.

As indicated in Section III.C, it is now possible to combine conventional scanning electron microscopy, transmission scanning electron microscopy, and X-ray microanalysis effectively in a single electron-probe instrument. All of the manufacturers of standard electron probes are working in this direction so that it may be invidious to mention separately those who have especially stressed the combination of X-ray microanalysis with high-resolution scanning microscopy (resolution in the electron image of approx. 100 Å); nonetheless out of the list above they are CSI, Hitachi, JEOL, and MAC.

The other major direction of development of electron-probe X-ray microanalysis is the combination of microanalysis with conventional transmission electron microscopy. In this field we should note

1. Makers of combination electron microscope-microanalyzers: AEI, Siemens.

2. Makers of a probe microanalyzer with an attachment for conventional electron microscopy: Cameca.

3. Makers of electron microscopes with attachments for microanalysis: Siemens and Philips. (Address of Philips: N. V. Philips' Gloeilampenfabrieken, Scientific Equipment Department, Eindhoven, The Netherlands.)

B. Apparatus for Electron Energy-Loss Microanalysis

Microprobe apparatus for electron energy-loss analysis does not seem to be available yet commercially. Castaing's system for producing energy-selected electron images is offered commercially by the firm Sopelem, 102 Rue Chaptal, 92-Levallois, Perret, France.

C. Apparatus for Ion-Beam Microanalysis

Ion-probe instruments are made by ARL. Instruments for ion-beam microanalysis by the output-focusing rather than the probe method are offered by Cameca. (Both addresses are listed above.)

D. Laser-Probe Microanalysis

This apparatus is offered by the Jarrell-Ash Company, 590 Lincoln Street, Route 128, Waltham, Massachusetts 02154, U.S.A.

Appendix II. Bibliography of Biological Applications of Electron-Probe X-Ray Microanalysis

In this chapter, we have not discussed the extensive biological work that has been done with the microprobe analyzer. The following bibliography can serve as a guide. The papers are listed under these headings: survey articles; the fate of foreign bodies; the fate of injected or ingested substances; metabolic abnormalities in soft tissues; physiology of soft tissues; prestages of mineralization in arteries, bone, and teeth; mineralized tissues (mainly bone and teeth); and studies of histological stains.

When the bibliography is used to judge what can be done in a particular field, it is important to keep one point in mind—the articles of the earlier years do not give an accurate indication, because of the big recent advances in the instruments and in operating techniques.

The bibliography is certainly not complete. However there is only one type of paper which I have generally deliberately omitted—conference reports whose contents seem to be entirely contained in other papers in the journals.

Many of the entries are papers which have been presented at the annual national (U.S.A.) conferences on electron microprobe analysis, now under the auspices of The Electron Probe Analysis Society of America. Readers who want material from one of these conferences should write to the treasurer of the Society, Dr. A. A. Chodos, Department of Geology, California Institute of Technology, Pasadena, California 91109, U.S.A.

A. Survey Articles

Andersen, C. A. (1967). An introduction to the electron probe microanalyzer and its application to biochemistry. *Methods Biochem. Anal.* **15,** 147–270.

Andersen, C. A., and Hasler, M. F. (1966). Extension of electron microprobe techniques to biochemistry by the use of long wave-length X-rays. In "X-ray Optics and Microanalysis" (R. Castaing, P. Deschamps, and J. Philibert, eds.), pp. 310–327. Hermann, Paris.

Carroll, K. G. (1967). Biological applications of the electron probe analyser. In "In Vivo Techniques in Histology" (G. H. Bourne, ed.), pp. 69–79. Williams & Wilkins, Baltimore, Maryland.

Cosslett, V. E., and Switsur, V. R. (1963). Some biological applications of the scanning microanalyzer. In "X-ray Optics and X-ray Microanalysis" (H. Pattee, V. E. Cosslett, and A. Engström, eds.), pp. 507–512. Academic Press, New York.

Galle, P. (1964). Mise au point d'une méthode de microanalyse des tissus biologiques au moyen de la microsonde de Castaing. Rev. Fr. Etud. Clin. Biol. 8, 203.

Hall, T. A. (1968). Some aspects of the microprobe analysis of biological specimens. In "Quantitative Electron Probe Microanalysis" (K. F. J. Heinrich, ed.), pp. 269–299. Nat. Bur. Stand. Spec. Publ. 298.

Hall, T. A. (1970). The applicability of electron probe X-ray microanalysis to trace element studies. In "Trace Element Metabolism in Animals" (C. F. Mills, ed.), pp. 504–512. E. & S. Livingstone, London.

Hall, T. A., and Höhling, H. (1969). The application of microprobe analysis to biology. In "X-Ray Optics and Microanalysis" (G. Möllenstedt and K. H. Gaukler, eds.), pp. 582–591. Springer, Berlin.

Hall, T. A., Hale, A. J., and Switsur, V. R. (1966). Some applications of microprobe analysis in biology and medicine. In "The Electron Microprobe" (T. D. McKinley, K. F. J. Heinrich, and D. B. Wittry, eds.), pp. 805–833. Wiley, New York.

Hall, T. A., Röckert. H. O. E., and Saunders, R. L. deC. H. (in press). "X-Ray Microscopy in Clinical and Experimental Medicine." Thomas, Fort Lauderdale, Florida, to be published.

Höhling, H. J., and Hall, T. A. (1969). Elektronenstrahlmikroanalyse als quantitative histologische Methode. Naturwissenschaften 56, 622–629.

Ingram, M. J., and Hogben, C. A. M. (1968). Procedures for the study of biological soft tissue with the electron microprobe. Develop. Appl. Spectrosc. 6, 43–64.

Läuchli, A. (1967). Zur Technik der Herstellung biologisches Präparate für die histochemische Untersuchung mit der Röntgen-Mikrosonde. Histochem. 11, 286–288.

Rasmussen, H. P., Shull, V. E., and Dryer, H. T. (1968). Determination of element localization in plant tissue with the microprobe. Develop. Appl. Spectrosc. 6, 29–42.

Robertson, A. J. (1968). The electron probe microanalyser and its applications in medicine. Phys. Med. Biol. 13, 505–522.

Tousimis, A. J. (1963). Electron-probe microanalysis of biological specimens. In "X-ray Optics and X-ray Microanalysis" (H. Pattee, V. E. Cosslett, and A. Engström, eds.), pp. 539–557. Academic Press, New York.

Tousimis, A. J. (1966). Applications of the electron probe X-ray microanalyzer in biology and medicine. Amer. J. Med. Electron. 5, 15–23.

Tousimis, A. J. (1969). Electron probe microanalysis of biological structures. Progr. Anal. Chem. 3, 87–103.

Tousimis, A. J. (1969). "A combined scanning electron microscopy and electron probe microanalysis of biological soft tissues." *Proc. Ann. Sympos. Scanning Electron Microscope, 2nd Chicago, 1969.* Illinois Institute of Technology Research Institute.

Tousimis, A. J., Brooks, E. J., and Birks, L. S. (1959). Possible biological applications of the electron probe. *In* "X-Ray Spectrochemical Analysis," p. 122. Wiley (Interscience), New York.

Wyckoff, R. W. G., Laidley, R. A., and Hoffmann, V. J. (1963). The probe analysis of non-conducting samples. *Norelco Rep.* **10,** 123–126.

Yasuzumi, G., Nagahara, T., and Nakai, Y. (1961). Application of X-ray microanalyser to biological specimens. *Nara Igaku Zasshi* **12,** 323.

B. THE FATE OF FOREIGN BODIES

Andersen, C. A. (1967). (See Section A of this list.)

Banfield, W. G., Tousimis, A. J., Hagerty, J. C., and Padden, T. R. (1969). Electron probe analysis of human lung tissues. *Progr. Anal. Chem.* **3,** 9–34.

Berkley, C., Langer, A. M., and Baden, V. (1967). Instrumental analysis of inspired fibrous pulmonary particulates. *Trans. N.Y. Acad. Sci.* **30,** 331–350.

Berkley, C., Langer, A. M. and Rubin, I. (1970). Electron microprobe analysis of particles in tissue. *Proc. Nat. Conf. Electron Probe Anal., 5th, New York City, 1970,* Paper 29.

Galle, P., Vivier, E., and Petitprez, A. (1968). "Étude par microscopie électronique couplee à la spectrographie des rayons X d'inclusions de matière inerte chez Spirostomum Ambiguum. *Electron Microsc., 1968, Proc. Eur. Reg. Conf., 4th, 1968,* **2,** pp. 439–440. Tipografia Poliglotta Vaticana, Rome.

Kaufman, T. B., Bigelow, W. C., Petering, L. B., and Drogos, Z. B. (1969). Silica in developing epidermal cells of Avena internodes: Electron probe analysis. *Science* **166,** 1015–1017.

Laroche, J. (1967). Localisation de la silice par le microanalyseur à sonde électronique. *C. R. Acad. Sci.* **D265,** 1695–1697.

Laroche, J. (1968). Contribution à l'étude de l'Equisetum arvense L. III. Recherches sur la nature et la localisation de la silice chez le sporophyle. Thèsis de doctorat, Librairie Générale de l'enseignement, 4 Rue Dante, Paris 4.

Mayer, P. B., Stumphius, J., von Rosenstiel, A. P., and Zeedijk, H. B. (1967). Elektronenmikroskopische und Röntgenmikroanalytische Untersuchungen von 'Asbestnadeln' in Sputum und Lunge eines Mesotheliomapatienten. *Conf. Deutsche Ges. Elektronenmikroskopie e.V., Marburg, 1967.*

Mellors, R. C., and Carroll, K. G. (1961). A new method for local chemical analysis of human tissue. *Nature (London)* **192,** 1090–1092.

Robertson, A. J., Rivers, D., Nagelschmidt, G., and Duncumb, P. (1961). Stannosis, benign pneumoconiosis due to tin oxide. *Lancet* 1089–1093 (1961).

von Rosenstiel, A. P., and Zeedijk, H. B. (1968). An electron probe and electron microscope investigation of asbestos bodies in lung sputum. *Proc. Nat. Conf. Electron Microprobe Anal., 3rd, Chicago, 1968,* Paper 43.

C. THE FATE OF INJECTED OR INGESTED SUBSTANCES

Banfield, W. G. *et al.* (1969). (See Section B of this list.)

Carroll, K. G., Spinelli, F. R., and Goyer, R. A. (1970). Electron probe microanalyser localization of lead in kidney tissue of poisoned rats. *Nature* **227,** 1056.

Galle, P. (1964). Étude comparée au microscope électronique et à la microsonde du Rein Pathologique. *J. Microsc. (Paris)* **3,** 347.

Galle, P. (1964). Oxalose rénale expérimentale. Étude au microscope électronique et par spectrographie des rayons X. *Nephron* **1**, 158.

Galle, P. (1967). Microanalyse des inclusions minérales du rein. *Proc. Int. Congr. Nephrol. 3rd, Washington, D.C., 1966,* **2**, pp. 306–319. Karger, Basel.

Galle, P. (1967). Les néphrocalcinoses: Nouvelles données d'ultrastructure et de microanalyse. *Actual. Nephrol. Hop. Necker* pp. 303–315.

Galle, P., and Morel-Maroger, L. (1965). Les lésions rénales du saturnisme humain et experimental. *Nephron* **2**, 273–286.

Galle, P., Berry, J. P., and Stuve, J. (1968). New applications of electron microprobe analysis in renal pathology. *Proc. Nat. Conf. Electron Microprobe Anal., 3rd, Chicago, 1968,* Paper 41.

Gueft, B., Kikkawa, Y., and Moskal, J. (1964). Electron-probe studies of tissues in Wilson's disease, lead poisoning and bunamiodyl nephropathy. *J. Appl. Phys.* **35**, 3077.

Johnson, F. B., and Zimmerman, L. E. (1965). Barium sulphate and zinc sulphide deposits resulting from golf-ball injury to the conjunctiva and eyelid. *Amer. J. Clin. Pathol.* **44**, 533–538.

Kaufman, T. B. *et al.* (1969). (See Section B of this list.)

Laroche, J. (1967). (See Section B of this list.)

Laroche, J. (1968). (See Section B of this list.)

Rasmussen, H. P. (1969). Entry and distribution of aluminum in Zea Mays. *Planta* **81**, 28–37.

Russ, J. C. and McNatt, E. (1969). Copper localization in cirrhotic rat liver by scanning electron microscopy. *Proc. Electron Microscop. Soc. Amer.* **27**, 38.

Waisel, Y., Hoffen, A., and Eshel, A. (1970). The localization of aluminum in the cortex cells of bean and barley roots by X-ray microanalysis. *Physiologia Plantarum* **23**, 75–79.

D. METABOLIC ABNORMALITIES IN SOFT TISSUES

Beaman, D. R., Nishiyama, R. H., and Penner, J. A. (1969). The analysis of blood diseases with the electron microprobe. *Blood* **34**, 401–413.

Carroll, K. G., and Tullis, J. L. (1968). Observations on the presence of titanium and zinc in human leucocytes. *Nature (London)* **217**, 1172–1173.

Duprez, A. (1969). Histochimie par Spectrographie des RX. *In* "X-Ray Optics and Microanalysis" (G. Möllenstedt and K. H. Gaukler, eds.), pp. 579–581. Springer, Berlin.

Gabbian, G., Badonnell, M. C., and Baud, C. A. (1969). The role of iron in experimental cutaneous calcinosis. *Fed. Proc. (Fed. Amer. Soc. Exp. Biol.)* **28**, 552.

Galle, P. (1964). (See Section C of this list.)

Galle, P. (1964). Oxalose rénale expérimentale, étude au microscope électronique et par spectrographie des rayons X. *Nephron* **1**, 158.

Galle, P. (1967). (See Section C of this list.)

Galle, P. (1967). Les néphrocalcinoses: Nouvelles données d'ultrastructure et de microanalyse. *Actual. Nephrol. Hop. Necker* pp. 303–315.

Galle, P., and Morel-Maroger, L. (1965). (See Section C of this list.)

Galle, P., Berry, J. P., and Stuve, J. (1968). (See Section C of this list.)

Goldfischer, S., and Moskal, J. (1966). Electron probe microanalysis of liver in Wilson's disease. *Amer. J. Pathol.* **48**, 305–315.

Gueft, B., Kikkawa, Y., and Moskal, J. (1964). (See Section C of this list.)

Ingram, M. J., and Hogben, C. A. M. (1967). Electrolyte analysis of biological fluids with the electron microprobe. *Anal. Biochem.* **18**, 54–57.

Marx, A. J., Gueft, B., and Moskal, J. F. (1965). Prostatic corpora amylacea. *Arch. Pathol.* **80**, 487–494.

Morel, F., and Roinel, N. (1969). Application de la microsonde électronique à l'analyse élémentaire quantitative d'échantillons liquides d'un volume inférieur à 10^{-9} l. *J. Chim. Phys.* **66**, 1084–1091.

Morel, F., Roinel, N., and Le Grimellec, C. (1969). Electron probe analysis of tubular fluid composition. *Nephron* **6**, 350–364.

Russ, J. C., and McNatt, E. (1969). (See Section C of this list.)

Sims, R. T., and Hall, T. A. (1968). X-ray emission microanalysis of the density of hair proteins in kwashiorkor. *Brit. J. Dermatol.* **80**, 35–38.

Terry, R. D., and Peña, C. (1965). Experimental production of neurofibrillary degeneration. II. Electron microscopy, phosphatase histochemistry and electron probe analysis. *J. Neuropathol. Exp. Neurol.* **24**, 200–210.

Tousimis, A. J., and Adler, I. (1963). Electron probe X-ray microanalyzer study of copper within Descemet's membrane in Wilson's disease. *J. Histochem. Cytochem.* **11**, 40.

E. PHYSIOLOGY OF SOFT TISSUES

Andersen, C. A. (1967). (See Section A of this list.)

Carroll, K. G. (1967). Chemical constitution of bacterial spores. *Proc. Nat. Conf. Electron Microprobe Anal., 2nd, Boston, 1967,* Paper 40.

Coleman, J. R., and Terepka, A. R. (1969). Electron probe demonstration of the distribution of calcium in the chick chorioallantoic membrane. *Biophys. J.* **9**, A-72 (abstracts).

Coleman, J. R., and Terepka, A. R. (1970). Correlated electron probe and electron microscope analysis of the distribution of calcium in the chick chorioallantoic membrane. *Ann. Meeting Biophys. Soc., 14th, Baltimore, 1970.* Abstract WPM-H8.

Hall, T. A. (1966). The microprobe analysis of zinc in mammalian sperm cells. *In* "X-Ray Optics and Microanalysis" (R. Castaing, P. Deschamps, and J. Philibert, eds.), pp. 679–685. Hermann, Paris.

Höhling, H. J., Kriz, W., Schnermann, J., and von Rosenstiel, A. P. (1970). Messungen von Elektrolyten in Nierenschnitten mit der Mikrosonde: Methodik und Vorläufige Ergebnisse. *65. Versamml. Anat. Ges., Wurzburg, 1970.* Papers 39 and 40.

Ingram, M. J., and Hogben, C. A. M. (1968). (See Section A of this list.)

Jones, W. C., and James, D. W. F. (1969). An investigation of some calcareous sponge spicules by means of electron probe micro-analysis. *Micron* **1**, 34–39.

Läuchli, A. (1967). Untersuchungen über Verteilung und Transport von Ionen in Pflanzengeweben mit der Röntgen-Mikrosonde. I. Versuche an vegetativen Organen von Zea Mays. *Planta* **75**, 185–206.

Läuchli, A. (1967). Nachweis von Calcium-Strontiumablagerungen im Fruchstiel von Pisum sativum mit der Röntgen-Mikrosonde. *Planta* **73**, 221–227.

Läuchli, A. (1968). Untersuchungen mit der Röntgen-Mikrosonde über Verteilung und Transport von Ionen in Pflanzengeweben. II. Ionentransport nach den Früchten von Pisum sativum. *Planta* **83**, 137–149.

Läuchli, A. (1968). Untersuchung des Stofftransports in der Pflanze mit der Röntgen-Mikrosonde. *Vorträge Gesamtgeb. Bot.*, **N. F. Nr. 2**, 58–65.

Läuchli, A., and Schwander, H. (1966). X-ray microanalyzer study on the localization of minerals in native plant tissue sections. *Experientia* **22**, 503–505.

Läuchli, A., and Lüttge, U. (1968). Untersuchung der Kinetik der Ionenaufnahme in das Cytoplasma von Mnium-Blattzellen mit Hilfe der Mikroautoradiographie und der Röntgen-Mikrosonde. *Planta* **83**, 80–98.

Lechene, C. (1970). The use of the electron microprobe to analyse very minute amounts of liquid samples. *Proc. Nat. Conf. Electron Probe Anal., 5th, New York City, 1970*, Paper 32.

Lever, J. D., and Duncumb, P. (1961). The detection of intracellular iron in rat duodenal epithelium. *In* "Electron Microscopy in Anatomy," pp. 278–286. Arnold, London.

Libanati, C. M., and Tandler, C. J. (1969). The distribution of the water-soluble inorganic orthophosphate ions within the cell: Accumulation in the nucleus. *J. Cell Biol.* **42**, 754–765.

Maroudas, A. (1970). Electron probe microanalysis of uncalcified articular cartilage. *Proc. Royal Microscop. Soc.* **5, part 5** (**Micro 70 issue**), 232.

Morel, F., and Roinel, N. (1969). (See Section D of this list.)

Morel, F. *et al.* (1969). (See Section D of this list.)

Noeske, O., Läuchli, A., Lange, O. L., Vieweg, G. H., and Ziegler, H. (1970). Konzentration und Lokalisierung von Schwermetallen in Flechten der Erzschlackenhalden des Harzes. *Dtsch. Bot. Ges.* **Neue Folge, Nr. 4**, 67–79.

Pautard, F. G. E., and Zola, H. (1968). The location of onuphic acid in hyalinoecia tubicola. *J. Histochem. Cytochem.* **15**, 737–744.

Podolsky, R. J., Hall, T., and Hatchett, S. L. (1970). Identification of oxalate precipitates in striated muscle fibers. *J. Cell Biol.* **44**, 699–702.

Robison, W. L., and Davis, B. (1969). Determination of iodine concentration and distribution in rat thyroid follicles by electron probe microanalysis. *J. Cell Biol.* **43**, 115–122.

Rosenstiel, A. P. von, Höhling, H. J., Schnermann, J., and Kriz, W. (1970). Electron probe microanalysis of electrolytes in kidney slices. *Proc. Nat. Conf. Electron Probe Anal., 5th, New York City, 1970*, Paper 33.

Russ, J. C., and McNatt, E. (1969). (See Section C of this list.)

Saffir, A. J. and Ogilvie, R. E. (1966). Analysis of oral tissues by X-ray absorption microanalysis with the electron microanalyzer. *Proc. Nat. Conf. Electron Probe Microanal., 1st, Washington, D.C., 1966*.

Tousimis, A. J., Hagerty, J. C., Padden, T. R., and Laster, L. (1969). The application of electron probe microanalysis to the study of amino acid transport in the small intestine. *Progr. Anal. Chem.* **3**, 75–86.

Sims, R. T., and Hall, T. A. (1968). X-ray emission microanalysis of proteins and sulphur in rat plantar epidermis. *J. Cell Sci.* **3**, 563–572.

Yasuzumi, G. (1962). X-ray scanning microanalysis of elemental iron localized in testicular nutritive cells of *Cipangopaludina Malleata* Reeve. *J. Cell Biol.* **14**, 496–498.

Yasuzumi, G. (1962). Electron microscopy and X-ray scanning microanalysis of needle biopsy material from human liver. *J. Cell Biol.* **14**, 421–431.

Yasuzumi, G., Tsubo, I., Okada, K., and Takahashi, H. (1969). Electron microscopy, electron probe X-ray microanalysis and microdiffraction of biopsy material from human liver. *Okajimas Folia Anat. Jap.* **45**, 279–307.

Zeigler, D. C., Zeigler, W. H., Harclerode, J. E., and White, E. W. (1969). Electron microprobe analysis of thyroidal iodine in the white-throated sparrow. *Proc. Nat. Conf. on Electron Microprobe Anal., 4th, Pasadena, 1969,* Paper 37.

Zeitz, L., and Andersen, C. A. (1966). X-ray microprobe analysis for zinc in the rat prostate. *In* "X-Ray Optics and Microanalysis" (R. Castaing, P. Deschamps, and J. Philibert, eds.), pp. 691–698. Hermann, Paris.

F. PRESTAGES OF MINERALIZATION IN ARTERIES, BONE, AND TEETH

Andersen, C. A. (1967). (See Section A of this list.)

Brooks, E. J., Tousimis, A. J., and Birks, L. S. (1962). The distribution of calcium in the epiphyseal cartilage of the rat tibia measured with the electron probe X-ray microanalyzer. *J. Ultrastruct. Res.* **7,** 56–60.

Carlisle, E. M. (1969). Silicon localization and calcification in developing bone. *Fed. Proc. Fed. Amer. Soc. Exp. Biol.* **28,** 374.

Carlisle, E. M. (1970). Silicon: A possible factor in bone calcification. *Science* **167,** 279–280.

Hagerty, J. C., and Tousimis, A. J. (1968). Electron probe morphological and chemical analyses of the rat epiphyseal plate. *Proc. Electron Microscop. Soc. Amer. 26.*

Hale, A. J., Hall, T. A., and Curran, R. C. (1967). Electron-microprobe analysis of calcium, phosphorus and sulphur in human arteries. *J. Pathol. Bacteriol.* **93.** 1–17.

Höhling, H. J., Hall, T. A., and Fearnhead, R. W. (1967). Electron Probe X-ray Microanalysis als Quantitative Histologische Methode zur Analyse von Vorstadien der Altersbedingten Aorten-Mineralisierung. *Naturwissenschaften* **54,** 93–94.

Höhling, H. J., Hall, T. A., Boothroyd, B., Cooke, C. J., Duncumb, P., and Fitton-Jackson, S. (1967). Untersuchungen der Vorstadien der Knochenbildung mit Hilfe der Normalen und Elektronenmikroskopischen Electron Probe X-ray Microanalysis. *Naturwissenschaften* **54,** 142–143.

Höhling, H. J., Hall, T. A., Boothroyd, B., Cooke, C. J., Duncumb, P., Fearnhead, R. W., and Fitton-Jackson, S. (1968). Electron probe studies of prestages of apatite formation in bone and aorta. *In* "Les Tissus Calcifiés" (G. Milhaud, M. Owen, and H. J. J. Blackwood, eds.), pp. 323–328. Societé D'Édition D'Enseignement Supérieur, Paris.

Höhling, H. J., Hall, T. A., and Boyde, A. (1967). Electron probe X-ray microanalysis of mineralization in rat incisor peripheral dentine. *Naturwissenschaften* **54,** 617–618.

Höhling, H. J., Hall, T. A., Boyde, A., and von Rosenstiel, A. P. (1968). Combined electron probe and electron diffraction analysis of prestages and early stages of dentine formation in rat incisors. *Calcif. Tissue Res. August Suppl.* **2,** 5.

Höhling, H. J., Schöpfer, H., Höhling, R. A., Hall, T. A., and Gieseking R. (1970). The organic matrix of developing tibia and femur, and macromolecular deliberations. *Die Naturwissenschaften* **57,** 357.

G. MINERALIZED TISSUES (MAINLY BONE AND TEETH)

Andersen, C. A. (1967). (See Section A of this list.)

Baud, C. A., Kimoto, S., and Hashimoto, H. (1963). Étude de la distribution du calcium dans l'os Haversien avec le radioanlyseur à microsonde électronique. *Experientia* **19,** 524–526.

Baud, C. A., and Lobjoie, D. P. (1966). L' emploi du radioanalyseur a microsonde électronique pour l'étude de la distribution du fluor dans les coupes d'email dentaire. *Ann. Histochim.* **11,** 151.

Baud, C. A., and Lobjoie, D. P. (1966). Biophysical investigations on the mineral phase in the superficial layers of human dental enamel. *Helvet. Odontol. Acta* **10,** 40.

Baud, C. A., Bang, S., Lee, H. S., and Baud, J. P. (1968). X-ray studies of strontium incorporation into bone mineral in vivo, *Calcif. Tissue Res. August Suppl.* **2,** 6.

Bax, D., and van der Linden, L. W. J. (1969). Electron probe microanalysis of filled human teeth. *In* "X-Ray Optics and Microanalysis" (G. Möllenstedt and K. H. Gaukler, eds.), pp. 597–600. Springer, Berlin.

Besic, F. C., Knowles, C. R., Wiemann, M. R., Jr., and Keller, O. (1969). Electron probe microanalysis of noncarious enamel and dentin and calcified tissues in mottled teeth. *J. Dental Res.* **48,** 131–139.

Boyde, A., and Switsur, V. R. (1963). Problems associated with the preparation of biological specimens for microanalysis. *In* "X-ray Optics and X-ray Microanalysis" (H. Pattee, V. E. Cosslett, and A. Engström, eds.), pp. 499–506. Academic Press, New York.

Boyde, A., Switsur, V. R., and Fearnhead, R. W. (1961). Application of the scanning electron-probe X-ray microanalyser to dental tissue. *J. Ultrastruct. Res.* **5,** 201–207.

Burny, F., Wollast, R., Balimont, P., and de Marneffe, R. (1968). Etude de la mineralisation osseuse à l'échelle de la microstructure au moyen de la microsonde électronique. *Calcif. Tissue Res. August Suppl.* **2,** 3.

Carlisle, E. M. (1969). (See Section F of this list.)

Carlisle, E. M. (1970). (See Section F of this list.)

Eick, J. D., Miller, |F|. A., Neiders, M. E., Leitner, J. W., and Anderson, C. H. (1969). Electron micro-probe analysis and scanning electron microscopy of human teeth. *Proc. Nat. Conf. Electron Microprobe Anal., 4th, Pasadena, 1969,* Paper 39.

Frank, R. M., Capitant, M., and Goni, J. (1966). Electron probe studies of human enamel. *J. Dental Res.* **45,** 672–682.

Frazier, P. D. (1966). Electron probe analysis of human teeth: Some problems in sample preparation. *Norelco Rep.* **13,** 25–27.

Frazier, P. D. (1967). Electron probe analysis of human teeth (Ca/P ratios in incipient carious lesions). *Arch. Oral Biol.* **12,** 25–33.

Fromme, H. G., Vahl, J., and Riedel, H. (1968). Elektronenmikroskopische und element-analytische Untersuchungen zur Differenzierung der Odontoblasten und zur Dentinbildung. *Deut. Zahnaerztl. Z.* **23,** 547.

Hoerman, K. C., Klima, J. E., Birks, L. S., Nagel, D. J., Ludwick, W. E., and Lyon, H. W. (1966). Tin and fluoride uptake in human enamel in situ: Electron probe and chemical microanalysis. *J. Amer. Dental Assoc.* **73,** 1301.

Höhling H. J., Hall, T. A., Mutschelknauss, R., and Niem, J. (1969). Kombinierte Elektronenbeugungs- und Electron Probe X-Ray Microanalysis Untersuchungen an Frühstadien der Zahnsteinbildung. *Z. Naturforsch.* B **24,** 58–60.

Johnson, A. R. (1969). The distribution of strontium in the rat femur as determined by electron microprobe analysis. *Proc. Nat Conf. Electron Microprobe Anal., 4th, Pasadena, 1969,* Paper 38.

Jones, W. C., and James, D. W. F. (1969). An investigation of some calcareous sponge spicules by means of electron probe micro-analysis. *Micron* **1,** 34–39.

Mellors, R. C. (1964). Electron probe microanalysis. I. Calcium and phosphorus in normal human cortical bone. *Lab. Invest.* **13,** 183–195.

Mellors, R. C. (1966). Electron microprobe analysis of human trabecular bone. *Clin. Orthopaedics* **45,** 157–167.

Remagen, W., Höhling, H. J., Hall, T. A., and Caesar, R. (1969). Electron microscopical and microprobe observations on the cell sheath of stimulated osteocytes. *Calcif. Tissue Res.* **4,** 60–68.

von Rosenstiel, A. P., Vahl, J., and Kyselova, J. (1968). Röntgenmikroanalytische Untersuchungen von Ag-Sn-Cu Legierungen. *Proc. Int. Congr. X-Ray Optics and Microanalysis, 5th, Tübingen, 1968.* (This paper is in the conference abstracts but not in the conference volume published by Springer-Verlag in 1969. Dr. von Rosenstiel's address: Metaalinstituut, T. N. O., Delft, The Netherlands.)

Rosser, H., Boyde, A., and Stewart, A. D. G. (1967). Preliminary observations of the calcium concentration in developing enamel assessed by scanning electron-probe X-ray emission microanalysis. *Arch. Oral Biol.* **12,** 431–440.

Saffir, A. J., and Ogilvie, R. E. (1966). (See Section E of this list.)

Saffir, A. J., and Ogilvie, R. E. (1967). The effects of diet on the microstructure of teeth. *Proc. Nat. Conf. Electron Microprobe Anal., 2nd, Boston, 1967,* Paper **38.**

Saffir, A. J., Ogilvie, R. E., and Harris, R. S. (1969). Electron microprobe analysis of fluorine in teeth. *Proc. Intern. Assoc. Dental Res. 1969.* Abstract 538, p. 175.

Saffir, A. J., Ogilvie, R. E., Alvarez, C. J., and Harris, R. S. (1970). Electron microprobe study of Ca/P ratio and fluorine distribution in rat teeth following feeding of Na fluoride and Na trimetaphosphate. *Proc. Intern. Assoc. Dental Res., 1970.* (In press.)

Söremark, R., and Grøn, P. (1966). Chlorine distribution in human dental enamel as determined by electron probe microanalysis. *Arch. Oral Biol.* **11,** 861–866.

Suga, S. (1967). Application of X-ray microanalysis in the study of biological tissues, especially hard tissues, *Shika Igaku* **55,** 217–224.

Suga, S. (1969). Electron microprobe studies on the mineralization process of tooth and bone. *In* "X-Ray Optics and Microanalysis" (G. Möllenstedt and K. H. Gaukler, eds.), pp. 592–596. Springer, Berlin.

Wei, S. H. Y., and Ingram, M. J. (1968). Electron microprobe analysis of the silver amalgam-tooth interface. *Proc. Nat. Conf. Electron Microprobe Anal., 3rd, Chicago, 1968,* Paper **44.**

Wörner, H., and Wizgall, H. (1969). Zerstörungsfreie Analyse von Zahnhartsubstanzen mit der Elektronenstrahlmikrosonde. *In* "X-Ray Optics and Microanalysis" (G. Möllenstedt and K. H. Gaukler, eds.), pp. 601–607. Springer, Berlin.

H. Studies of Histological Stains

Andersen, C. A. (1967). (See Section A of this list.)

Engel, W. K., Resnick, J. S., and Martin, A. E. (1968). The electron probe in enzyme histochemistry. *J. Histochem. Cytochem.* **16,** 273–275.

Gardner, D. L., and Hall, T. A. (1969). Electron-microprobe analysis of sites of silver deposition in avian bone stained by the v. Kossa technique. *J. Pathol.* **98,** 105–109.

Hale, A. J. (1962). Identification of cytochemical reaction products by scanning X-ray emission microanalysis. *J. Cell Biol.* **15,** 427–435.

Lane, B. P., and Martin, E. (1969). Electron probe analysis of cationic species in pyroantimonate precipitates in epon-embedded tissue. *J. Histochem. Cytochem.* **17**, 102–106.

Scheuer, P. J., Thorpe, M. E. C., and Marriott, P. (1967). A method for the demonstration of copper under the electron microscope. *J. Histochem. Cytochem.* **15**, 300–301.

Sims, R. T., and Marshall, D. J. (1966). Location of nucleic acids by electron probe X-ray microanalysis. *Nature (London)* **212**, 1359.

Tandler, C. J., Libanati, C. M. and Sanchis, C. A. (1970). The intracellular localization of inorganic cations with potassium pyroantimonate. Electron microscope and electron microprobe analysis. *J. Cell Biol.* **45**, 355–366.

CHAPTER 4

Infrared and Raman Spectroscopy

GEORGE J. THOMAS, JR.

I. Introduction

A. GENERAL NATURE OF VIBRATIONAL SPECTRA

1. Molecular Energy Levels

Molecular spectroscopy is the study of the absorption or emission of electromagnetic radiation by molecules. These processes, as described by the laws of quantum mechanics, are the result of a transition of the molecule from one energy state to another, and the total energy of the molecule is increased or decreased by one or more quanta depending, respectively, upon, whether absorption or emission takes place.

It is convenient and to a good approximation accurate to represent the total energy of a free molecule as a sum of contributions from electronic, vibrational, rotational, and translational parts, as in Eq. (1). The

$$E_{\text{total}} = E_e + E_v + E_r + E_t \tag{1}$$

electronic energy (E_e) is associated with the forces which hold the molecule itself together, the vibrational energy (E_v) with periodic displacements of the atoms with respect to one another, the rotational energy (E_r) with the periodic motions of the molecule as a whole, and the translational energy (E_t) with the displacement of the molecular center-of-mass. For most molecules, changes in E_e are brought about by radiation of relatively short wavelengths, i.e., in the ultraviolet and visible regions of the electromagnetic spectrum. These are discussed in the first edition of this volume. Changes in E_v are due to the absorption or emission of quanta in the mid-infrared region of the spectrum. These lead to infrared vibrational spectra discussed in this chapter. Changes in rotational and translational energies of free molecules correspond to the longer wavelengths of the far-infrared and microwave regions and are subject to quantization only for the gas phase. For biological applications, only condensed phases are encountered in which molecular rotations and translations are transformed into oscillations of the molecule as a whole and can usually be neglected.

The change in molecular energy, ΔE, accompanying the absorption or emission of a light quantum of frequency ν (sec^{-1}) or wave number $\bar{\nu}$ (cm^{-1}) is given by the Bohr frequency rule

$$\Delta E = h\nu = hc\bar{\nu} \qquad (2)$$

where h is Planck's constant (6.625×10^{-27} erg-sec) and c is the velocity of light (3.0×10^{10} cm/sec). When vibrational transitions are considered, it is convenient to describe the radiation by its wave number, $\bar{\nu}$, rather than by its frequency, ν. Vibrational energies are also expressed in wave number units and vibrational spectra are usually recorded on a linear wave number scale. The wave number is sometimes referred to loosely as a "frequency" although its proper unit as the reciprocal of a wavelength should be kept in mind.

2. Normal Vibrations and Vibrational Energies

A free molecule of N atoms has $3N$ types of motion (degrees of freedom). Three are translational, and, if the molecule is nonlinear, three are rotational. If all the atoms of a molecule lie on a single axis, only two degrees of rotational freedom are allowed. There are therefore $3N - 6$ additional degrees of freedom for a nonlinear molecule ($3N - 5$ for a linear molecule). These are all vibrational and associated with each is a vibrational frequency. When the atomic displacements from the equilibrium molecular configuration are sufficiently small, the vibrations are essentially harmonic. Any actual vibrational motion of a molecule may be quantitatively described as the superposition of simpler motions called the normal vibrations or *normal modes* of the molecule.

There are $3N - 6$ normal modes for a nonlinear molecule and associated with each is a *fundamental frequency*. These frequencies may all be different, or there may be more than one normal mode with the same fundamental frequency. Different normal vibrations having the same fundamental frequency are said to be degenerate.

The forms of the normal vibrations, their degeneracies, and their interactions with electromagnetic radiation (selection rules) all depend strongly on the symmetry of the molecule.

The actual atomic displacements which occur during normal vibrations (i.e., the *normal coordinates*) and their corresponding frequencies have been calculated for a large number of small molecules (Herzberg, 1945; Wilson *et al.*, 1955). The normal modes of some simple molecules are shown in Fig. 1. Calculation of normal modes is greatly simplified if the molecule possesses high symmetry. Most biomolecules are very complex and contain little or no symmetry and thus the calculation of their normal coordinates is difficult and generally impractical. Other approximate methods are therefore required to correlate vibrational frequencies with atomic displacements for complex polyatomic molecules. The concept of group frequencies, discussed below, is very useful in this regard.

The energy of a harmonic vibration with fundamental frequency ν is, according to the results of quantum mechanics,

$$E_\mathrm{v} = (v + \tfrac{1}{2})h\nu = (v + \tfrac{1}{2})hc\bar{\nu} \tag{3}$$

where v is the vibrational quantum number (0, 1, 2, . . .), h is Planck's constant, and c is the velocity of light. The energy levels for each degree of vibrational freedom are thus equally spaced by the amount $hc\bar{\nu}$. A more detailed discussion is given by Herzberg (1945).

In the harmonic-oscillator formulation, the fundamental frequency, ν is related to the reduced mass μ and the Hooke's law force constant k by the relation

$$\nu = c\bar{\nu} = \frac{1}{2\pi}\left(\frac{k}{\mu}\right)^{1/2} \tag{4}$$

One consequence of this relation is that vibrations of heavy atoms occur at lower frequencies than those of light atoms. For example, the carbon-hydrogen oscillator C-H having approximately half the reduced mass of the carbon deuterium oscillator C-D (but the same force constant), vibrates at approximately 1.4 times the frequency. Another consequence is that multiple bonds, which have higher force constants than single bonds, vibrate at higher frequencies.

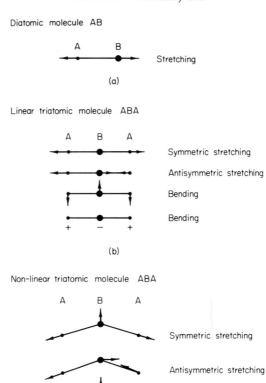

FIG. 1. Approximate normal vibrations of some simple molecules. Atomic displacements are depicted by arrows (in plane of page) and by + and − symbols (out of plane of page).

3. Vibrations of Complex Molecules

The relatively small biomolecule, ATP, has **43** atoms and **123** normal vibrations. Since ATP lacks any molecular symmetry, each of these is expected to appear both in the infrared and in the Raman spectrum. Other bands may also appear, corresponding to simple multiples of the fundamental frequencies (overtones) and to combinations of their sums and differences. Coupling of the overtones or combination frequencies with fundamentals could also occur. Thus a very complicated vibrational spectrum can result from ATP, a molecule which is still small by standards of biochemistry.

In order to simplify the interpretation of spectra of such complexity, a number of approximations have been introduced. One of these is the concept of characteristic group frequencies (Bellamy, 1958, 1968). This concept rests upon the experimental fact known since the 1880's, that certain groups of atoms give rise to vibrational bands at or near the same frequency irrespective of the particular molecule in which the group occurs. The particular frequency is said to be characteristic of the chemical group. For example, the methyl group —CH$_3$ usually gives rise to an intense infrared absorption band near 1460 cm^{-1} regardless of the molecule in which the group occurs. It must be kept in mind that this concept is an approximation which varies in its validity in different parts of the vibrational spectrum.

It is useful to classify the characteristic vibrations of covalently bonded atoms into two types: *bond stretching,* which involves a periodic extension and contraction of the bond, and *bending,* which involves a periodic increase and decrease in the angle formed by two adjacent bonds. More complex types of motions involving the simultaneous stretching or bending of groups of bound atoms also occur frequently. Examples are given in Fig. 2. Other examples are given by Jones and Sandorfy (1956) for acyclic molecules and by Katritzky and Ambler (1963) for cyclic molecules.

From studies of vibrational spectra of many molecules, charts of large numbers   iled. One

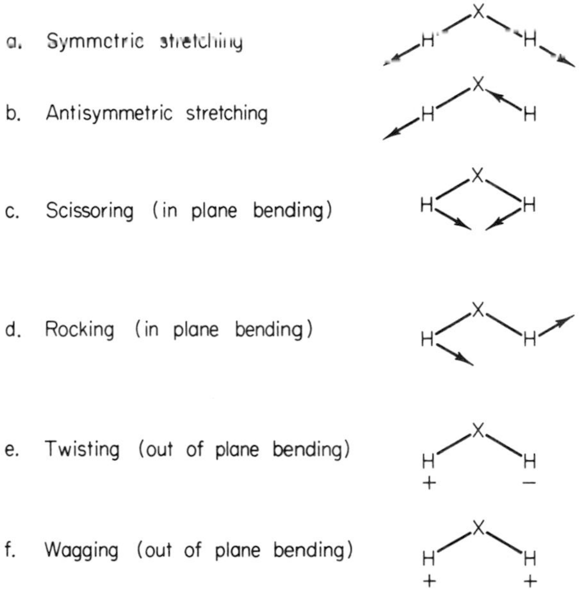

a. Symmetric stretching

b. Antisymmetric stretching

c. Scissoring (in plane bending)

d. Rocking (in plane bending)

e. Twisting (out of plane bending)

f. Wagging (out of plane bending)

FIG. 2.   tly bonded
at atom X. Examples are the primary amino (—NH$_2$) and methylene groups ($\diagdown$CH$_2$).
Atomic displacements depicted as in Fig. 1.

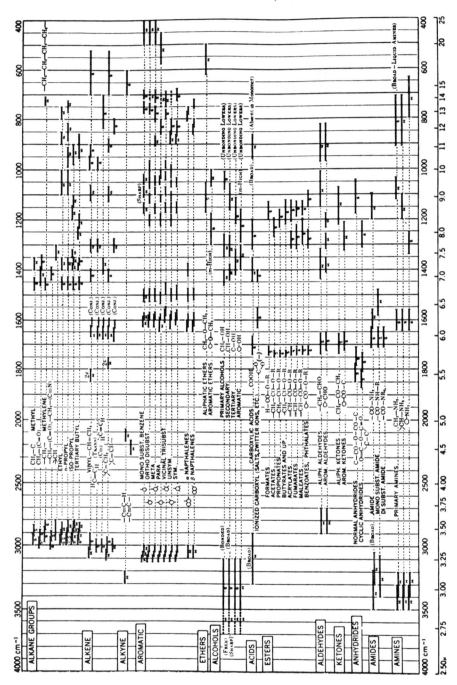

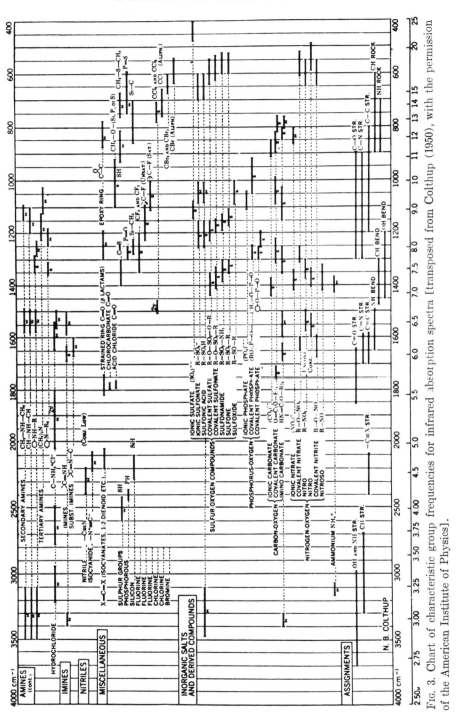

Fig. 3. Chart of characteristic group frequencies for infrared absorption spectra [transposed from Colthup (1950), with the permission of the American Institute of Physics].

covering most of the mid-infrared region, due to Colthup (1950), is reproduced in Fig. 3. Although group frequencies have also been recognized for many years in the Raman effect (Kohlrausch, 1943; Hibben, 1939), no comparable chart has been compiled for Raman frequencies. For molecules lacking symmetry, however, infrared and Raman frequencies are not expected to differ appreciably, though their relative intensities in the two spectra may be quite different.

A second simplification in vibrational spectra of complex molecules results from the fact that coupling between different group vibrations often does not occur. It is observed, for example, that certain vibrations characteristic of the methylene and methyl groups are so localized that coupling is negligible even when the groups occupy adjacent positions in a hydrocarbon chain.

It has also been pointed out by Jones and Sandorfy (1956, pp. 252, 287) that spectra of complex molecules can frequently be simplified by considering the intensities of vibrational bands. A practical application of this fact is in the interpretation of Raman and infrared spectra of aqueous purine and pyrimidine nucleotides, where it is found that vibrations of the sugar residues occur only very weakly in the Raman effect by comparison with their high infrared intensities. The same is not true of many vibrations localized in the heterocyclic bases, which are intense in both the Raman and infrared spectra (Lord and Thomas, (1967b).

More detailed discussions of characteristic group frequencies are given by Bellamy (1958, 1968), Jones and Sandorfy (1956), and Katritzky and Ambler (1963).

4. Factors Affecting Group Frequencies

(a) *Intramolecular Effects.* Intramolecular factors which make a characteristic group frequency vary slightly from one compound to another include atomic mass in adjacent groups, mechanical coupling between similar groups, symmetry, electronegativity, conjugation and the like. The effect of atomic mass can be seen from Eq. (4).

The pure group frequency is an abstraction which is sometimes approximated quite well in practice. Most molecular vibrations involve a mixing of several kinds of group motions which are said to be mechanically coupled. A classic example of mechanical coupling occurs in secondary amides, such as polypeptides. The group R—CO—NH—R shows infrared absorption bands at about 1650 and 1550 cm^{-1}. The first of these is known to arise predominantly from the C=O stretching vibration. The 1550 cm^{-1} band, however, can neither be ascribed solely to N—H bending nor to C—N stretching vibrations. Instead, actual calculation of the normal modes (Miyazawa *et al.*, 1958) shows the 1550 cm^{-1} band to result from simultaneous motions of the C, N, and H atoms.

Two group vibrations must have the same symmetry in order for them to couple with one another. Although this is an important restriction for highly symmetric molecules, such as benzene, methane, and the like, the requirement is easily satisfied for biomolecules, which have very low symmetry. This and other effects of symmetry on group frequencies are discussed by Herzberg (1945).

Electronegative substituents attached to functional groups have a pronounced effect on characteristic group frequencies. Several examples are discussed by Lord and Miller (1956) and by Hadzi (1963). Still more pronounced effects can result from ionizations. Purine and purimidine nucleosides, for example, show striking differences in their infrared (Shimanouchi *et al.*, 1964) and Raman spectra (Lord and Thomas, 1967b), particularly in bands due to double-bond vibrations, when the ionization of ring nitrogen or imino groups takes place at extremes of pH.

Further discussion of intramolecular perturbations of group frequencies is given by Lord and Miller (1956).

(*b*) *Intermolecular Effects.* One of the most important factors affecting group frequencies of biomolecules is that of hydrogen-bonding interaction. Many examples of hydrogen bonding in biological molecules are actually intramolecular, e.g., in the α-helix; however the effects on group frequencies are similar for both intra- and intermolecular hydrogen bonds. For aqueous solutions, hydrogen bonding between solvent and solute molecules will always occur.

The bonding interaction between a donor X—H group (e.g., O—H or N—H) and an acceptor atom, Y (e.g., O, N, or F), will affect all vibrational modes involving the X—H group. The X—H stretching vibration suffers the largest perturbation, usually a decrease in frequency corresponding to several hundred wave numbers. The infrared bands and Raman lines are also very much broader than those of nonhydrogen-bonded groups and are thus easily recognized. The magnitude of the wave number shift in the X—H stretching band ($\Delta\bar{\nu}$) is a rough measure of the energy of interaction (Bellamy and Pace, 1969; Bellamy and Owen, 1969). When the distinct vibrational bands of hydrogen-bonded and nonhydrogen-bonded species are sufficiently separated from one another, it is possible to determine association constants governing formation of hydrogen-bonded complexes. Various techniques for determining the association constants in such cases have been described (Tsuboi, 1951; Lord and Porro, 1960). Several applications to purine-pyrimidine associations are discussed below (Section II.C.2).

Quantitative determinations of association constants at several temperatures allow calculation of the thermodynamic parameters of hydrogen-bond formation, ΔG, ΔH, and ΔS. Since the molar absorptivities of monomers and complex may be temperature dependent, a technique such as that of Lord and Porro (1960), which circumvents the need for an assumption to the contrary, should be followed.

Bending vibrations of the X—H group are also perturbed by hydrogen-bond formation. However, these are frequently coupled with other vibrational modes and are less reliable as indicators of hydrogen-bond formation. In most cases, it is observed (Hallam, 1963) that hydrogen bonding causes the X—H bending frequencies to be increased. The relative upward shifts $(\Delta \bar{\nu}/\bar{\nu})$ are smaller than the relative downward shifts of the corresponding X—H stretching modes. A more detailed discussion of the effects of hydrogen bonding on donor group frequencies is given by Pimentel and McClellan (1960).

In addition to perturbations of the X—H frequencies, hydrogen bond formation will also influence the vibrations within the acceptor group Y. A widely studied example is the carbonyl acceptor group, C=O. In most cases, the C=O stretching frequency is shifted downwards by 5 to 25 cm^{-1}.

A number of other intermolecular factors can also affect the vibrational frequencies of molecular groups as well as the shapes of infrared absorption bands or Raman lines. These perturbations, present in condensed phases, may result variously from van der Waals, dispersion, electrostatic, or other forces. Spectra of crystals are particularly sensitive to coupling between individual molecules in the unit cell and possibly also to combination with low frequency lattice vibrations. A fuller discussion of these topics is given by Jones and Sandorfy (1965, pp. 298ff) and by Hallam (1963).

B. Interaction of Infrared Radiation with Vibrating Molecules— Infrared Absorption

1. Absorption Process for the Harmonic Oscillator

Infrared absorption spectra arise when a molecule absorbs one or more quanta of infrared radiation in the transition from one vibrational energy level to another level of higher energy. In order for a quantum of radiation to be absorbed, the Bohr frequency rule [Eq. (2)] must be satisfied. For a harmonic vibration, the change in molecular energy is equal to the energy difference between the levels given by Eq. (3).

Not every transition, however, can occur with the absorption of radiation. The rules determining which transitions will give rise to infrared absorption are called *selection rules*. For a harmonic vibration, absorption will not occur unless the radiation-induced transition is between *adjacent* vibrational levels, i.e., a harmonic oscillator will undergo electric dipole transitions only between vibrational levels which differ by a unit in their quantum numbers. For absorption, this selection rule is expressed as

$$\Delta v \equiv v' - v'' = 1 \tag{5}$$

where v' and v'' are vibrational quantum numbers [Eq. (3)] of the final and initial vibrational energy levels, respectively. For most molecular vibrations, the Boltzmann distribution law indicates that only the ground vibrational energy levels ($v'' = 0$) are appreciably populated at room temperatures. Therefore absorption transitions other than from the state $v'' = 0$ to $v' = 1$, for a given harmonic vibration, occur with low probability. Transitions between these states, represented as $1 \leftarrow 0$, are referred to as fundamentals and the corresponding spectral bands are called fundamental absorption bands.

Since real molecules do not vibrate harmonically, the harmonic oscillator selection rule is not obeyed strictly. Transitions involving the absorption of more than one quantum of radiation, e.g., the transitions $2 \leftarrow 0$ and $3 \leftarrow 0$, are sometimes observed. These are referred to as overtones and are said to result from anharmonicity.

2. Intensities of Infrared Transitions

In addition to the selection rule $\Delta v = +1$ for the absorption of radiation by a harmonic oscillator, another selection rule which depends upon the symmetry of the oscillator (or molecule) must be satisfied. The process of infrared absorption is a dipole-induced transition, i.e., the vibrating molecule procedures an alternating electric field which interacts with the electric field (dipole) of the infrared radiation. To a first approximation, the dipole moment of a vibrating molecule is

$$\mu = \mu_0 + (d\mu/dQ)Q \tag{6}$$

where μ_0 is the permanent dipole moment and Q is the normal coordinate, i.e., the coordinate which defines the positions of the nuclei during the vibration, and $(d\mu/dQ)Q$ is the change in the dipole moment produced by the vibration. The intensity of a fundamental absorption band depends upon the square of the integral containing the dipole moment of Eq. (6) and the wave functions of the vibrational states corresponding to v' and v'', i.e.,

$$\text{Intensity} = \text{constant} \cdot |M_{v'v''}|^2 = \text{constant} \cdot |\int \psi_{v'} \mu \psi_{v''} \, dQ|^2 \qquad (7)$$

The quantity $M_{v'v''}$ is called the transition moment between the states v' and v''. It is found (Herzberg, 1945) that the integral, and therefore $|M_{v'v''}|^2$, vanish unless $(d\mu/dQ)$ is nonzero. It follows that a quantum of infrared radiation can be absorbed in a harmonic vibration only if there is a dipole moment change accompanying the vibration. This statement is often referred to as a symmetry selection rule, since the symmetry of the molecule will determine whether a changing dipole moment is associated with the given vibration. When this condition is satisfied, the vibration is said to be *infrared-active*. It follows also from Eq. (7) that the larger the dipole moment change accompanying an infrared-active vibration, the more intense will be the resulting absorption band. Other factors also influence absorption band intensities (Herzberg, 1945).

A more detailed discussion of selection rules based on molecular symmetry, independent of specific mechanical or electrical models, can be found in various places (Herzberg, 1945; Duncan, 1956; King, 1964).

3. Examples

(a) *The CO_2 Molecule.* The carbon dioxide molecule is an example of the linear triatomic molecule ABA, the normal modes of which are shown in Fig. 1b, above. Both the antisymmetric stretching mode and the bending mode involve changes in the dipole moment during vibration and are infrared-active. The symmetric stretching vibration, on the other hand, occurs with no dipole moment change and is consequently infrared-inactive. The two forms of the bending vibration shown in Fig. 1b differ from one another only in the orientation of the plane (with respect to viewer) in which the vibration occurs. It is therefore expected that both forms of the vibration will have the same frequency; i.e., they are degenerate. The doubly degenerate bending mode of CO_2 gives rise to a strong infrared absorption band near 667 cm^{-1}.

(b) *The H_2O Molecule.* Another molecule of interest is the nonlinear ABA molecule, such as H_2O. The forms of the normal modes are shown in Fig. 1c. All three fundamentals involve a change of dipole moment and are hence infrared-active. Because of the large difference in electronegativities of O and H, the water molecule is highly polar and the dipole moment changes associated with the valence bond stretching vibrations give rise to intense infrared absorption bands. These properties make water a difficult solvent for use in infrared spectroscopy of solutions (Section II.B.2, below).

C. INTERACTION OF VISIBLE LIGHT WITH VIBRATING MOLECULES— RAMAN SCATTERING

1. Raman Scattering for the Harmonic Oscillator

When a beam of light traverses a medium containing molecules, some of the light is transmitted, some absorbed, and some diffused or "scattered." The fraction scattered is usually very small and most of this fraction is scattered elastically, i.e., without change in frequency and is referred to as *Rayleigh scattering*. Another type of elastic scattering is *Tyndall scattering*, which results not from individual molecules or atoms as such, but from scattering by larger particles randomly distributed, such as colloidal particles and dust particles. A portion of the scattered light (called *Raman scattering* after its discoverer) is scattered inelastically and differs in frequency from the incident light. These differences are produced by quantized vibrations of the molecules which give rise to the scattering.

When scattered monochromatic radiation is examined spectrophotometrically, it is no longer monochromatic but consists of the original frequency (Rayleigh and Tyndall scattering) plus the new frequencies (Raman scattering). The original frequency is referred to as the *"exciting line."* The new frequencies or "lines," as well as their intensities and polarizations, are characteristic of the scattering substance and constitute its Raman spectrum. If the inelastic scattering process results in a net transfer of energy from the incident light to the molecule, the energy of the light quantum and hence its frequency are lowered. Conversely, the incident light quantum has its frequency increased if energy is transferred from the molecule to the light quantum in the scattering process. Thus when the Raman line occurs at a higher frequency than the exciting line, a net loss by the molecule of one or more vibrational quanta occurs (anti-Stokes transition). Conversely, the Raman lines resulting from the net gain of vibrational energy by the molecules produce radiation of frequency below that of the exciting line (Stokes transitions). The Boltzmann distribution law quantitatively explains the observation that Stokes lines are more intense than anti-Stokes lines. As a practical matter, therefore, Raman spectrometers are operated so as to scan the frequency region below that of the exciting line.

The vibrational transitions in a polyatomic molecule that can occur by Raman scattering are severely limited. For harmonic vibrations, only those transitions are permitted that follow the selection rule $\Delta v = +1$ for one vibration at a time. Thus the most likely allowed Raman transition is the Stokes transition $1 \leftarrow 0$ for a single vibrational mode. Which modes can appear are further restricted if the molecule possesses symmetry. Finally it should be mentioned that the selection rules can be altered by anharmonicity, but experience has shown that the combination tones permitted under such alteration are relatively rare in the Raman effect.

2. Intensities of Raman Transitions

A light wave of frequency ν may be represented as an oscillating dipole with electric vector E given by

$$E = E_0 \cos 2\pi\nu t \tag{8}$$

When such a wave is incident upon a molecule, the electrons are induced to oscillate with the same frequency. The induced dipole moment, P, is proportional to the electric field

$$P = \alpha E = \alpha E_0 \cos 2\pi\nu t \tag{9}$$

where the proportionality constant α is the *polarizability* of the molecule. The polarizability is a symmetric tensor quantity with components α_{ij} ($i, j = x, y,$ or z). It is this property which gives rise to light scattering by molecules.

The intensity of a Raman transition is proportional to the square of the integral containing the induced dipole moment and the wave functions of the vibrational states coresponding to the transition $v' \leftarrow v''$,

$$\text{Intensity} = \text{constant} \cdot |P_{v'v''}|^2 = \text{constant} \cdot |\int \psi_{v'} P \psi_{v''} \, dQ|^2 \tag{10}$$

In order for the transition moment to be nonzero, it is necessary that the integral of Eq. (10) be nonzero. This integral actually has six components, one for each of the quantities α_{ij} given by Eq. (9), and therefore at least one of the six integral, $\int \psi_{v'} (\alpha_{ij}) \psi_{v''} \, dQ$, must not vanish.

In order to evaluate the integrals $\int \psi_{v'} (\alpha_{ij}) \psi_{v''} \, dQ$, for a harmonic vibration, use is made of the expansion

$$\alpha_{ij} = (\alpha_{ij})_0 + (d\alpha_{ij}/dQ)Q \tag{11}$$

for the polarizability, where $(\alpha_{ij})_0$ is the polarizability in the equilibrium configuration of the molecule and Q is the normal coordinate. The vibrational Raman effect therefore requires that $(d\alpha_{ij}/dQ)$ be nonzero, i.e., there must be a change in at least one component of the molecular polarizability in order for a vibration to be Raman-active. It is further required that $\Delta v = +1$.

It follows from Eq. (10) that the greater the change in polarizability (or induced dipole moment) during the vibration, the more intense will be the corresponding Raman spectral line. A number of other electrical and mechanical factors may also influence Raman line intensities (Herzberg, 1945; Evans, 1963).

The Raman scattering intensity (Stokes line), in terms of experimentally attainable quantities, was derived by Placzek (1934) as,

$$I_{v'v''} = \text{constant} \cdot \frac{(\nu_0 - \nu_{v'v''})^4}{\nu_{v'v''}} \cdot \frac{NI_0}{[1 - \exp{(h\nu_{v'v''}/kT)}]} \cdot [45(\alpha_s')^2_{v'v''} + 7(\alpha_a')^2_{v'v''}] \quad (12)$$

where ν_0 is the frequency of the exciting line with intensity I_0, N is the number of molecules in the initial state v'', $\nu_{v'v''}$ is the vibrational frequency, $(\alpha_s')_{v'v''}$ is the integral $\int\psi_{v'}(d\alpha_s/dQ)\psi_{v''}\,dQ$, where α_s is the symmetric or isotropic part of the polarizability, and $(\alpha_a')_{v'v''}$ is a similar integral for the anisotropic part of the polarizability. The polarizability tensor components α_{ij} are combined into the components α_s and α_a because these latter quantities are invariants of the tensor and simplify the discussion. The constant coefficients 45 and 7 arise from averaging over all orientations of the scattering molecules in the calculation.

3. Depolarization of Raman Lines

Knowledge of the polarization properties of Raman scattering can provide information about the symmetry of the molecular vibrations which give rise to the scattering. The depolarization ratio ρ, which provides this information, is defined as follows for the arrangement in which the scattered light is observed at right angles to the direction of the incident beam, e.g., the laser beam.

$$\rho = I_\perp/I_\| \quad (13)$$

$I_\perp$ is the intensity of scattered light polarized perpendicular to a plane which is in turn perpendicular to the laser beam, and $I_\|$ is the intensity of scattered light polarized parallel to this plane. For the case when the laser beam is polarized perpendicular to the direction of observation,

$$\rho = \frac{I_\perp}{I_\|} = \frac{3(\alpha_a')^2_{v'v''}}{45(\alpha_s')^2_{v'v''} + 4(\alpha_a')^2_{v'v''}} \quad (14)$$

When a vibration causes the polarizability of the molecule to change unsymmetrically, i.e., anisotropically, then $(\alpha_s')_{v'v''} = 0$, and $\rho = \frac{3}{4}$. The resulting Raman line is said to be depolarized. For totally symmetric vibrations, however, the anisotropy in the polarizability $(\alpha_a')_{v'v''}$ usually is small and $\rho < \frac{3}{4}$. The value of ρ is thus useful in distinguishing Raman lines of symmetric vibrations from those due to unsymmetric vibrations. Further discussion of this subject is given by Jones and Sandorfy (1956) and by Evans (1963).

4. General Features of Raman Spectra

(a) *Measurement of Raman Intensities.* There is no expression, analogous to the Beer-Lambert relation for light absorption (first edition) in terms of which the absolute scattering intensity can be expressed. The intensity at the band maximum of a given vibrational Raman line, however, is generally assumed to be proportional to the volume concentration of molecular species. The proportionality constant is referred to as the *scattering coefficient*. Different conventions for determining scattering coefficients have been described (Jones and Sandorfy, 1956, pp. 290–293; Jones *et al.*, 1965).

(b) *Relative Intensities of Group Vibrations.* Because the Raman intensity depends upon the polarizability change associated with a molecular vibration, it is not surprising that vibrations of multiple bonds are more intense than those of single bonds. In purine and pyrimidine nucleosides, for example, the C=C, C=N, and C=O stretching vibrations are considerably more intense in the Raman effect than are C—C, C—N, C—O, or even C—H stretching modes (Lord and Thomas, 1967b). Roughly speaking, the more electrons which are localized in a valence bond, the more intense will be the Raman line resulting from its stretching vibration, intramolecular coupling or other perturbing effects being absent. For example, the S—S stretching mode is more intense in the Raman effect than the O—O stretching mode.

Bending vibrations, even those of polar groups, such as X—O—H, X—N—H etc., . . . also give rise to Raman lines which are relatively weak. The bending vibrations of liquid H_2O and D_2O are extremely weak in the Raman effect. This fact explains the very weak background scattering of water in the Raman effect over the frequency range 200–3000 cm^{-1} and accounts for its use as a valuable solvent for Raman spectroscopy.

II. Infrared Methods

A. INSTRUMENTATION

1. General Features of Infrared Spectrometers

All spectrometers are constructed of the following basic components: a source of radiation, an optical system for condensing and focusing the radiation, a monochromator, a detector of radiation, and an electronic system for amplifying the detected radiation and recording the signal. The operation of these components is discussed below.

It is necessary to distinguish between single, and double-beam spectrometers. The former measures, as a function of wavelength (or wave number), the energy of radiation transmitted by the sample, while the latter measures the *difference* in intensity between radiation incident on the sample (I_0) and radiation transmitted by the sample (I). The double-beam spectrometer usually records the energy difference as transmittance (I/I_0) or absorbance [log (I_0/I)]. Most laboratories today are equipped with double-beam infrared spectrometers and these only shall be considered here. A discussion of single beam spectrometers is given by Potts (1963).

The block diagram of a typical double-beam infrared spectrometer is shown in Fig. 4.

(*a*) *Sources of Infrared Radiation.* The infrared source is an object that is heated to high temperature so that it emits radiation. The intensity of radiation emitted as a function of wavelength approximates that of a "blackbody" which is a perfect emitter of radiation. Emission curves for a blackbody at several temperatures are shown in Fig. 5. In order to obtain sufficient intensity over the entire mid-infrared region, a blackbody of about 1500 to 1800°K is required. Three common types of sources which radiate in this temperature range are used in infrared spectrometers: the Globar, the Nernst glower, and a nichrome wire coil.

(*b*) *Condensing and Focusing Optics.* The basic features of the condensing system of a double-beam (optical-null) spectrometer are shown in Fig. 6. Light from the source *a* is collected on the two mirrors *b* and *b'* and reflected through the exit windows *c* and *c'* to form the

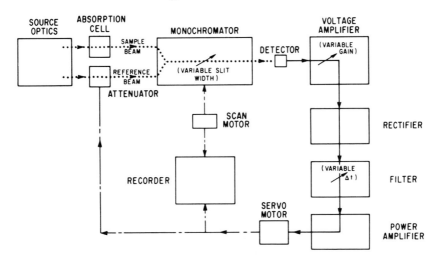

Fig. 4. Block diagram of the optical-null double-beam spectrometer. From Potts (1963), with the permission of Wiley (Interscience).

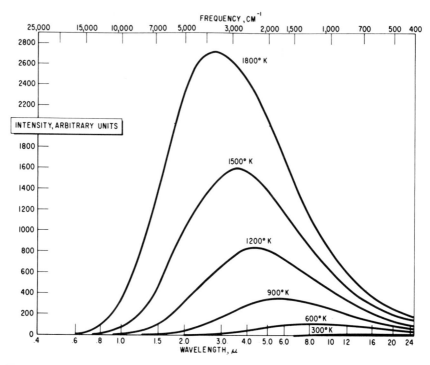

FIG. 5. Energy distribution of radiation from blackbody at various temperatures. From Potts (1963), with the permission of Wiley (Interscience).

sample and reference beams, respectively. The sample compartment is between the windows c and d. The windows d and d' are entrance slits to the photometer section. The sample beam is chopped by the motor-driven chopping mirror e and the reference beam is reflected at the plane mirror f in such manner that the sample and reference beams are focused alternately onto the monochromator entrance slit g. The attenuator h determines the amount of energy in the reference beam which reaches e. As the spectrometer scans through an absorption band, the attenuator is automatically driven into the reference beam so that the reference intensity will be equal to the sample beam light intensity.

(c) *Monochromators.* The purpose of the monochromator is to isolate a narrow band of spectral wavelengths. This is achieved by use of either a diffraction grating or prism. Both types are widely used at the present time. A typical prism monochromator is shown in Fig. 7. Light entering the monochromator entrance slit g from the photometer section (see Fig. 6, above) reaches the collimating mirror m. The collimator is an off-axis paraboloid so designed to render the light entering the slit g

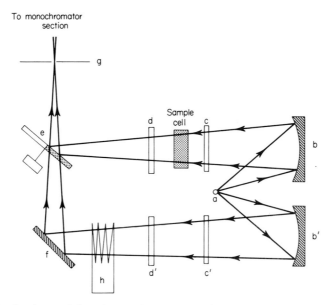

FIG. 6. Condensing and focusing optics of a double-beam spectrometer. See text.

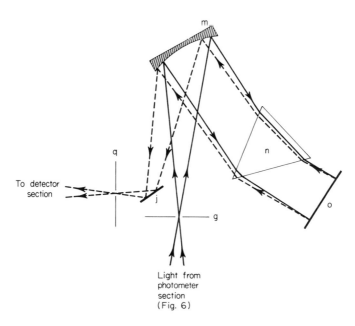

FIG. 7. Schematic view of a monochromator employing a prism as dispersing element and a Littrow mirror mount.

strictly parallel (Potts, 1963, p. 32). The parallel light is passed through the prism n which acts as the dispersing element to separate the continuous blackbody radiation into narrow wavelength intervals. The light is reflected at the plane mirror o and passes through the prism a second time to the exit slit q, via the collimator and plane mirror p. The spectrum is scanned by varying the position of the Littrow mirror about a vertical axis through its front surface.

Besides the Littrow system, other optical arrangements are commonly employed in infrared monochromators. A more comprehensive discussion is given by Harrison *et al.* (1948).

(*d*) *Detection and Amplification of Infrared Signals.* In order to detect infrared radiation and determine its intensity as a function of wavelength so that the spectrum can be obtained, the radiation signal is usually converted to an electrical voltage signal which is more easily recorded. There are several devices which perform this function. A full discussion is given by Martin (1963) and by Potts (1963, Chapter 4).

The usual procedure for mid-infrared radiation is to collect the light exiting at the slit, q in Fig. 7 on an ellipsoidal mirror which in turn focuses the beam onto a very small area of the detector. The detector itself is a thermocouple i.e., a junction of two unlike metals which heats up when it receives the infrared radiation. The potential difference which exists between the detector junction and another junction kept at room temperature is proportional to the number of infrared photons falling on the detector.

The signal reaching the detector is suitably amplified and rectified to a dc voltage ($\sim$0.01 V) which is used to drive the recorder (Potts, 1963, Chapter 4).

The motion of the dispersive element (e.g., Littrow mirror or diffraction grating) is coupled to the motion of chart paper and the detector signal drives a recording pen so that a continuous plot of the radiation intensity transmitted by the sample at each frequency (spectrum) is obtained.

2. Examples of Infrared Spectrometers

(*a*) *Perkin-Elmer Corp., Model 457.* This instrument is a double-beam, optical null, grating spectrophotometer which operates in the range 4000–250 cm^{-1}, recording linear transmittance versus linear wave number. The optics consist of a single pass Littrow system employing two plane gratings as the dispersive element (Fig. 8a). The optimum resolving power is 2 cm^{-1} at 3000 cm^{-1} or 1 cm^{-1} at 1000 cm^{-1}, i.e., absorption bands near 3000 cm^{-1} spaced by less than 2 cm^{-1} cannot be distinguished

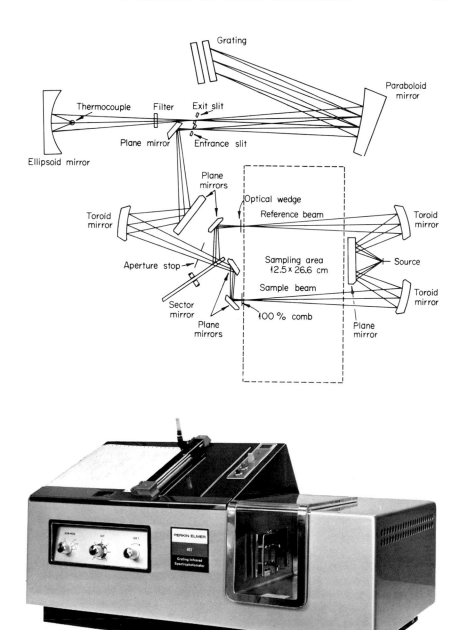

FIG. 8. Perkin-Elmer Model 457 Infrared Spectrometer. Top: Schematic view of optics; bottom: actual photograph.

from one another. The instrument is compact, as shown in Fig. 8b, and suitable for most biological applications. Like the Beckman IR-10, it combines low cost with moderate performance.

(b) *Beckman Instruments, Inc., Model IR-12.* This instrument is a higher performance spectrophotometer suitable for high resolution (gas-phase) studies as well as for ordinary applications. The IR-12 employs a single monochromator with four gratings for dispersion, each operating in a specified spectral range (Fig. 9). The presentation is linear transmittance versus linear wave number from 4000 to 200 cm^{-1}. Optimum resolving power is 0.25 cm^{-1} at 1000 cm^{-1}. The chopper motor is placed before the sample area so that operation in either single-beam

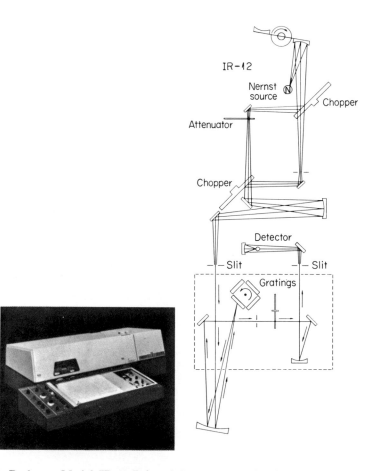

FIG. 9. Beckman Model IR-12 Infrared Spectrometer. A schematic view of optics and an actual photograph.

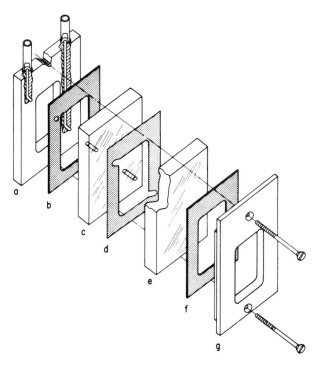

FIG. 10. Diagram of fixed thickness infrared absorption cell for liquid samples: a, back plate; b, gasket; c, window; d, spacer; e, window; f, gasket; g, front plate. From Potts (1963), with the permission of Wiley (Interscience).

or double-beam modes is possible. Like the Perkin-Elmer Model 621, the IR-12 is a high cost, high performance instrument.

3. Cells and Accessories

(a) *The Fixed Thickness Cell.* Different types of sealed cells for liquids have been described (Lord *et al.*, 1952; Martin, 1963; Potts, 1963). An example is shown in Fig. 10. In a cell of this type, the optical path is fixed and hence must be suitably chosen beforehand so that the transmittance of the sample falls in a convenient range (usually <0.9). A disadvantage of this cell for biological applications is the relatively large volume of sample required to fill it (0.5 to 1.0 ml depending on cell thickness). Microcells of similar design are available but these require special beam condensing systems discussed below (Section II.B.4.f).

(b) *The Variable Thickness Cell.* A cell of adjustable optical path which is continuously variable over a wide range is extremely useful

for many applications. For example, in differential spectrophotometry
(Section II.B.1,2) when compensating a spectrum for solvent absorption,
the amount of solvent in both beams must be the same. With a variable
thickness cell in the reference beam, it is a simple matter to compensate
solvent absorption quite accurately. A cell of this type is very convenient
for infrared spectroscopy of deuterium oxide solutions of biomolecules.

(c) *The Variable Temperature Cell.* It is often required to obtain
infrared spectra of biological materials at different temperatures, e.g.,
in denaturation studies of proteins or nucleic acids. A leak-proof thermostatable cell is then required. A typical unit consists of a fixed thickness
cell mounted in a suitable heat sink (e.g., metal frame). The cell is
then mounted in a large jacket through which heating or cooling liquid
can be circulated.

A number of manufacturers of infrared spectrometers provide cells
of various types. Examples of commercially available, fixed thickness,
variable thickness, and variable temperature cells are shown in Fig. 11.

Other accessories, including devices for internal reflection spectroscopy
and microspectrophotometry, are also available from a number of commercial sources. Further details of these accessories are discussed in
the section on sample-handling (II.B.4.e,g).

		Variable temperature
Fixed thickness	Variable thickness	cell assembly
cell	cell	Upper : cell
		Lower : heating jacket
(a)	(b)	(c)

FIG. 11. Commercially available infrared absorption cells for liquid samples. See
text. (a) Fixed thickness cell (Beckman Instruments, Inc.); (b) variable thickness
cell (R.I.I.C., Model XL); (c) thermostatable cell with jacket (R.I.I.C., Models
FH-01 and WJ-1).

B. Sample-Handling Procedures

An important practical advantage of infrared spectroscopy as an analytical technique is that it can be applied to most materials regardless of such gross physical and chemical properties of the sample as phase, color, turbidity, molecular weight, and the like. With the proper, usually simple, procedures, nearly any material can be prepared for infrared study, a versatility that has been a factor in the widespread use of infrared spectroscopy in the study of biological materials. The same is not true of the Raman method, which requires more specialized sample-handling techniques (Section III.B).

We shall consider here procedures for obtaining infrared spectra of pure liquids, liquid solutions (nonaqueous and aqueous), and solids, since these are most frequently encountered in biological research. Choice between spectroscopy of solutions and that of pure materials is usually dictated by the type of molecular information sought. However, there are certain advantages to solution spectroscopy which favor its use in many cases. First, an appropriate solvent provides a constancy of molecular environment for correlating vibrational frequencies with molecular structure when spectra of many materials are to be compared. Second, the perturbations of vibrational frequencies caused by interactions between solute and solvent molecules are usually less severe than the interactions between neighboring molecules in a pure liquid or crystal lattice (Hallam, 1963). Third, infrared absorption bands are often too intense to permit quantitative analyses to be made even of thin layers of pure materials. Solution spectroscopy is by far the most commonly employed in quantitative infrared analyses.

For solution spectroscopy, it is of course required that the solvent be transparent to radiation in the spectral regions of interest. An effective solvent which is transparent throughout the mid-infrared region (4000–200 cm^{-1}) does not exist and certain compromises must be made. For example, infrared studies of biomolecules are frequently restricted to narrow spectral regions of solvent transparency and some "vibrational information" is thereby sacrificed.

Water is the most important solvent for biological research because the chemistry of living material takes place in an aqueous medium. However, it is often desirable to remove the relatively large effects of hydrogen bonding between water and its solutes so that bonding between solute molecules themselves may be studied. To do this, a solvent is needed which lacks both donor and acceptor sites for hydrogen-bond formation. Therefore, nonpolar and weakly polar solvents which lack such sites also find use in infrared studies of biological materials. Sample-handling procedures for infrared spectroscopy of nonaqueous and aqueous solutions differ appreciably.

1. *Nonaqueous Solutions*

(a) *Solvents.* Details of the use of organic solvents are given in text-books on infrared techniques (Jones and Sandorfy, 1956; Brugel, 1962; Potts, 1963). Transmission properties of a number of common solvents are summarized in Fig. 12. Small symmetrical molecules containing little or no hydrogen, like CCl_4, $CHCl_3$, and CS_2, are transparent over large parts of the mid-infrared region and are therefore suitable as solvents, though the chemical reactivity of the last named should be kept in mind. Chloroform is of particular usefulness in biochemical studies because of its ability to dissolve polar molecules (nucleosides, antibiotics, etc., . . .) without forming strong intermolecular hydrogen bonds.

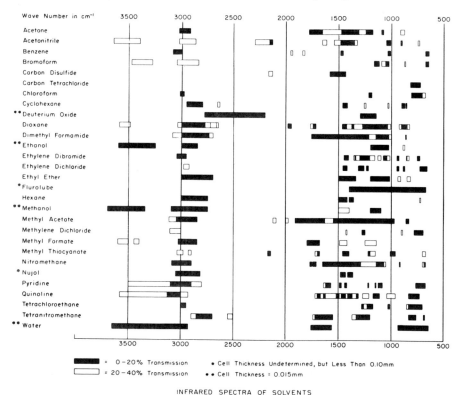

FIG. 12. Transmission properties of common solvents in the mid-infrared region of the spectrum. Chart prepared by Spectroscopy Laboratory, MIT, Cambridge, Massachusetts.

When absorption cells of moderate thicknesses (up to about 0.01 cm) are used, solutions of chloroform are transparent *except* in the intervals 3100–2980 cm⁻¹ (CH stretching), 1560–1400 cm⁻¹ (CH deformation), and below 1300 cm⁻¹. Interfering absorption in the wave number regions of the hydrogen vibrations may be avoided by the use of deuterochloroform, $CDCl_3$, because the vibrations of the CD group appear at much lower frequencies than the corresponding vibrations of the CH group. This can be seen in Fig. 13, where infrared spectra of $CHCl_3$ and $CDCl_3$ are compared. Thus useful information can often be obtained from absorption bands of a solute dissolved in $CDCl_3$ which is obscured in corresponding spectra of $CHCl_3$ solutions. The use of a solvent in nondeuterated and deuterated forms is a common technique in infrared solution spectroscopy. It is of particular advantage when, as is usually the case for $CHCl_3$ or $CDCl_3$, the proton or deuteron does not readily exchange with protons of the solute molecules. Many deuterated solvents are available commercially.

For quantitative studies, the purity of the solvent is especially important. "Spectral grade" solvents can be purchased which are free of impurities with interfering absorption bands in the region of solvent transparency. Solvents should also be kept free of contaminants or decomposition products which may interact with the solute. A solvent can be checked before use by comparing its infrared spectrum with a published spectrum of reliably pure material. Even better, if such pure material is available, a sample of the same material may be compared with it by differential spectroscopy (Section II.B.1.c). Polar contaminants such as alcohols, ketones, etc. . . . , which may be added as preservatives to certain solvents can be satisfactorily removed by passing the liquid slowly over a column of dry alumina gel before use. In Fig. 14, spectra of $CHCl_3$ before and after alumina gel treatment are compared. Alcohols absorbing near 3600 cm⁻¹ have been effectively removed by the column. Solvents contaminated by moisture, detectable by broad bands at 3400 and 1620 cm⁻¹, can be dried by the use of a suitable nonreactive dessicant.

(b) *Cells.* Techniques for the preparation, use, and care of infrared absorption cells have been described in various places (Lord *et al.*, 1952; Jones and Sandorfy, 1956; Potts, 1963, Chapter 5). A suitable window material for solution cells must be inert to both solute and solvent and transparent to infrared radiation in the spectral range of interest. Table I lists the ranges of transparency of a number of commercially available infrared window materials. In studies of hydrogen bond formation, NH and OH stretching vibrations that appear near 3400 cm⁻¹ can be observed with cells of fused silica windows ("near IR silica").

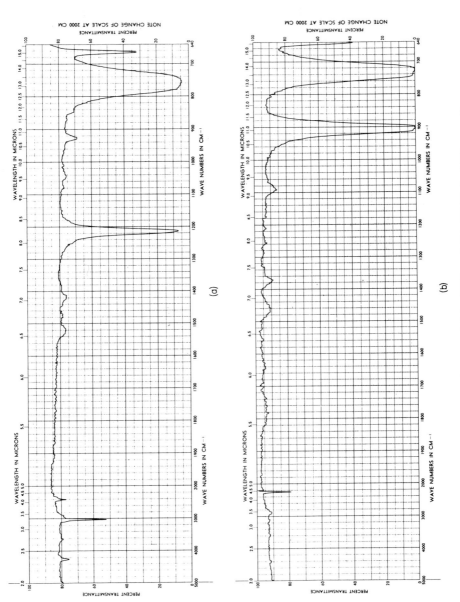

Fig. 13. Infrared absorption spectra of (a) CHCl₃ and (b) CDCl₃ in 0.001 cm cells.

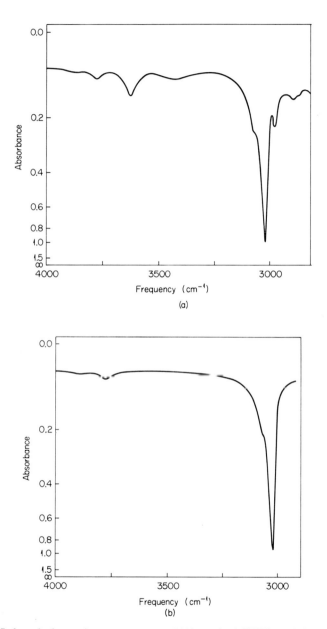

Fig. 14. Infrared absorption spectra near 3500 cm^{-1} of CHCl$_3$. (a) Reagent grade CHCl$_3$; (b) CHCl$_3$ after passage over column of alumina gel.

TABLE I
Infrared Window Materials

Material	Useful limit (cm^{-1})	Comments
Glass	>3700, <185	Not suitable for narrow path cells.
Fused silica	>2500	Not suitable for narrow path cells.
Sapphire (Al$_2$O$_3$)	>1800	Excellent thermal and mechanical properties. Fusible to glass.
LiF	>1400	
CaF$_2$	>1000	
As$_2$S$_3$	>900	Red glass. High n. Unstable >200°C.
BaF$_2$	>800	
Irtran-2 (ZnS)	>700	Yellow. High n.
InSb	>650	High n.
NaCl	>550	[a] Most common cell window.
Si	700–500, <300	Very high n. High m.p. Low frequency limit depends on purity, thickness, temperature.
Ge	5200–400	Same comments as for Si.
KCl	>400	[a]
AgCl	>400	High n. Soft and malleable. Darkens in light. Reacts with brass, etc.
KBr	>330	[a]
KI	>280	[a] Very hygroscopic, very soft.
CsBr	>250	[a] Hygroscopic, soft.
KRS-5	>230	TlBr-TlI. Red. High n. Toxic. Easily deformed.
CsI	>185	[a] Soft, hygroscopic.
Polyethylene	700–30	Flexible, weak, low melting.

[a] Not usable with aqueous solutions. High n means high refractive index (large reflection losses).

In quantitative analyses, absorption cell path lengths often must be known precisely, and when they are less than about 0.1 cm they are usually measured by the interference fringe method (Potts, 1963, pp. 117–122). The choice of path length (often called *thickness*) depends upon the concentration of the absorbing species and on its molar absorptivity a_v according to the Beer–Lambert law. For a typical infrared absorption band, a_v is approximately 10^3 liter-mole^{-1}-cm^{-1} so that the product of solute concentration in moles-liter^{-1} and absorption path in centimeters should have a magnitude about 10^{-3}. Thus 0.01 M solutions will probably require 0.1-cm cells, 0.1 M solutions 0.01-cm cells, and so on.

(c) *Compensated Spectra.* It is often required to obtain the spectrum of a solute in the presence of a solvent whose own absorption overlaps or obscures the absorption bands of the solute. *With due precautions,* the interfering absorption may be compensated by placing an equivalent amount of the solvent in the reference beam of a double-beam spectrometer. The resultant spectrum will then be that of the solute only, since the double beam spectrometer will measure the difference in energy of the two beams. A full discussion of the use of differential spectroscopy and the necessary precautions is given by Potts (1963, Chapter 7). In particular, attention is called to the importance of working only in regions of moderate solvent absorption (Potts, 1963, Fig. 5-5, p. 106).

This technique is important for the spectroscopy of biological samples in aqueous solution and will be discussed below. Another application is the use of deuterated chloroform solutions in studies of hydrogen-bond formation between solute molecules. The stretching vibrations of hydrogen-bonded OH and NH groups occur in a spectral range (ca. 3400 cm^{-1}) where $CDCl_3$ has little interfering absorption. However it is often required to examine very dilute solutions of materials so that intermolecular associations of solute molecules will be minimized. In order to do this, long absorption paths are required. In such cases solvent "background" absorption becomes a serious problem and accurate reference-beam compensation is required. It is important to keep in mind that compensation of a spectrum for solvent absorption requires that the amount of solvent in each beam be the same. The need for checking the purity of the solvent before use is obvious, since interaction of impurities with the solute could occur.

2. Aqueous Solutions

(a) *The Solvent.* Because of its opacity to infrared radiation, water is a difficult solvent to use in mid-infrared solution spectroscopy. Figure 15a shows the intense infrared absorption bands of liquid H_2O which obscure most of the mid-infrared region. However, water is of central importance in biological systems and with special techniques it can be employed as a spectroscopic solvent.

One such technique is the use of deuterium oxide, D_2O, as a companion solvent (Gore *et al.*, 1949). The range of infrared transparency of aqueous solutions is widened considerably by the complementary use of both H_2O and D_2O. This can be seen by comparing the spectra of H_2O and D_2O (Fig. 15). The usable regions are 2850–1800 cm^{-1} and 1500–950 cm^{-1} for H_2O solutions, and 4000–2850 cm^{-1}, 2130–1250 cm^{-1}, and 1100–750 cm^{-1} for D_2O solutions.

Blout and Lenormant (1953, 1955) were the first to apply D_2O solutions to the study of the spectra of biomolecules, and the technique

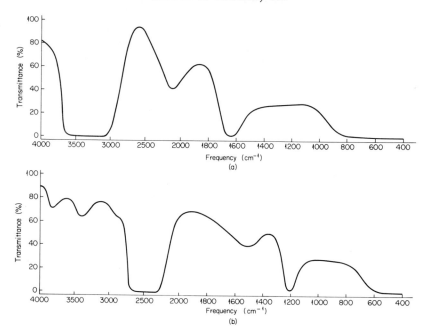

Fɪɢ. 15. Infrared absorption spectra of liquid water in the region 4000–400 cm⁻¹. Optical path is 25 μm, AgCl plates. (a) H_2O; (b) D_2O.

has since become widely used for this purpose (Section II.C). Many investigators, taking advantage of the transparency of D_2O in the 1800–1500 cm⁻¹ range (the so-called double-bond region), have examined absorption bands due to stretching vibrations of $C=O$, $C=N$, and $C=C$ groups which are common to many types of biological molecules. The infrared spectroscopy of nucleotides, polynucleotides, and nucleic acids, for example, has been developed almost entirely with D_2O as the solvent.

An important consequence of aqueous solution spectroscopy is the loss of information because of proton exchange with the solvent. Absorption bands due to vibrations of chemical groups which may be involved in hydrogen-bond formation, such as N—H and O—H stretching and deformation modes, cannot be observed in D_2O solutions. The resultant N—D and O—D bands are also obscured by intense D_2O absorption. Therefore absorption bands due to vibrations of hydrogen-bonding donor groups of solute molecules often cannot be observed in either H_2O or D_2O solutions.

The usual requirements of solvent purity must be met for aqueous solutions. Spectral grade D_2O (>99%) is available at low cost from a number of commercial suppliers. D_2O is easily contaminated with H_2O through contact with the atmosphere or with the walls of vessels in which it is stored or transferred. When small quantities of H_2O become dissolved in D_2O, most of the contaminant is converted to HDO through exchange. HDO in turn gives rise to infrared absorption bands near 3400 cm^{-1}, 2500 cm^{-1}, and 1500 cm^{-1} which may obscure solute absorption. Care should therefore be exercised in handling and storing D_2O to prevent such contaminations.

(b) *Cells.* Many of the usual window materials (Table I) are not suitable for construction of aqueous-solution cells because of their solubility in water. Fluorite (CaF_2), transparent down to 1000 cm^{-1}, and the somewhat more brittle barium fluoride (BaF_2) transparent to 800 cm^{-1}, are commonly employed. Silver chloride (AgCl) is transparent to 400 cm^{-1} but is too soft to provide cells of accurate and uniform thickness. AgCl is also somewhat unstable to visible light. Lithium fluoride is slightly more water-soluble than CaF_2 or BaF_2. The remaining water-insoluble materials of Table I are usually less satisfactory than CaF_2 or BaF_2 for reasons listed there.

Cells for aqueous solution spectroscopy are constructed in the usual manner (Jones and Sandorfy, 1956). Ordinarily they range in thickness from 0.0075 to 0.0025 cm (75–25 μm). Cells thicker than 75 μm do not transmit sufficient energy, while those thinner than 25 μm are cumbersome to construct and difficult to fill.

The present author has used cells of fixed thickness, Model FH, manufactured by Research and Industrial Instrument Co. of London (obtainable in U.S.A. through Beckman Instruments, Inc.), with CaF_2 or BaF_2 windows and Teflon spacers. The advantages of these cells for infrared spectroscopy of nucleic acids and related molecules, and techniques for their maintenance have been described by Miles (1968).

(c) *Compensated Spectra.* Compensation for solvent absorption is virtually a necessity for infrared spectroscopy of aqueous solutions. This is best achieved by use of a variable path cell in the reference beam. For example, D_2O solution spectra in the double-bond region may be recorded with a nearly flat baseline provided the background of solvent absorption is compensated by an equivalent thickness of D_2O in the reference beam cell. The R.I.I.C. Model XL variable path cell is well suited for this purpose.

(d) *Preparation of Solutions.* In order to approximate physiological conditions, it is often necessary to add certain reagents to aqueous solutions of biomolecules. For example, biological systems are in general strongly buffered and a buffer may be required *in vitro* so that the solution pH (or pD) remains within certain narrow limits. For infrared spectroscopy, it is helpful if the buffer is transparent in the spectral range of interest; the spectrum of a solution of buffer should be obtained beforehand to make sure that absorption bands of the buffer do not obscure the region of interest. Since the concentration of buffer in solutions will usually exceed that of other solute, it is important that these bands do not overlap those of the solute. Compensation for the absorption of buffer by addition of an equivalent concentration to the reference cell is not necessarily feasible, since this may only result in removing energy from both beams so that no meaningful spectrum can be obtained.

Many common buffers, including the phosphates, sodium cacodylate, and "tris" (tris-hydroxymethyl-aminomethane), are sufficiently transparent for use in the double-bond region. Even more suitable for this region is the sodium salt of trichloromethylphosphonic acid, $Na_2CCl_3PO_3$ (Hartman and Rich, 1965). Ethylenediaminetetraacetic acid (EDTA), frequently used in solutions of biomolecules, is also largely transparent in the double-bond region. All of the above reagents, on the other hand, have absorption bands in the 1500–950 cm^{-1} interval and consequently would be less suitable for use in infrared studies of H_2O solutions.

For D_2O solution spectroscopy, solutes should be lyophilized (freeze-dried) from excess D_2O whenever possible to minimize proton contamination. Otherwise the exchangeable hydrogen present in most biomolecules—in bound chemical groups and in water of crystallization—will introduce substantial HDO into the solvent. Buffers and salts to be added to D_2O solutions should also be lyophilized from pure D_2O for the same reason.

3. Pure Liquids

Infrared spectra of pure liquids are frequently employed in determinations of molecular structure, identification of specific functional groups, detection of trace impurities, and similar qualitative investigations. Apart from the lipids, few biological materials exist as liquids, and most infrared studies of pure liquids have been confined to the lipid class (Section II.C.3.a).

Infrared absorption of pure liquids is generally so intense that cell thicknesses of about 10 μm are required. Since most biological materials are too viscous to be inserted conveniently in such thin cells, spectra of such materials are best obtained by forming a capillary film of the liquid ($\sim$10 μm in thickness). The capillary film is made by placing a drop of liquid on a flat plate of infrared window material, covering this with another flat plate, and then squeezing the resulting sandwich so that the liquid spreads to cover the entire area of the plates. Techniques for mounting such cells and for maintenance of the plates have been described (Potts, 1963, Chapter 5). When considerable water is present in the liquid film, appropriate water-insoluble plates (Table I) should be used.

Another procedure for obtaining infrared spectra of liquid samples is that of internal reflection spectroscopy[1] (Fahrenfort, 1961; C.I.C. Newsletter, 1961). In this technique, often abbreviated IRS, the sample is examined by reflection of the infrared radiation from the interface between the sample and a material of high refraction index rather than by transmission through a thin layer. The type of spectrum obtained by IRS, and the instrumentation and techniques required, differ somewhat from those of absorption spectroscopy. A full discussion of the principles involved is given by Harrick (1967). Sample-handling techniques for IRS are discussed below (Section II.B.4.e).

4. Solids

The selection of a suitable technique for obtaining infrared spectra of crystalline and amorphous solids will depend upon the sample's morphology and stability as well as on the type of information desired. Crystalline powders, for example, require techniques which will minimize the losses due to multiple reflections from the solid particles. This is accomplished by keeping the particles to a size well below that of the shortest wavelength of light to be used. i.e., $<<3$ μm, and by suspending or imbedding them in a matrix of about the same refractive index as that of the particles themselves. A thin, single crystal (~ 10 μm) would also produce a satisfactory spectrum, but in practice it is difficult to produce a crystal of sufficient cross section ($\sim 2 \times 0.5$ cm) to fill the sample area of an infrared spectrophotometer. Microsampling techniques or the use of an infrared microscope may be required for small crystallites (Section II.B.4.f).

Special consideration should be given to the crystalline form of the sample, since vibrational spectra are sensitive to the arrangement of molecules in the crystal lattice (space group symmetry) and to intermolecular forces. Often a single chemical material can exist in two or even more polymorphic crystalline forms, each giving rise to a characteristic infrared spectrum. Transformations between different forms, or to an amorphouse form, may result from changes in humidity or temperature. Transfering a material from ambient conditions to an infrared beam, for example, will usually result in an increase of 5 to 10°C in sample temperature. Some of the techniques described below are suitable

---

[1] The method is often referred to as attenuated total reflectance (ATR).

only for small, highly stable biomolecules since the mechanical stresses involved, such as grinding, can easily degrade many biopolymers.

(a) *The Oil-Mull Technique.* This is a common and simple method of preparing solids for spectroscopic study. A small amount of the solid material ($\sim$5 mg) is ground in a mortar to a fine particle size ($\sim$.1 μm) and a drop of mineral oil (alkane hydrocarbons in the C_{20}–C_{30} range) is added. The grinding is continued until a slurry is formed. A drop of the slurry is then squeezed between two flat plates of infrared window material to form a cell containing a capillary film. The cell is then mounted directly in the spectrophotometer.

The hydrocarbon mulling agent, often referred to by the trade name "Nujol," is transparent throughout most of the chemical infrared region, but has intense absorption near 2900 cm^{-1} due to CH stretching vibrations and bands of lesser intensity near 1460 and 1375 cm^{-1} due to CH deformation vibrations. If one wished to observe absorption bands of the solid near these spectral regions, it would be necessary to use a nonhydrogenic oil as the mulling agent, such as the perhalocarbon "Fluorolube." Fluorolube is transparent in the region above 1400 cm^{-1}. Thus, if the spectrum of a solid is obtained from 4000 to 1400 cm^{-1} with Fluorolube as a mulling agent and from 1400 to 200 cm^{-1} with Nujol, the entire mid-infrared region may be scanned free of interfering absorption bands of the mulling agents (Fig. 16). Hexachlorobutadiene can be employed instead of Fluorolube but is less satisfactory because of its higher volatility and lower viscosity.

Properly prepared mulls yield excellent spectra. Much care should be exercised in grinding the solid, with a mortar and pestle of a hard, smooth material such as agate, in keeping the mull free of moisture, and in using properly polished cell windows. The art of proper mull-making is described in detail by Potts (1963, pp. 141–145).

The best mulls are made from relatively hard crystalline substances. Many biological materials are waxy, fibrous, or rubbery and do not form suitable mulls. Others, like DNA, may be easily degraded by the grinding process.

(b) *The Pressed-Pellet Technique.* In the pressed-pellet or pressed-disc technique, a few milligrams of the solid are ground to a very fine powder with a large excess of very dry (dessicated) KBr. The mixture is then placed in a specially designed die and compressed at about 500 to 1000 atm into a thin disc which can be mounted directly in the spectrophotometer (Jones and Sandorfy, 1956).

This technique, if properly employed, can yield spectra of a quality comparable to those obtainable from mulls. Advantages over the mull technique are: (1) rubbery or plastic samples are more easily pressed

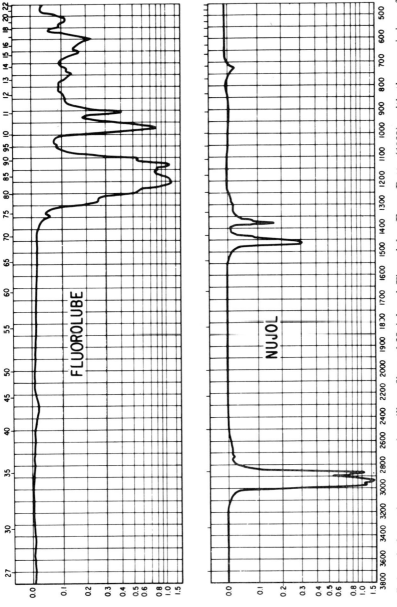

FIG. 16. Infrared absorption spectra of capillary films of Nujol and Fluorolube. From Potts (1963), with the permission of Wiley (Interscience).

into a pellet than ground into a mull; (2) the halides of potassium have no interfering absorption in the 4000–400 cm^{-1} region; (3) quantitative reproducibility is more easily achieved; and (4) the discs can be stored for future reference. Disadvantages of the pressed pellet technique are: (1) pellets nearly always contain contaminating moisture with interfering absorption in the 3400 cm^{-1} region; (2) the high pressures required for pellet formation may alter the crystal structure of the material being investigated; and (3) the potassium halides are not chemically inert and may react with the sample. Because of the severe shear forces attendant with grinding and compressing, the pellet technique is also not suitable for fragile macromolecules. Acceptable pellets of small biomolecules, however, such as amino acid or nucleotide monomer residues, can usually be made with little difficulty.

(c) *Polymer Films.* Many polymeric materials can be formed into thin, coherent films. For convenience in handling, the film can be mounted with adhesive tape on a slotted cardboard holder and placed directly in the spectrophotometer beam.

A related technique is the casting of a thin film of the polymer from solution onto a plate of infrared window material, which may then be mounted in the spectrophotometer. Spectra of polynucleotides and polypeptides, for example, may be obtained by casting films from aqueous solution onto AgCl, CaF$_2$, or BaF$_2$ plates. A high water content in the films may be required to stabilize the secondary structure of the biopolymer, and consequently water bands (Fig. 15) may appear in the spectrum. Numerous infrared studies have been made of nucleic acid and polynucleotide films cast from H$_2$O and D$_2$O solutions (Section II.C.2).

In casting films of biopolymers, the correct amount of aqueous solution of the material is spread over the window and solvent is evaporated by dessication or evacuation, so that a film of suitable thickness ($\sim$10 μm) is obtained. The plate may then be mounted as one window of a hygrostatic cell in which the relative humidity can be controlled with an appropriate saturated salt solution. A cell of this type, first described by Sutherland and Tsuboi (1957), is shown in Fig. 17.

(d) *Dichroism and Oriented Films.* The structural information derivable from the infrared spectrum of a biopolymer film is increased if the molecules in the film can first be oriented in a preferred direction, and then the spectrum of the film determined with polarized infrared radiation. The sample is mounted so that the molecules of the sample have one specific orientation (say parallel) with respect to the plane of polarization of the radiation and its spectrum is recorded. The orientation of the sample is then changed by 90° and a second spectrum is

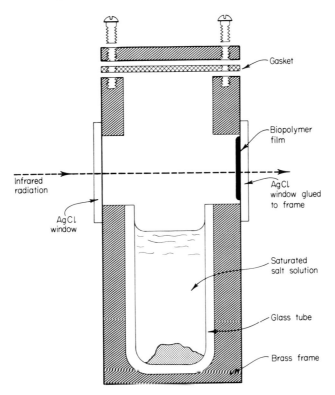

FIG. 17. Infrared absorption cell for examination of biopolymer films at constant humidity. With the permission of Wiley (Interscience).

obtained. If a particular absorption band of the sample shows different absorbance in the two orientations, it is said to be "dichroic." The ratio of the two absorbances is called the *dichroic ratio*. With help of simple models for infrared absorption by functional groups, the dichroic ratios can provide information about the orientation of these groups with respect to the direction of orientation of the film. Quantitative discussions of infrared dichroism and its uses are given by Krimm (1963) and Zbinden (1964).

Oriented films of nucleic acids and polynucleotides are made by repeatedly stroking the wet fibers unidirectionally until dry. A high degree of orientation is rather easily achieved. Similar techniques may be used for polypeptides. The oriented film may be mounted in a cell like that of Fig. 17. Infrared radiation is efficiently polarized by use of a simple polarizer based on the principle of Brewster's angle. The customary procedure is to rotate the sample rather than the polarizer in recording the two spectra because polarization effects in grating spectrophotometers vary seriously with wave number.

Infrared dichroism of biopolymers will in general depend on the temperature and relative humidity of the films since the geometrical conformations of most chemical groups in biopolymers will vary with changes in the secondary structure. It is therefore important to control carefully the temperature and humidity of films while dichroic studies are being made on them. A number of salt solutions have been described for relative humidity control in the 0–100% range (Falk *et al.*, 1962). Temperature control is more difficult because of heating caused by the beam of infrared radiation. However the temperature at the film can usually be monitored with a thermocouple junction imbedded in the salt plate.

(e) *Internal Reflection Spectroscopy (IRS).* This technique is suitable for the investigation of liquids, including solutions, as well as for solids (Wilks and Hirschfeld, 1967). A full discussion of the many applications of the method is given by Harrick (1967).

IRS is a useful supplement to transmission spectroscopy but is not ordinarily employed in cases where conventional transmission methods work satisfactorily. It is of greatest value for investigation of materials that are difficult or impossible to study by transmission spectroscopy. Biological and medical applications, for example, have included the reflection spectra of skin, tissue, membranes, and whole cells.

In IRS, the sample is placed in contact with a transparent medium of greater refractive index (internal reflection element) and the reflectance, single or multiple, from the interface between sample and element is measured, generally at an angle of incidence greater than the critical angle. When the reflectance is less than unity, as a result of absorption of incident radiation, the process is termed "attenuated total reflection" (ATR). A comparison of transmission and reflection is given in Fig. 18. In the former case (Fig. 18a), incident light strikes the sample normal to its surface and is transmitted through it. In the latter (Fig. 18b), the incident beam impinges on the reflection element at an angle θ to the normal and is reflected at the interface.

The reflection may be single (Fig. 18b) or multiple (Fig. 18c) according to the design of the reflection cell.

Although transmission and reflection spectra have a similar appearance, there are important differences. Reflection spectra often have bands which are more intense and displaced towards lower wave number than those of transmission spectra. The bands of reflection spectra are also sensitive to the angle of incidence of the infrared radiation (θ, Fig. 18) and to the refractive indices of sample and reflective element.

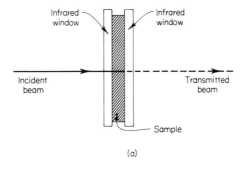

(a)

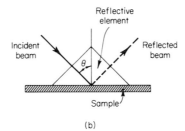

(b)

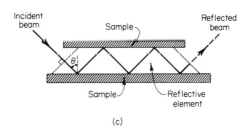

(c)

FIG. 18. Comparison of conventional transmission and reflection techniques. (a) Transmission; (b) single reflection; (c) multiple reflection.

To obtain a spectrum by IRS, the sample and internal reflection element are brought into satisfactory physical contact with one another. For liquids, pastes, and flexible solids, this is easily achieved. For hard solids, special sample-handling may be required (Harrick, 1967, Chapter 7). For quantitative measurements, the sample should have a smooth surface capable of contact over a fixed area of the element. For qualitative applications, a rough surface will provide sufficient contact with the reflective element.

The design and fabrication of the element will depend upon the chemical composition of the sample and the information sought. For most infrared applications, a flat plate of inert material of high refractive index is used (Table I). IRS accessories for standard spectrophotometers are available commercially.

IRS may be employed with greater ease than transmission spectroscopy in the quantitative analysis of liquids and in the investigation of viscous solutions or gels (aqueous solutions of biopolymers, for example) since the difficulties associated with filling thin cells are eliminated.

When the use of IRS is being considered, it must be kept in mind that only a thin film of the sample near the surface is observed, particularly at large angles of incidence. Thus, vibrational information is obtained only from that part of the sample near the surface. The need for avoiding contamination of the sample surface is obvious.

(f) *Microsampling Techniques.* Microsampling techniques are often necessary in the spectroscopic study of biological materials. The micro approach generally involves a reduction in the area of illumination (cross section) in the sample beam so that correspondingly smaller samples may be used without reduction of sample thickness. This can be accomplished with a beam condenser, as shown in Fig. 19 (top). The device is

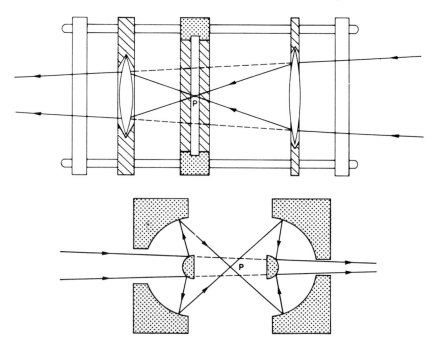

Fig. 19. Top: A simple beam condensing system of KBr or AgCl lenses. P is a frame in which microcells or other microsample devices may be mounted. Bottom: The Schwarzchild microscope principle used as a beam condensing system. Microsample is placed at point P. With the permission of Wiley (Interscience).

mounted as a unit in the sample compartment of a spectrophotometer. A condenser of this type employs either KBr or AgCl lenses and therefore may either fog in moist surroundings or darken on exposure to light. Alternatively, a mirror system may be employed like that of the Schwartzchild microscope, shown in Fig. 19 (bottom). Revices of both types are available commercially.

A disadvantage of beam condensers is the increased temperature of the sample which results from the more intense radiant flux at the focus. Reduction of the sample beam area can also be achieved by masking, but this results in a serious loss of energy and is much less satisfactory than the use of a condensing system.

A greater reduction in acceptable sample area than is obtainable from a beam condenser can be achieved with use of a reflecting microscope, often called an infrared microscope. Several instruments of this type have been described (Blout and Abbate, 1955; Duecker and Lippincott, 1964). The major disadvantage of the infrared microscope is the severe heating of sample which can result.

A number of liquid microcells, requiring as little as 0.1 μl of sample, are available commercially for use with beam condensing units (Mason, 1958; Jones and Nadeau, 1958). The preparation and mounting of micro mulls and micro pellets, requiring about 20 μg of solid sample, have also been described (Lohr and Kaier, 1960; Dinsmore and Edmondson, 1959). Further discussion of infrared microspectroscopy is given by Blout (1960).

C. Biological Applications of Infrared Spectroscopy

Investigations of various classes of biomolecules using methods of infrared spectroscopy have been reviewed frequently (Sutherland, 1952; Bauman, 1957; Fraser, 1960; Timasheff and Gorbunoff, 1967; Felsenfeld and Miles, 1967; Susi, 1969). Discussed here are some recent applications which demonstrate the scope of the method and the kinds of information obtainable.

1. Proteins, Polypeptides, and Amino Acids

Infrared studies of proteins and other biopolymers have the general objective of determining molecular conformation from band frequencies, intensities, and polarizations. Ideally the conformation is uniquely related to these properties, but for molecules as complex as biopolymers it is rare that sufficient spectroscopic data are available to draw rigorous conclusions about polymer geometry. It is usually necessary to resort to approximate methods and assumptions, the nature of which depends on the data obtainable and the kind of structural information which is useful. The commonest of these assumptions is that a chemically distinct group of atoms, such as the amide group in polypeptides, has its own characteristic absorption frequencies in the infrared. This approximation, called the "group-frequency" assumption (Bellamy, 1958, 1968) varies in its validity in different parts of the mid-infrared spectrum, and has been discussed (Section I.A.3).

One region of great usefulness for study of proteins is the interval 1700–1500 cm^{-1}, where certain bands due to vibration of the amide group (—CO—NH—) appear (Sutherland, 1952). The nature of these vibrations, the so-called amide I and amide II motions at ca. 1650 and 1525 cm^{-1}, respectively, has been determined by calculation of the normal coordinates, i.e., the actual atomic displacements, of trans-N-methyl-acetamide and its deuterated derivative (Miyazawa 1955, 1956; Miyazawa et al., 1958). Thus amide I is known to be due mainly to the C=O stretching vibration, and amide II to simultaneous N—H bending and C—N stretching vibrations of the peptide groups.

Miyazawa has used these findings to correlate the amide I and amide II bands of fibrous proteins and polypeptides with their conformations. The frequencies and polarizations (dichroic ratios) of the amide bands were interpreted in terms of the different intra- and inter-molecular interactions which occur in various conformations. Separate band assignments were made for the α-helical, β-pleated sheet, and random-coil conformations (Miyazawa, 1960; Miyazawa and Blout, 1961; Miyazawa, 1962). Krimm (1962) has extended this work to include the parallel-chain pleated-sheet conformation. Further discussion of this subject is given by Schellman and Schellman (1964) and by Susi (1969), and a number of applications of the Miyazawa–Krimm results are reviewed by Timasheff and Gorbunoff (1967). More recently, these results have been found useful in characterizing the conformations of such materials as glucagon (Gratzer et al., 1967, 1968), ribosomal protein (Cotter and Gratzer, 1969b), and poly-γ-benzyl-DL-glutamate (Masuda and Miyazawa, 1969).

Numerous other bands appear in spectra of proteins and polypeptides. Those characteristic of the peptide grouping, amide A ($\sim$330 cm^{-1}) and B ($\sim$3100 cm^{-1}) and amide III ($\sim$1300 cm^{-1}), IV ($\sim$625 cm^{-1}), V ($\sim$725 cm^{-1}), VI ($\sim$600 cm^{-1}), and VII ($\sim$200 cm^{-1}) (Bellamy, 1958; Miyazawa, 1962), have found only limited application in conformational studies (Miyazawa et al., 1962; Masuda and Miyazawa, 1967; Miyazawa et al., 1967). This is due mainly to their complex origins and frequent overlap with bands of residues in the side chains.

Absorption bands originating from vibrations of the side-chain residues and additional vibrations of the principal chain itself have also been investigated. A discussion of these for several fibrous proteins is given by Beer *et al.* (1959). A detailed analysis of the complete mid-infrared spectrum of a polypeptide is that of Tsuboi (1962), who investigated films of poly-γ-benzyl-L-glutamate in the α-helical conformation. Measurements of dichroism for bands in the interval 3800–500 cm^{-1} made assignments possible for both main-chain and side-chain groups, as well as determination of the orientations of the latter with respect to the former (fiber direction). Interchain interactions and their effects on the spectrum of poly-γ-benzyl-L-glutamate have also been explored (Tsuboi, 1964a; Fraser *et al.*, 1967).

Detailed spectral studies have been made of polyglycine I (Fukushima *et al.*, 1963), polyglycine II (Suzuki *et al.*, 1966; Krimm, 1967), L-proline oligopeptides (Isemura *et al.*, 1968), and acetylglycine N-methylamide (Koyama and Shimanouchi, 1968). Infrared evidence for C—H · · · O $=$ C hydrogen bonds in polyglycine II has been discussed (Krimm *et al.*, 1967; Krimm and Kuroiwa, 1968). Itoh *et al.* (1968, 1969) have obtained far infrared spectra (below 200 cm^{-1}) of polyalanines in α- and β-structural conformations.

Most of the above studies were made on films of fibrous materials containing relatively little water. Globular proteins, on the other hand, must be investigated in aqueous solution or in films of rather high water content. The experimental difficulties are accordingly greater (Section II.B, above). A further complication in spectra of globular proteins is the presence of a multiplicity of chain conformations which give rise to broad, diffuse absorption bands. Early attempts at infrared spectroscopy of such materials in H_2O solution were of limited success [see, for example, Elliott *et al.* (1957)]. More useful has been the examination of such proteins in D_2O solution, since there is then little solvent absorption in the amide region. Spectra of a number of proteins in D_2O have been reported (Timasheff and Susi, 1966). The amide II band is at least partially obscured by absorption due to residual HOD present in the D_2O. Studies of proteins in D_2O are therefore concerned with the amide I band, which originates from a C $=$ O stretching motion and is consequently only slightly (~ 10 cm^{-1}) shifted by deuteration. A systematic study of amide I for H_2O and D_2O solutions, and for mulls and films of proteins has been made by Timasheff and co-workers (Timasheff *et al.*, 1967a,b; Susi *et al.*, 1967). A recent application has been the study of protein conformation in reconstituted ribosomes (Cotter and Gratzer, 1969b).

A number of other nucleic acid-protein complexes have been examined by infrared methods. Nucleohistone and nucleoprotamine studies are reviewed by Bradbury *et al.* (1967). In addition to the usual correlations of band position with conformation, structural information on these materials has been obtained from the rate of deuterium exchange. In nucleohistone, for example, there is a slowly exchanging fraction of protons which can be attributed to amide groups shielded by secondary structure from the D_2O.

Infrared spectroscopy, including the near infrared region (Section II.C.3.d), is now often used to study exchange kinetics in proteins and related biopolymers. The subject has been reviewed recently (Harrington et al., 1966; Hvidt and Nielson, 1966).

Other quantitative analytical studies on proteins and related molecules by infrared spectroscopy are few in number [see, for example, Bendit (1967)] due to the diffuseness and overlap of absorption bands which make applications of Beer's law rather difficult. Molecular interactions involving the amino acid monomer residues, on the other hand, might be investigated by infrared methods without the difficulties usually encountered with high molecular weight proteins.

2. Nucleic Acids, Polynucleotides, Purines, and Pyrimidines

Infrared studies of nucleic acids have been reviewed by Shimanouchi et al. (1964) and by Susi (1969). Most applications before 1960 paralleled work done previously on proteins. For example, polarized infrared spectra of films of oriented DNA fibers were obtained by several workers (Fraser and Fraser, 1951; Sutherland and Tsuboi, 1957; Wilkinson et al., 1959; Bradbury et al., 1961). Studies of this type are useful in determining relative directions of base, sugar, and phosphate residues with respect to the fiber axis. Recently, double helical viral RNAs have been investigated using similar techniques [see, for example, Sato et al. (1966)].

In other infrared studies of nucleic acid films, Falk et al. (1963) determined the sites on DNA which become hydrated as the relative humidity is increased and showed that the helix-coil transition can be followed by the dichroic behavior of the 1660 cm^{-1} band which originates in vibrations of the base residues. Kyogoku et al. (1961a,b, 1962) observed helix-coil transitions of DNA and RNA in D_2O solutions. Infrared studies of the denaturation of nucleic acid helices are based upon the markedly different spectra which result in the double-bond region from the paired and unpaired bases occurring in native and denatured structures, respectively. Recently, Fritzsche (1967) examined D_2O solutions of DNAs with different base compositions and Tsuboi (1968) investigated the effects of isotopic substitution on spectra of DNA films.

The purine and pyrimidine bases, nucleosides, nucleotides, and related monomeric derivatives of nucleic acids have been extensively studied by infrared spectroscopy. The prevalent tautomeric structures of the heterocycles in the solid state and in aqueous solutions were established largely through infrared spectra. Reviews are given by Shimanouchi et al. (1964) and Susi (1969).

Structures of polynucleotides have also been elucidated from comparisons of their infrared spectra with spectra of the corresponding mononucleotides obtained under the same conditions. Thus Hartman and Rich (1965) observed the formation of double helical hemiprotonated polyribocytidylic acid (poly C) in solutions of low pD. Miles and collaborators have conducted studies of the structures and helix-strandedness of polynucleotide complexes in D_2O solution. Poly A, poly U, poly A + U, poly A + 2U (Miles, 1958, 1960; Miles and Frazier, 1964a), poly I, poly C, poly I + C (Miles, 1959, 1961), and poly G (Miles and Frazier, 1964b) have been investigated. Helical complexes between monomers (nucleosides, nucleotides, and the like) and complementary polynucleotides have also been investigated. Reviews are given by Felsenfeld and Miles (1967) and Miles (1968). Tsuboi and co-workers have investigated deuterated films of polynucleotides (Shimanouchi et al., 1964, pp. 454–460; Tsuboi, 1964b).

Several quantitative infrared studies of nucleic acids and their monomer derivatives have been made recently. Hamlin et al. (1965) demonstrated the formation of a 1:1 hydrogen bonded complex between an adenine and a uracil derivative in chloroform solution, while Kuechler and Derkosch (1965) made a similar study on carbon tetrachloride solutions. Kyogoku et al. (1966, 1967a,b, 1968, 1969) extended the chloroform solution investigations to include a number of other purine–pyrimidine interactions and calculated association constants for several complexes from the infrared data. This work is reviewed by Lord and Thomas (1968). Interactions of riboflavin with adenine derivatives have been investigated by similar methods (Kyogoku and Yu, 1968, 1969).

Infrared spectroscopy has also been used to determine the percentages of AU and GC base pairs in RNAs containing both single-stranded and double-helical regions (Thomas, 1969). Applications to D_2O solutions of yeast rRNA (Thomas and Spencer, 1969), E. coli. rRNA (Cotter and Gratzer, 1969a; Hartman and Thomas, 1970), and to deuterated films of several specific tRNAs (Tsuboi et al., 1969) have been described. The anticodon loop of formyl methionine tRNA has also been investigated by this method (Morikawa et al., 1969).

Up to the present, there has been no vibrational analysis of nucleic acids which is comparable to the Miyazawa treatment of polypeptides and proteins. Except for the phosphate groups (Shimanouchi et al., 1964, pp. 467–493), very little is understood about the normal vibrations of nucleic acid constituents. The effects of base stacking on vibrational spectra of nucleic acids, for example, are not well understood, although such effects appear to be negligible by comparison with effects of inter-chain base pairing (Thomas, 1969). A great deal more work on model compounds will be required before infrared spectra of nucleic acids can yield the type of secondary structural (conformational) information now derivable from protein spectra.

3. Other Biomolecules

(a) *Lipids*. The lipids and lipoidal derivatives have been extensively examined. A review of applications is given by Chapman (1965).

Systematic study of fatty acids in CCl_4 and CS_2 solutions by Jones (1962a) has permitted a number of bands to be assigned. Infrared spectra of fats have also been used to investigate hydrogen bonding of the car-boxyl groups, polymorphism in crystalline samples, extent of branching or unsaturation in side chains, and quantitative analysis of fat mixtures and natural products.

The steroids have been the object of very detailed spectroscopic study (Rosenkrantz, 1955, 1957). Several thousand steroid spectra have been catalogued by Dobriner *et al.* (1953), Roberts *et al.* (1958), and Neudert and Ropke (1965). The spectral range 1350–650 cm^{-1} has been found useful as a fingerprint region for distinguishing one steroid from another. Applications to the study of carotenoid structures are discussed by Zech-meister (1962).

Infrared studies of tissue lipids (Schwarz *et al.*, 1957) and serum lipids (Freeman, 1957) have been described. A number of phospholipids have been investigated by Rouser *et al.* (1963) and by Chapman and collaborators (Byrne and Chapman, 1964; Chapman and Collin, 1965; Chapman *et al.*, 1966). Reviews of recent infrared studies of phospho-lipids and sphingolipids are given by Chapman and Wallach (1968) and by Chapman (1969a). The study of lipid–protein interactions by infrared spectroscopy has also been reviewed by Chapman (1969b).

(b) *Carbohydrates*. Early infrared studies of carbohydrates are re-viewed by Clark (1955). The use of infrared spectroscopy for the de-termination of carbohydrate structure is described in detail by Barker *et al.* (1956). Parker (1960) has made a systematic infrared study of the kinetics of mutarotation of α-D-glucose, β-D-glucose, and β-D-man-nose in H_2O solutions. Spectra of several other sugars have been pub-lished by Goulden (1959). A number of more recent infrared studies of carbohydrates are reviewed by Parker (1968).

(c) *Cells and Tissues*. Various living materials, including different types of bacteria, have been investigated by infrared spectroscopy. Several applications are reviewed by Clark (1955). Viruses (Benedict, 1956) and tissues (May and Grenell, 1956) have also been examined. The use of infrared spectra for characterizing mycobacterial strains has been described by Smith *et al.* (1956). Chapman (1966) has investigated liquid crystals derived from cell membranes. Norris (1961) has described techniques for the infrared study of microorganisms with several applications.

(*d*) *Special Applications.* 1. *Internal reflection spectroscopy.* As mentioned above (Section II.B.4.e), this technique is useful for samples which are not easily examined by transmission spectroscopy. Aqueous solutions of biomolecules (Parker, 1963; Katlafsky and Keller, 1963; Goulden and Manning, 1964), bacterial cultures (Johnson, 1965), biological tissues (Herman, 1965), blood (Kapany and Silbertrust, 1964), and bone (Furedi and Walton, 1968) are among the many materials investigated using IRS. Other applications are reviewed by Harrick (1967) and Wilks and Hirschfeld (1967).

2. *Near-infrared spectroscopy.* The spectral region referred to as the "near-infrared" is generally identified as the region between 4,000 and 12,000 cm^{-1}. In this region, overtone and combination bands (Section I, above) occur. The near-infrared region has not been widely employed for the study of biological materials. Its use for selected structural problems in biochemistry is reviewed by Hanlon and Klotz (1968).

3. *Far infrared spectroscopy* The "far-infrared" region is that occurring below 200 cm^{-1}. In this region, only low energy vibrations can occur for molecules which are in a condensed phase. Oscillators of either large reduced mass or small force constant can give rise to vibrations in the far-infrared (Section I, above). Stretching and bending vibrations of hydrogen bonds, for example, generally occur in this range.

Far-infrared spectra of polyalanines have been published by Itoh *et al.* (1968). Harada and Lord (1970) recently studied far-infrared spectra of nucleic acid derivatives.

III. Raman Methods

A. INSTRUMENTATION

1. General Features of Raman Spectrometers

Raman scattering is extremely weak, usually a factor of 10^3 weaker than Rayleigh scattering. In Raman spectrometers, therefore, every effort is made to increase the intensity of Raman scattering which reaches the detector. To this end powerful sources, efficient collection optics, and sensitive detectors must be employed. Consequently, high perfor-

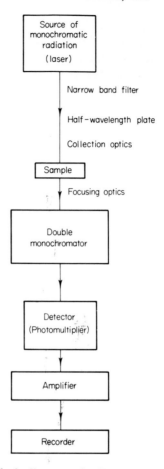

FIG. 20. Block diagram of a Raman spectrometer.

mance Raman spectrometers are far more costly today than are infrared spectrometers.

The basic components of a Raman spectrometer are shown in Fig. 20. Many of these have been discussed in detail elsewhere (Harrison *et al.*, 1948; Evans, 1963; Koningstein, 1967).

(a) *Light Sources for Raman Spectroscopy.* A suitable light source for exciting Raman spectra should be monochromatic and of high and continuous intensity. The rare gas laser is the best such source available and has been incorporated into all commercial spectrometers (Kogelnik and Porto, 1963; Leite and Porto, 1964; Hawes *et al.*, 1966).

Some of the more commonly used lasers and their wavelengths of emission are: helium-neon (6328Å), argon ion (4880 Å and 5145 Å), and krypton ion (5208 Å, 5682 Å, and 6471 Å). Of these the helium-neon laser is the most stable in output power. Its emission in the red (6328 Å) has the further advantage that most molecules do not undergo electronic transitions in this region. Consequently absorption of the radiation is less likely to occur. However, the use of red light to excite Raman spectra is disadvantageous because the intensity of molecular scattering decreases as the fourth power of the exciting wavelength [i.e., decreases with decreasing frequency ν_0 of Eq. (12)]. A further, often more serious, disadvantage of red excitation is the unavailability today of low-cost efficient detectors, operable throughout the red region.

An advantage of the ion lasers is the versatility provided by having more than one wavelength for exciting the Raman effect. Ar+ and Kr+ lasers with power stability suitable for most Raman applications are now available. Lasers containing a mixture of Ar+ and Kr+ are also available. These provide at least four different exciting lines (one in each of the blue, green, yellow, and red regions) which can be tuned selectively by rotating a prism at the exit window of the laser cavity.

(b) Collection Optics. Optical systems to collect the scattered radiation and focus it on the entrance slit of the monochromator vary from one instrument to another. Various lens systems and image slicers have been described (Cary, 1955; Hawes et al., 1967). Two such systems are considered below (Section III.A.2).

(c) Monochromators. Most commercial Raman spectrometers employ a double monochromator, which greatly reduces the stray light (i.e., internally scattered light of frequencies other than the desired frequency) inherent in single monochromator instruments. A further advantage of the double monochromator is that it has twice the dispersion of a single monochromator. The instruments discussed below both employ modified Czerny monochromator systems.

(d) Detectors. For Raman spectroscopy, the scattered light (in the visible region of the spectrum) is detected by the use of a phototube, such as the FW-130 photomultiplier. Response of the phototube to red light however is considerably weaker than for light of shorter wavelengths.

Radiation exiting the monochromator is focused onto the cathode of the phototube. After photomultiplication, the signal from the anode of the tube is then amplified and fed to a strip chart recorder.

2. Examples of Raman Spectrometers

(a) Varian Associates, Cary Model-81. The design and operation of the basic instrument have been described (Evans, 1963). Conversion of the Model-81 to laser excitation does not alter the basic operation

(Hawes *et al.*, 1966). The optical arrangement of the laser Raman instrument is shown in Fig. 21a and the coaxial excitation assembly is shown in Fig. 21b.

Light entering the capillary cell (Fig. 21b) undergoes multiple internal reflections off the cell walls. This can be a serious disadvantage for solutions which are not optically homogeneous, since the effective path of the scattered radiation is quite long. Tyndall scattering may then obscure much of the Raman scattering.

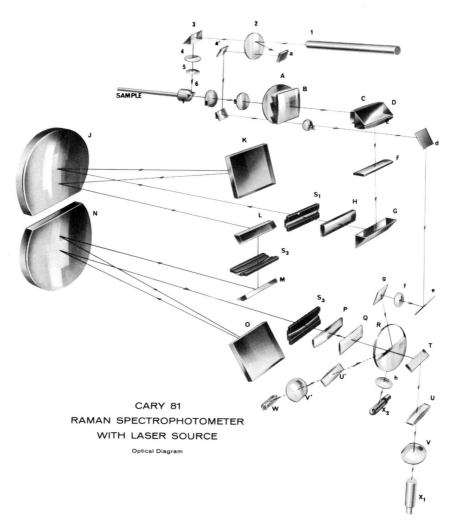

CARY 81
RAMAN SPECTROPHOTOMETER
WITH LASER SOURCE
Optical Diagram

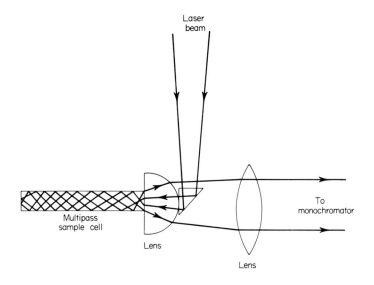

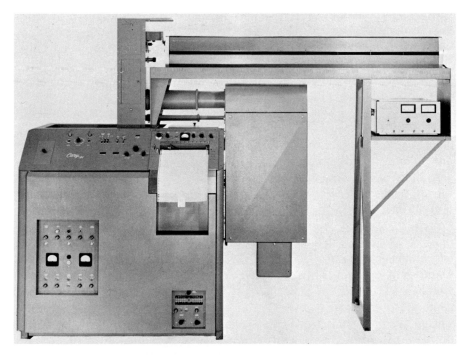

FIG. 21. Varian Associates Cary-81 Raman spectrometer. Opposite page: Diagram of optics; top: coaxial excitation assembly; bottom: actual photograph.

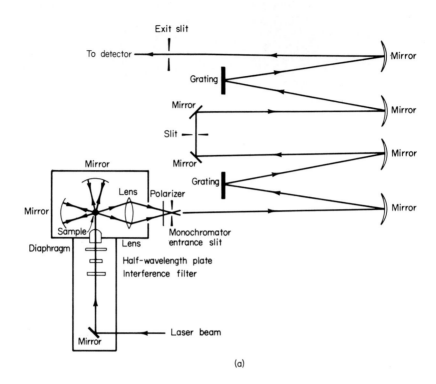

(a)

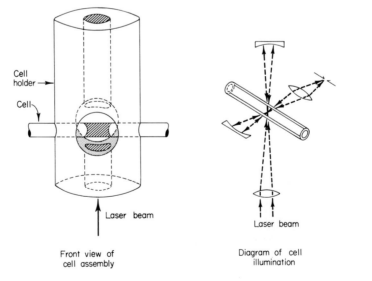

Front view of
cell assembly

Diagram of cell
illumination

(b)

(b) *Spex Industries, Inc., Model 1401-Ramalog.* This instrument differs considerably from the Model-81 in the geometry of sample illumination as well as in the design of the optical system. As shown in Fig. 22b, the laser beam enters the cell at a right angle to the cell axis and is scattered again perpendicular to this axis and to the direction of incidence. This transverse excitation assembly requires a much smaller volume of sample since only a small element of the capillary cell is irradiated. Larger cells which are mounted coaxially with the laser beam may also be employed (Freeman and Landon, 1968). An instrument similar to the Model 1401 in design and performance characteristics is manufactured by the Jarrell-Ash Co.

3. Sampling Devices

For Raman spectroscopy, sample cells for liquids are ordinarily made of glass. Most commercial spectrometers are designed to accommodate 1.0 mm (inner diameter) capillary tubes, e.g., Kimax #34507 "melting point tubes." Powder samples or small crystallites may also be packed into capillary cells to yield good quality spectra. Single crystals are

Fig. 22. Spex Industries, Inc., Model 1401-Ramalog Raman spectrometer. Opposite page: (a) Diagram of optics; (b) transverse excitation assembly. This page: actual photograph.

conveniently mounted in a goniometer for precise orientation of the crystallographic axes with respect to the direction of polarization of the laser beam.

Methods and devices for mounting a Raman cell in the sample illuminator will vary from one instrument to another. Details of procedure for optimization of the Raman scattering are usually described for each instrument. The proper positioning of a Raman cell in the focus of the laser beam may be the decisive factor in obtaining a spectrum of satisfactory quality.

B. SAMPLE-HANDLING PROCEDURES

Although infrared and Raman spectra yield the same kind of molecular information, the methods differ radically in experimental approach and consequently in sample-handling procedures. Frequently the chemical purity and homogeneity of the sample must be more carefully controlled for Raman spectroscopy. Physical properties of the sample, such as color, molecular weight, and the like, must also be taken into account in preparing a sample so that its Raman spectrum can be recorded.

In other respects, sample-handling for Raman spectroscopy is simpler than for the infrared. Sample cells, for example, are ordinarily made of glass, thus greatly simplifying the handling of aqueous samples. Solid samples, moreover, need not be dispersed in a mull or pressed into a pellet and therefore may be recovered intact and unchanged after spectra have been recorded.

Details of sample preparation for Raman spectroscopy may also depend on the intensity of the exciting radiation and on its wavelength. The following discussion refers primarily to laser-Raman spectroscopy. Experimental techniques for Raman spectroscopy in general are discussed in several places (Harrison *et al.*, 1948; Evans, 1963; Jones *et al.*, 1965; Ferraro, 1967).

1. General Requirements

In order to obtain a satisfactory Raman spectrum of a particular material, it is necessary that interaction of the radiation with the sample by such other mechanisms as absorption, fluorescence, and Tyndall scattering be minimized.

Absorption of the incident radiation is eliminated by the choice of a wavelength for exciting the Raman spectrum which is removed from any absorption bands of the sample. Fluorescence may also be avoided by appropriate choice of the excitation wavelength. Usually, the fluorescence observed in Raman spectra of biological materials is not due to the biomolecules themselves but to organic contaminants. Such contaminants are often "burned up" after prolonged exposure to the source.

In other cases, the addition of a "fluorescence quencher," such as potassium iodide, may help to reduce fluorescence.

In many cases only a single wavelength, such as HeNe 6328 Å, is available to excite a Raman spectrum. If a sample is encountered which absorbs or fluoresces near this wavelength, then it may not be possible to obtain a satisfactory Raman spectrum under these circumstances.

Tyndall scattering is due to the presence in the sample cell of suspended particles, such as air bubbles, dust, colloids, or other undissolved matter, with particle size comparable to or greater than that of the excitation wavelength. This type of scattering is mostly eliminated by careful filtration or centrifugation of the sample.

Because the Raman effect is intrinsically very weak, every effort should be made to prepare the sample for optimum Raman scattering. Several excitation geometries have been described which yield efficient Raman scattering for relatively small quantities of sample (Hawes *et al.*, 1967; Bailey *et al.*, 1967).

In contrast to modern infrared spectrophotometers, which operate as double-beam instruments, Raman spectrophotometers operate essentially in a single-beam mode. Consequently, it is not possible to compensate for Raman scattering of solvents, impurities, and the like, by differential spectroscopy. Spectral interference from a solvent is compensated by running a spectrum of the pure solvent separately and then subtracting its spectrum from that of the solution. For quantitative applications, it is usually necessary to make use of an internal standard (Jones *et al.*, 1965) or to normalize all Raman intensities to a single Raman line of reliably constant intensity (Lord and Thomas, 1967a). Quantitative Raman spectroscopy is therefore somewhat more cumbersome than quantitative infrared spectroscopy.

The intensity of absorption of infrared radiation is known to follow the Beer–Lambert law. The intensity of Raman scattering, on the other hand, is not simply related to the optical path [Eq. (12)]. For a given cell assembly and excitation geometry, however, the intensity of Raman scattering is usually found to be proportional to the volume concentration of scattering molecules. This proportionality factor is often referred to as the "scattering coefficient." A more detailed discussion of the subject of quantitative Raman intensities is given by several authors (Bernstein and Allen, 1955; Rea, 1959, 1962; Tunnicliff and Jones, 1962; Jones *et al.*, 1965).

2. Aqueous Solutions

One of the truly great advantages of Raman spectroscopy for the study of biological molecules is that aqueous solutions may be investigated with little or no solvent interference over most of the vibrational

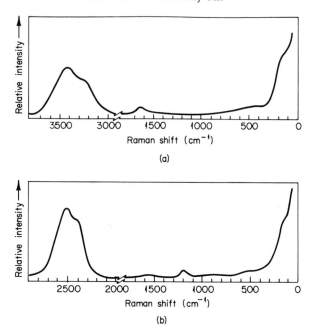

FIG. 23. Raman spectra of liquid H_2O (a) and D_2O (b). With the permission of the American Institute of Physics.

spectrum. This is shown in Fig. 23 where Raman spectra of H_2O and D_2O are presented. Liquid water gives rise to only weak Raman scattering in the region 2000–200 cm^{-1}. Thus one can obtain considerably more information regarding the effects of isotopic exchange than is possible by infrared absorption spectroscopy. For aqueous solutions, therefore, the Raman effect would appear to be a more powerful method for investigating molecular structure and associative equilibria.

3. Solutions of Low-Molecular Weight Solutes

We are concerned here with aqueous and nonaqueous solutions of monomeric constituents of polynucleotides, polypeptides, or polysacharrides as well as steroids, vitamins, hormones, and the like. Molecular weights of such solutes may range up to 1000 daltons. However, solutions of such materials are generally of low viscosity and are far simpler to handle for purposes of Raman spectroscopy than are solutions of polymers, oligomers, or ordered aggregates of monomers which are discussed later.

The concentration of solute required to obtain a satisfactory spectrum will depend upon the intensity of the Raman lines associated with the vibrational transitions being observed (i.e., on the magnitude of the polarizability change or "scattering coefficient" associated with each Raman line). Vibrations of multiply bonded groups of atoms (e.g., $C=C$, $C=N$, $C\equiv C$, etc., . . .) generally give rise to more intense Raman lines than those of singly bonded groups. As a general rule, most Raman transitions of solute molecules can be observed if a solution containing 0.5 $M\%$ solute is prepared. For example, a solution of uridine, approximately 0.25 M in H_2O, will be of sufficient concentration to allow most Raman lines of uridine to be detected with a tolerable signal-to-noise ratio $(S:N)$ under normal operating conditions of spectral slit width, scan speed, and amplifier response time. More concentrated solutions will generally permit a "better" spectrum to be recorded (greater $S:N$), while more dilute solutions may result in the failure to detect the weaker Raman lines over the level of background noise. Experience must be the spectroscopist's ultimate guide in these matters.

The total volume of solution required to fill a glass capillary cell (1-mm bore) for most laser-Raman spectrometers is about 0.05 ml (50 μl). The use of cells of larger capacity does not appreciably increase the quality of Raman spectra except when very high spectral resolution is required. Cells smaller than the 1-mm glass capillary tube may give adequate spectra for pure liquids but not for dilute solutions in most cases. A further discussion of microsampling techniques is given by Bailey et al. (1967) and Freeman and Landon (1969).

Before filling the sample cell, the solution should be carefully filtered or centrifuged to remove any suspended particles. Millipore membrane filters, adaptable to syringes, offer a convenient filtration-loading method. For spectrometers employing axial excitation and transverse viewing of cells (Bailey et al., 1967), considerable care must be exercised to prevent trapping of air bubbles throughout the length of the cell. For spectrometers employing transverse excitation and transverse viewing of the cell, only the volume of liquid on which the laser beam impinges must be kept free of trapped air (Fig. 22b).

A clear solution will not be appreciably heated by the laser beam. However in practice solutions which are optically homogeneous are difficult to prepare and load into the cell, and some heating of the sample may result from scattering by suspended particles. The actual temperature of a sample which is being irradiated by the laser beam is difficult to measure accurately, particularly while the spectrum is being recorded. The approximate temperature can usually be deduced from repeated trial measurements with a thermocouple placed adjacent to the focus of the laser beam.

Further details of sample handling for aqueous solutions of low-molecular weight solutes are described by Lord and Thomas (1967b). General techniques for solution spectroscopy are described by Jones *et al.* (1965).

4. Solutions of High-Molecular Weight Solutes

Polymers and other materials which form solutions of high viscosity pose formidable problems in sample-handling for Raman spectroscopy. The major difficulty is that of loading a cell with a viscous fluid or gel without loss of optical homogeneity. The high molecular weight solutes rarely give clear solutions, so that Tyndall scattering becomes a severe problem.

Prior to the advent of laser-Raman spectroscopy, it was virtually impossible to obtain spectra of viscous solutions, e.g., like those of aqueous native-DNA. Only little progress has been made to date in obtaining spectra of such solutions and very few Raman studies of viscous biopolymer solutions have been reported (Section III.C).

Many commercially available biopolymers are of inadequate purity for Raman spectroscopy of their aqueous solutions. In some cases, chromatographic purification may be required.

Sample-handling procedures for the preparation of selected proteins for Raman study are discussed by Yu (1969). A general discussion of laser-Raman spectroscopy of polymers is given by Schaufele (1969).

5. Solids

In the early days of Raman spectroscopy, few solids could be investigated with ease. The more intense laser sources now available permit the routine examination of solids of many types, including high molecular weight (hydrocarbon) polymers. Unlike infrared spectroscopy of solids, grinding of the sample and its preparation as a mull or pellet are not required. Crystalline and amorphous solids containing particles of varied sizes may be investigated directly. The sample is mounted on a suitable holder which is placed directly in the laser beam.

Special sample-handling techniques for several types of solid materials are discussed by Schaufele (1967, 1969). In the case of biopolymer samples, care must be taken to prevent overheating or photodecomposition of the solid by the laser beam. Polynucleotides, for example, are quickly "burned up" by direct exposure to a 70 mW HeNe 6328 Å laser beam. Oriented films of polynucleotides and polypeptides would also be difficult to examine by Raman spectroscopy.

C. BIOLOGICAL APPLICATIONS OF RAMAN SPECTROSCOPY

In the period immediately following the discovery of the Raman effect in 1928, Raman spectroscopy was very widely used by chemists because of the simplicity of photographic spectroscopy compared to the infrared procedures of that time. When automatically recording infrared instruments became generally available in the late 1940's, the situation was reversed and the Raman effect was relatively little used. Today, however, technical advances in spectrometers and sources, particularly the advent of lasers, have brought Raman spectroscopy once more into competition with infrared methods. Consequently, most chemical applications of Raman spectroscopy have been carried out in the periods pre-1950 and post-1965.

1. Early Applications

One of the first large collections of Raman spectra was compiled by Kohlrausch in his classic "Ramanspektren" (1943). Spectra of many heterocycles and other bio-organic molecules are included in this work. Unfortunately, however, by today's standards, many of the Raman spectra must be considered incomplete, with possible inaccuracies in the vibrational frequencies reported. An early collection of Raman spectral data by Hibben (1939) suffers from the same limitations. A review of these and other very early applications of Raman spectroscopy is given by Loofbourow (1940), who discussed the suitability of the Raman method for the study of biophysical problems.

The first detailed Raman investigations of biological molecules were those of Edsall and collaborators, who used the Raman method as a probe of amino acid structures and interactions in aqueous media. In a series of papers, they studied the ionization of amino acid carboxyl groups (Edsall, 1936, 1937c) and amino groups (Edsall, 1937b), structures of guanidine and urea derivatives (Edsall, 1937a, 1938), spectral effects of deuterium substitution in amino acid functional groups (Edsall and Scheinberg, 1940), and the structures of a number of specific amino acids (Edsall, 1943; Edsall et al., 1950).

Detailed Raman studies of histidine and related imidazoles (Garfinkel and Edsall, 1958a), of the ionization and deuteration of glycine and alanine (Takeda et al., 1958), of oligopeptides and lysozyme (Garfinkel and Edsall, 1958c), and of several other amino acids derived from proteins (Garfinkel, 1958) were also reported. Considerable attention is given in these publications to experimental technique. The excellent quality of the spectra is apparent in the large numbers of Raman lines detected. The study of lysozyme and polylysine (Garfinkel and Edsall, 1958b) represents the first published Raman study of biopolymers.

Raman spectra of several sugars, in the solid state and in aqueous solutions, were obtained by Spedding and Stamm (1942) and Raman spectra of fats by Yvernault (1946). A study of cis-trans isomerism in unsaturated fatty acids is given by Dupont and Yvernault (1945).

Goguadze (1950) applied Raman spectroscopy to the study of steroids. A large number of Raman studies of steroids, fats, and oils have been carried out by Russian investigators. Some of these applications are included in the bibliography of Jones and Jones (1965).

2. Recent Applications

A new era of Raman spectroscopy began to evolve circa 1960 with the introduction of a high performance, photoelectric-recording instrument by Applied Physics Corporation (Cary Model-81, Raman spectrophotometer). The subsequent development of laser sources promises the eventual use of Raman spectroscopy as a routine analytical technique, very likely on a par with modern infrared spectroscopy.

Some of the first Raman studies of this period were those of Jones and collaborators in steroids and related molecules. A number of steroids containing the unsaturated lactone ring were examined in CCl_4, $CHCl_3$, and CS_2 solutions (Jones et al., 1959). A comparison of the results of infrared and Raman studies of steroids (Jones et al., 1962) and a discussion of the experimental techniques for Raman applications (Jones, 1962b) indicate the usefulness of the Raman method for identification of steroids. Other bio-organic molecules have also been examined (Jones and Ripley, 1964).

Elson and Edsall (1962) have determined the ionization constants of cysteine and thioglycolic acid in a quantitative application of Raman spectroscopy. The effects of ionization of amino groups on the C—H stretching frequencies observed in the Raman effect are also discussed (Ghazanfar et al., 1964a). Glycine and its carbon-deuterated derivatives have been examined in detail (Ghazanfar et al., 1964b).

In other studies of amino acids and derivatives, Krishnan and co-workers have obtained Raman spectra of single crystals of asparagine monohydrate (Krishnan and Krishnan, 1962) and of single crystals of glycine (Balasubramanian et al., 1962). A Raman study of hydrogen bonding in amino acids and its effect on NH stretching frequencies has also been undertaken (Krishnan and Krishnan, 1964). Krishnan and Plane (1967) recently investigated the formation of complexes between glycine and divalent zinc, beryllium, and cadmium ions using Raman spectroscopy.

The feasibility of obtaining Raman spectra of nucleotides and derivatives in aqueous solution has been discussed by Malt (1966). Spectra of a large number of ribonucleotides, ribonucleosides, and related purine and pyrimidine derivatives in H_2O and D_2O were published by Lord and Thomas (1967b) who investigated tautomeric structures and ionizations of the bases. In a related study (Lord and Thomas, 1967a), Raman spectra of aqueous mixtures of complementary purine and pyrimidine nucleosides were examined and a quantitative study of the cytidine-mercury ion complex was made. A number of adenosine phosphates were also examined by Rimai *et al.* (1969).

The first Raman spectrum of a polynucleotide was obtained from aqueous solutions of polyriboadenylic acid in the neutral (single-stranded) form (Thomas, 1967). A more detailed spectrum obtained by laser excitation was later published by Lord and Thomas (1968). Yu (1969) investigated both neutral and acidic forms of poly A. Hirano (1968) made a preliminary study of the Raman spectrum of DNA; however only a few broad lines were detected and these may have been due to sample fluorescence. The first detailed Raman spectrum of a nucleic acid, ribosomal RNA, has recently been reported (Thomas, 1970).

The first detailed Raman study of a protein was that of Garfinkel and Edsall (1958b), who obtained spectra of lysozyme and polylysine in aqueous solution using Hg 4358 Å excitation and a photographic-recording spectrometer. More recently Tobin (1968) reported laser-excited (HeNe 6328 Å and Ar⁺ 5145 Å) Raman spectra of lysozyme, pepsin, and α-chymotrypsin in the solid state. Fluorescence is much less a problem in the laser excited Raman spectra. An investigation of the secondary structures of aqueous lysozyme, ribonuclease, and α-chymotrypsin using laser Raman spectroscopy has been made by Yu (1969) and by Lord and Yu (1970).

Raman spectral studies of biological molecules are now being carried out in several laboratories, including the author's. It seems likely that new techniques for obtaining high quality spectra of currently intractable biopolymers will soon be developed.

3. Applications of Special Methods

In addition to the general (vibrational and rotational) Raman effect described above (Section I), several types of special Raman and related scattering mechanisms have been described. A review is given by Porto (1968, 1969). Among these, Rayleigh and Brillouin scattering, anisotropy Raman scattering in liquids, and the resonance Raman effect have potentially useful biological applications.

The resonance Raman effect results from the use of an excitation frequency which is close to that of an upper electronic state of the scattering molecules. Some applications of the resonance Raman effect to biological systems are reviewed by Behringer (1967).

ACKNOWLEDGMENTS

I thank Prof. Richard C. Lord, Director, M.I.T. Spectroscopy Laboratory, for his many contributions to this chapter and Mrs. Marion Roberts for typing the original manuscript.

REFERENCES

Bailey, G. F., Kint, S., and Scherer, J. R. (1967). *Anal. Chem.* **39,** 1040.
Balasubramanian, K. A., Krishnan, R. S., and Iitaka, Y. (1962). *Bull. Chem. Soc. Jap.* **35,** 1303.
Barker, S. A., Bourne, E. J., and Whiffen, D. H. (1956). *Methods Biochem. Anal.* **3,** 213.
Bauman, R. P. ed. (1957). Biological applications of infrared spectroscopy. *Ann. N.Y. Acad. Sci.* **69,** 1.
Beer, M., Sutherland, G. B. B. M., Tanner, K. N., and Wood, D. L. (1959). *Proc. Roy. Soc. Ser. A* **249,** 147.
Behringer, J. (1967). *In* "Raman Spectroscopy" (H. A. Szymanski, ed.), Chapter 6. Plenum, New York.
Bellamy, L. J. (1958). "Infrared Spectra of Complex Molecules," 2nd ed. Methuen, London.
Bellamy, L. J. (1968). "Advances in Infrared Group Frequencies." Methuen, London.
Bellamy, L. J., and Owen, A. J. (1969). *Spectrochim. Acta Part A* **25,** 329.
Bellamy, L. J., and Pace, R. J. (1969). *Spectrochim. Acta Part A* **25,** 319.
Bendit, E. G. (1967). *Biopolymers* **5,** 525 (1967).
Benedict, A. A. (1956). *Ann. N.Y. Acad. Sci.* **69,** 158.
Bernstein, H. J., and Allen, G. (1955). *J. Opt. Soc. Amer.* **45,** 237.
Blout, E. R. (1960). *In* "Technique of Organic Chemistry (A. Weissburger, ed.), 3rd ed., Vol. I, Pt. II, pp. 1475–1519. Wiley (Interscience), New York.
Blout, E. R., and Abbate, M. J. (1955). *J. Opt. Soc. Amer.* **45,** 1028.
Blout, E. R., and Lenormant, H. (1953). *J. Opt. Soc. Amer.* **43,** 1093.
Blout, E. R., and Lenormant, H. (1955). *Biochim. Biophys. Acta* **17,** 325.
Bradbury, E. M., Price, W. C., and Wilkinson, G. R. (1961). *J. Mol. Biol.* **3,** 301.
Bradbury, E. M., Crane-Robinson, C., Rattle, H. W. E., and Stephens, R. M. (1967). *In* "Conformation of Biopolymers" (G. N. Ramachandran, ed.), Vol. 2, pp. 583–605. Academic Press, New York.
Brugel, W. (1962). "An Introduction to Infrared Spectroscopy" (translated by H. Katritzky). Methuen, London.
Byrne, P., and Chapman, D. (1964). *Nature (London)* **202,** 987.
Cary, H. H. (1955). Raman spectrophotometer. U.S. Patent No. 2,940, 355.
Chapman, D. (1956). "The Structure of Lipids," Chapter 4. Methuen, London.
Chapman, D. (1966). *Ann. N.Y. Acad. Sci.* **137,** 745.
Chapman, D. (1969a). Infrared spectroscopy of lipids. *Symp. Quantitative Methodology Lipid Res.* Amer. Oil Chemists Soc., Chicago, Illinois.
Chapman, D. (1969b). *Lipids* **4,** 251.
Chapman, D., and Collin, D. T. (1965). *Nature (London)* **206,** 189.
Chapman, D., and Wallach, D. F. H. (1968). *In* "Biological Membranes" (D. Chapman, ed.), Chapter 4. Academic Press, New York.
Chapman, D., Byrne, P., and Shipley, G. G. (1966). *Proc. Roy. Soc.* **290,** 115.
C.I.C. Newsletter (1961). Nos. 14 and 15. Connecticut Instrum. Div., Barnes Eng. Co., Stamford, Connecticut.

Clark, C. (1955). *In* "Physical Techniques in Biological Research" (G. Oster and A. W. Pollister, eds.), 1st ed., Vol. I, Chapter 5, pp. 308-ff. Academic Press, New York.

Colthup, J. N. (1950). *J. Opt. Soc. Amer.* **40**, 397.

Cotter, R. I., and Gratzer, W. B. (1969a). *Nature (London)* **221**, 154.

Cotter, R. I., and Gratzer, W. B. (1969b). *Eur. J. Biochem.* **8**, 352.

Dinsmore, H. L., and Edmondson, P. R. (1959). *Spectrochim. Acta* **15**, 1032.

Dobriner, K., Katzenellenbogen, E. R., and Jones, R. N. (1953). "Infrared Absorption Spectra of Steroids—An Atlas." Wiley (Interscience), New York.

Duecker, H., and Lippincott, E. R. (1964). *Rev. Sci. Instrum.* **35**, 1108.

Duncan, A. B. F. (1956). Technique of organic chemistry. *In* "Chemical Applications of Spectroscopy" (W. West, ed.), Vol. IX, pp. 187-245. Wiley (Interscience), New York.

Dupont, G., and Yvernault, F. (1945). *Bull. Soc. Chem.* **12**, 84.

Edsall, J. T. (1936). *J. Chem. Phys.* **4**, 1.

Edsall, J. T. (1937a). *J. Phys. Chem.* **41**, 133.

Edsall, J. T. (1937b). *J. Chem. Phys.* **5**, 225.

Edsall, J. T. (1937c). *J. Chem. Phys.* **5**, 508.

Edsall, J. T. (1938). *Cold Spring Harbor Symp. Quant. Biol.* **6**, 40.

Edsall, J. T. (1943). *J. Amer. Chem. Soc.* **65**, 1767.

Edsall, J. T., and Scheinberg, H. (1940). *J. Chem. Phys.* **8**, 520.

Edsall, J. T., Otvos, J. W., and Rich, A. (1950). *J. Amer. Chem. Soc.* **72**, 474.

Elliott, A. Hanby, W. E., and Malcolm, B. R. (1957). *Nature (London)* **180**, 1340.

Elson, E. L., and Edsall, J. T. (1962) *Biochemistry* **1**, 1.

Evans, J. C. (1963). *In* "Infrared Spectroscopy and Molecular Structure" (M. Davies, ed.), pp. 199-225. Amer. Elsevier, New York.

Fahrenfort, J. (1961). *Spectrochim. Acta* **17**, 698.

Falk, M., Hartman, Jr., K. A., and Lord, R. C. (1962). *J. Amer. Chem. Soc.* **84**, 3843.

Falk, M., Hartman, Jr., K. A., and Lord, R. C. (1963). *J. Amer. Chem. Soc.* **85**, 387, 391.

Felsenfeld, G., and Miles, H. T. (1967). *Ann. Rev. Biochem.* **37**, 407.

Ferraro, J. R. (1967). *In* "Raman Spectroscopy" (H. A. Szymanski, ed.), Chapter 2. Plenum, New York.

Fraser, M. J., and Fraser, R. D. B. (1951). *Nature (London)* **167**, 761.

Fraser, R. D. B. (1960). *In* "Analytical Methods of Protein Chemistry" (P. Alexander and R. J. Block, eds.), Vol. 2. Pergamon, New York.

Fraser, R. D. B., Harrop, B. S., Ledger, R., MacRae, T. P., Stewart, F. H. C., and Suzuki, E. (1967). *Biopolymers* **5**, 797.

Freeman, N. K. (1957). *Ann. N.Y. Acad. Sci.* **69**, 131.

Freeman, S. K., and Landon, D. O. (1969). *Anal. Chem.* **41**, 398.

Fritzsche, H. (1967). *Biopolymers* **7**, 863.

Fukushima, K., Ideguchi, Y., and Miyazawa, T. (1963). *Bull. Chem. Soc. Jap.* **36**, 1301.

Furedi, H., and Walton, A. G. (1968). *Appl. Spectrosc.* **22**, 23.

Garfinkel, D. (1958). *J. Amer. Chem. Soc.* **80**, 3827.

Garfinkel, D., and Edsall, J. T. (1958a). *J. Amer. Chem. Soc.* **80**, 3807.

Garfinkel, D., and Edsall, J. T. (1958b). *J. Amer. Chem. Soc.* **80**, 3818.

Garfinkel, D., and Edsall, J. T. (1958c). *J. Amer. Chem. Soc.* **80**, 3823.

Ghazanfar, S. A. S., Edsall, J. T., and Myers, D. V. (1964a). *J. Amer. Chem. Soc.* **86**, 559.

Ghazanfar, S. A. S., Myers, D. V., and Edsall, J. T. (1964b). *J. Amer. Chem. Soc.* **86**, 3439.

Goguadze, V. P. (1950). *Izv. Akad. Nauk SSSR Otd. Khim. Nauk* p. 185.

Gore, R. C., Barnes, R. B., and Peterson, E. (1949). *Anal. Chem.* **21**, 382.

Goulden, J. D. S. (1959). *Spectrochim. Acta* **15**, 657.

Goulden, J. D. S., and Manning, J. D. (1964). *Nature (London)* **203**, 403.

Gratzer, W. B., Bailey, E, and Beaven, G. H. (1967). *Biochem. Biophys. Res. Commun.* **28**, 914.

Gratzer, W. B., Beaven, G. H., Rattle, H. W. E., and Bradbury, E. M. (1968). *Eur. J. Biochem.* **3**, 276.

Hadzi, D. (1963). *In* "Infrared Spectroscopy and Molecular Structure" (M. Davies, ed.), pp. 226–269. Amer. Elsevier, New York.

Hallam, H. E. (1963). *In* "Infrared Spectroscopy and Molecular Structure" (M. Davies, ed.), pp. 405–440. Amer. Elsevier, New York.

Hamlin, Jr., R. M., Lord, R. C., and Rich, A. (1965). *Science* **148**, 1734.

Hanlon, S., and Koltz, I. M. (1968). *Develop. Appl. Spectrosc.* **6**, 219.

Harada, I., and Lord, R. C. (1970). *Spectrochim. Acta* (in press).

Harrick, N. J. (1967). "Internal Reflection Spectroscopy." Wiley (Interscience), New York.

Harrington, W. F., Josephs, R., and Segal, D. M. (1966). *Annu. Rev. Biochem.* **35**, 599.

Harrison, G. R., Lord, R. C., and Loofbourow, J. R. (1948). "Practical Spectroscopy." Prentice-Hall, Englewood Cliffs, New Jersey.

Hartman, Jr., K. A., and Rich, A. (1965). *J. Amer. Chem. Soc.* **87**, 2033.

Hartman, Jr., K. A., and Thomas, Jr., G. J. (1970). *Science* **170**, 740.

Hawes, R. C., George, K. P., Nelson, D. C., and Beckwith, R. (1966). *Anal. Chem.* **38**, 1842.

Hawes, R. C., Sloane, H. J., and Haber, H. S. (1967). *Eur. Congr. Mol. Spectrosc., 9th, Madrid, September 1967*. (Reprint available from Cary Instruments, Monrovia, California.)

Herman, T. S. (1965). *Anal. Biochem.* **12**, 406.

Herzberg, G. (1945). "Infrared and Raman Spectra of Polyatomic Molecules." Van Nostrand, Princeton, New Jersey.

Hibben, J. H. (1939). "The Raman Effect and Its Chemical Applications." Reinhold, New York.

Hirano, K. (1968). *Bull. Chem. Soc. Jap.* **41**, 731.

Hvdit, A., and Nielson, S. O. (1966). *Advan. Protein Chem.* **21**, 288.

Itoh, K., Nakahara, T., Shimanouchi, T., Oya, M. Uno, K., and Iwakura, Y. (1968). *Biopolymers* **6**, 1759.

Itoh, K., Shimanouchi, I., and Oya, M. (1969). *Biopolymers* **7**, 649.

Isemura, T., Okabayashi, H., and Sakakibara, S. (1968). *Biopolymers* **6**, 307.

Johnson, R. D. (1965). *Anal. Chem.* **38**, 160.

Jones, R. N. (1962a). *Can J. Chem.* **40**, 321.

Jones, R. N. (1962b). *Proc. Int. Symp. Mol. Struct. Spectrosc. 1962,* p. A218.

Jones, R. N., and Jones, M. K. (1965). *Anal. Chem.* **38**, 393R.

Jones, R. N., and Nadeau, A. (1958). *Spectrochim. Acta* **12**, 183.

Jones, R. N., and Ripley, R. A. (1964). *Can. J. Chem.* **42**, 305.

Jones, R. N., and Sandorfy, C. (1956). Technique of organic chemistry. *In* "Chemical Applications of Spectroscopy" (W. West, ed.), pp. 247–580. Wiley (Interscience), New York.

Jones, R. N., Angell, C. L., Ito, T., and Smith, R. J. D. (1959). *Can. J. Chem.* **37**, 2007.

Jones, R. N., DiGiorgio, J. B., Elliot, J. J., and Nonnenmacher, G. A. A. (1965). *J. Org. Chem.* **30**, 1822.

Jones, R. N., Krueger, P. J., Noack, K., Elliot, J. J., Ripley, R. A., Nonnenmacher, G. A. A., and DiGiorgio, J. B. (1962). *Proc. Colloq. Spectrosc. Int. 10th,* p. 461.

Kapany, N. S., and Silbertrust, N. (1964). *Nature (London)* **204**, 138.

Katlafsky, B., and Keller, R. E. (1963). *Anal. Chem.* **35**, 1665.

Katritzky, A. R., and Ambler, A. P. (1963). Physical methods in heterocyclic chemistry. "Spectroscopic Methods" (A. R. Katritzky, ed.), Vol. II, Chapter 10. Academic Press, New York.

King, G. W. (1964). "Spectroscopy and Molecular Structure." Holt, New York.

Kogelnik, H., and Porto, S. P. S. (1963). *J. Opt. Soc. Amer.* **53**, 1446.

Kohlrausch, K. W. F. (1943). "Ramanspektren." Beker and Erler, Leipzig.

Koningstein, J. A. (1967). *In* "Raman Spectroscopy" (H. S. Szymanski, ed.), Chapter 6. Plenum, New York.

Koyama, Y., and Shimanouchi, T. (1968). *Biopolymers* **6**, 1037.

Krimm, S. J. (1962). *J. Mol. Biol.* **4**, 528.

Krimm, S. (1963). *In* "Infrared Spectroscopy and Molecular Structure" (M. Davies, ed.), pp. 270–310. Amer. Elsevier, New York.

Krimm, S. (1967). *Nature (London)* **212**, 1482.

Krimm, S. and Kuroiwa, K. (1968). *Biopolymers* **6**, 401.

Krimm, S., Kuriowa, K., and Rebane, T. (1967). *In* "Conformation of Biopolymers" (G. N. Ramachandran, ed.), Vol. 2, pp. 439–447. Academic Press, New York.

Krishnan, K., and Krishnan, R. S. (1962). *Proc. Indian Acad. Sci. Sect. A* **55**, 153.

Krishnan, K., and Plane, R. A. (1967). *Inorg. Chem.* **6**, 55.

Krishnan, R. S., and Krishnan, K. (1964). *Proc. Indian Acad. Sci. Sect. A* **60**, 11.

Kuechler, E., and Derkosch, J. (1965). *Z. Naturforsch.* **21**, 209.

Kyogoku, Y., and Yu, B. S. (1968). *Bull. Chem. Soc. Jap.* **41**, 1742.

Kyogoku, Y., and Yu, B. S. (1959). *Bull. Chem. Soc. Jap.* **42**, 1387.

Kyogoku, Y., Tsuboi, M., Schimanouchi, T., and Watanabe, I. (1961a). *Nature (London)* **189**, 120.

Kyogoku, Y., Tsuboi, M., Shimanouchi, T., and Watanabe, I. (1961b). *J. Mol. Biol.* **3**, 741.

Kyogoku, Y., Tsuboi, M., Shimanouchi, T., and Watanabe, I. (1962). *Nature (London)* **195**, 459.

Kyogoku, Y., Lord, R. C., and Rich, A. (1966). *Science* **154**, 518.

Kyogoku, Y., Lord, R. C., and Rich, A. (1967a). *J. Amer. Chem. Soc.* **89**, 496.

Kyogoku, Y., Lord, R. C., and Rich, A. (1967b). *Proc. Nat. Acad. Sci. U.S.* **57**, 250.

Kyogoku, Y., Lord, R. C., and Rich, A. (1968). *Nature (London)* **218**, 69.

Kyogoku, Y., Lord, R. C., and Rich, A. (1969). *Biochim. Biophys. Acta* **179**, 10.

Leite, R. C. C., and Porto, S. P. S. (1964). *J. Opt. Soc. Amer.* **54**, 981.

Lohr, L. J., and Kaier, R. J. (1960). *Anal. Chem.* **32**, 301.

Loofbourow, J. R. (1940). *Rev. Mod. Phys.* **12**, 267.

Lord, R. C., and Miller, F. A. (1956). *Appl. Spectrosc.* **10**, 115.

Lord, R. C., and Porro, T. J. (1960). *Z. Elektrochem.* **64**, 672.

Lord, R. C., and Thomas, Jr., G. J. (1967a). *Biochim. Biophys. Acta* **142**, 1.

Lord, R. C., and Thomas, Jr., G. J. (1967b). *Spectrochim. Acta Part A* **23**, 2551.

Lord, R. C., and Thomas, Jr., G. J. (1968). *Develop. Appl. Spectrosc.* **6**, 179.

Lord, R. C., and Yu, N. (1970). *J. Mol. Biol.* **50**, 509.

Lord, R. C., McDonald, R. S., and Miller, F. A. (1952). *J. Opt. Soc. Amer.* **42**, 149.

Malt, R. A. (1966). *Biochim. Biophys. Acta* **120**, 461.

Martin, A. E. (1963). *In* "Infrared Spectroscopy and Molecular Structure" (M. Davies, ed.), pp. 22–84. Amer. Elsevier, New York.

Mason, W. B. (1958). "Infrared Microsempling in Bio-Medical Investigations." (Reprint available from Perkin-Elmer Corp., Norwalk, Connecticut.)

Masuda, Y., and Miyazawa, T. (1967). *Makromol. Chem.* **103**, 261.

Masuda, Y., and Miyazawa, T. (1969). *Bull. Chem. Soc. Jap.* **42**, 570.

May, L., and Grenell, R. G. (1956). *Ann. N.Y. Acad. Sci.* **69**, 171.

Miles, H. T. (1958). *Biochim. Biophys. Acta* **30**, 324.

Miles, H. T. (1959). *Biochim. Biophys. Acta* **35**, 274.

Miles, H. T. (1960). *Biochim. Biophys. Acta* **45**, 196.

Miles, H. T. (1961). *Proc. Nat. Acad. Sci. U.S.* **47**, 791.

Miles, H. T. (1968). *Methods Enzymol. Part B* **12**, 256–267.

Miles, H. T., and Frazier, J. (1964a). *Biochem. Biophys. Res. Commun.* **14**, 21, 129.

Miles, H. T., and Frazier, J. (1964b). *Biochim. Biophys. Acta* **7**, 216.

Miyazawa, T. (1955). *Nippon Kagaku Zasshi* **76**, 341.

Miyazawa, T. (1956). *Nippon Kagaku Zasshi* **77**, 171.

Miyazawa, T. (1960). *J. Chem. Phys.* **32**, 1647.

Miyazawa, T. (1962). *In* "Polyamino Acids, Polypeptides and Proteins" (M. A. Stahmann, ed.), pp. 201–217. Univ. of Wisconsin Press, Madison, Wisconsin.

Miyazawa, T., and Blout, E. R. (1961). *J. Amer. Chem. Soc.* **83**, 712.

Miyazawa, T., Shimanouchi, T., and Mizushima, S. (1958). *J. Chem. Phys.* **29**, 611.

Miyazawa, T., Masuda, Y., and Fukushima, K. (1962). *J. Polym. Sci.* **52**, S62.

Miyazawa, T., Fukushima, K., Sugano, S., and Masuda, Y. (1967). *In* "Conformation of Biopolymers" (G. N. Ramachandran, ed.), Vol. 2, pp. 557–568. Academic Press, New York.

Morikawa, K., Tsuboi, M., Kyogoku, Y., Seno, T., and Nishimura, S. (1969). *Bull. Chem. Soc. Jap.* **223**, 537.

Neudert, W., and Ropke, H. (1965). "Atlas of Steroid Spectra" (translated by J. B. Leane). Springer, New York.

Norris, K. P. (1961). *Advan. Spectrosc.* **2**, 293.

Parker, F. S. (1960). *Biochim. Biophys. Acta* **42**, 513.

Parker, F. S. (1963). *Nature (London)* **200**, 1093.

Parker, F. S. (1968). *Develop. Appl. Spectrosc.* **6**, 237.

Pimentel, G. C., and McClellan, A. L. (1960). "The Hydrogen Bond," Chapter 3. Reinhold, New York.

Placzek, G. (1934). "Handbuch der Radiologie," Vol. VI, Pt. 2. Akad. Verlagsges., Leipzig.

Porto, S. P. S. (1968). "The Spex Speaker," Vol. XIII, No. 2. Obtainable from Spex Ind., Inc., Metuchen, New Jersey.

Porto, S. P. S. (1969). "The Spex Speaker," Volume XIV, No. 2.

Potts, Jr., W. J. (1963). Techniques. "Chemical Infrared Spectroscopy," Vol. I. Wiley, New York.

Rea, D. G. (1959). *J. Opt. Soc. Amer.* **49**, 90.

Rea, D. G. (1962). *J. Mol. Spectrosc.* **4**, 507.

Rimai, L., Cole, T., Parsons, J. L., Hickmott, Jr., J. T, and Carew, E. B. (1969). *Biophys. J.* **9,** 320.

Roberts, G., Gallagher, B. S., and Jones, R. N. (1958). "Infrared Absorption Spectra. of Steroids—An Atlas." Wiley (Interscience), New York.

Rosenkrantz, H. (1955). *Methods Biochem. Anal.* **2,** 1.

Rosenkrantz, H. (1957). *Methods Biochem. Anal.* **5,** 407.

Rouser, G., Kritchevsky, G., Heller, D., and Lieber, E. (1963). *J. Amer. Oil Chem. Soc.* **40,** 425.

Sato, T., Kyogoku, Y., Higuchi, S., Mitsui, Y., Iitaka, Y., Tsuboi, M., and Miura, K. (1966). *J. Mol. Biol.* **16,** 180.

Schaufele, R. F. (1969). *Macromol. Rev.* **4,** to be published.

Schaufele, R. F. (1967). *Trans. N.Y. Acad. Sci.* **30,** 69.

Schellman, J. A., and Schellman, C. (1964). *In* "The Proteins" (H. Neurath, ed.), 2nd ed., Vol. II, Chapter 7. Academic Press, New York.

Schwarz, H. P., Dreisbach, L., Childs, R., and Mastrangelo, S. V. (1957). *Ann. N.Y. Acad. Sci.* **69,** 116.

Shimanouchi, T., Tsuboi, M., and Kyogoku, Y. (1964). *Advan. Chem. Phys.* **7,** 435.

Smith, D. W., Randall, H. M., Gastambide-Odier, M. M., and Koevoet, A. L. (1956). *Ann. N.Y. Acad. Sci.* **69,** 145.

Spedding, F. H., and Stamm, R. F. (1942). *J. Chem. Phys.* **10,** 176.

Susi, H. (1969). *In* "Structure and Stability of Biological Macromolecules" (S. N. Timasheff and G. D. Fasman, eds.), Chapter 7. Dekker, New York.

Susi, H., Timasheff, S. N., and Stevens, L. (1967). *J. Biol. Chem.* **242,** 5460.

Sutherland, G. B. B. M. (1952). *Advan. Protein Chem.* **7,** 291.

Sutherland, G. B. B. M., and Tsuboi, M. (1957). *Proc. Roy. Soc. Ser. A* **239,** 446.

Suzuki, S., Iwashita, Y., Shimanouchi, T., and Tsuboi, M. (1966). *Biopolymers* **4,** 337.

Takeda, M., Iavazzo, R. E. S., Garfinkel, D., Scheinberg, I. H., and Edsall, J. T. (1958). *J. Amer. Chem. Soc.* **80,** 3813.

Timasheff, S. N., and Gorbunoff, M. J. (1967). *Annu. Rev. Biochem.* **37,** 13.

Thomas, Jr., G. J. (1967). Ph.D. Thesis, Dept. of Chem., M.I.T., Cambridge, Massachusetts.

Thomas, Jr., G. J. (1969). *Biopolymers* **7,** 325.

Thomas, Jr., G. J. (1970). *Biochim. Biophys. Acta* **213,** 417.

Thomas, Jr., G. J., and Spencer, M. (1969). *Biochim. Biophys. Acta* **179,** 360.

Timasheff, S. N., and Susi, H. (1966). *J. Biol. Chem.* **241,** 249.

Timasheff, S. N., Susi, H., and Stevens, L. (1967a). *J. Biol. Chem.* **242,** 5467.

Timasheff, S. N., Susi, H., Townend, R., Stevens, L., Gorbunoff, M. J., and Kumosinski, T. F. (1967b). *In* "Conformation of Biopolymers" (G. N. Ramachandran, ed.), Vol. 1, pp 173–196. Academic Press, New York.

Tobin, M. C. (1968). *Science* **161,** 68.

Tsuboi, M. (1951). *Bull. Chem. Soc. Jap.* **24,** 75.

Tsuboi, M. (1962). *J. Polym. Sci.* **59,** 139.

Tsuboi, M. (1964a). *Biopolym. Symp.* **n1,** 527.

Tsuboi, M. (1964b). *J. Polym. Sci. Part C* **7,** 125.

Tsuboi, M. (1968). *Bull. Chem. Soc. Jap.* **41,** 1821 (1968).

Tsuboi, M., Shuto, K., Takemura, S., and Nishimura, S. (1969). *Bull. Chem. Soc. Jap.* **42,** 102.

Tunnicliff, D. D., and Jones, A. C. (1962). *Spectrochim. Acta* **18,** 579.

Walrafen, G. E. (1964). *J. Chem. Phys.* **40,** 3249.

Wilkinson, G. R., Price, W. C., and Bradbury, E. M. (1959). *Spectrochim. Acta* **14,** 284.

Wilks, P. A., and Hirschfeld, T. (1967). *Appl. Spectrosc. Rev.* **1,** 99.

Wilson, Jr., E. B., Decius, J. C., and Cross, P. C. (1955). "Molecular Vibrations." McGraw-Hill, New York.

Yu, N. (1969). Ph.D. Thesis, Dept. of Chem., M.I.T., Cambridge, Massachusetts.

Yvernault, T. (1946). *Oleagineux* **1,** 189.

Zbinden, R. (1964). "Infrared Spectroscopy of High Polymers." Academic Press, New York.

Zechmeister, L. (1962). "Cis-Trans Isomeric Carotenoids, Vitamins A and Aryl-polyenes." Academic Press, New York.

CHAPTER 5

Optical Rotatory Dispersion and Circular Dichroism

C. ALLEN BUSH

I. Introduction

A. BACKGROUND

1. Historical

Although it is a relative newcomer to research in biology, optical activity has a history stretching as far back as that of organic chemistry itself. Optical rotation and its dependence on wavelength were first described in 1815 to 1817 by Biot. It was Pasteur who, in 1848 to 1860, demonstrated the concept of molecular dissymetry as the origin of optical activity. Throughout the latter part of the century, the ideas of the tetrahedral carbon atom, on which much of modern organic chemistry is based, were developed with the aid of optical rotation. The chemists of that era owe a debt to Bunsen, who in 1866 introduced a simple source of monochromatic radiation in order to overcome the troublesome dependence of the rotation on the wavelength of the light. On the other hand, the confinement of measurements to the wavelength of the sodium emission at 5893 Å was a constraint in the development of optical rotatory dispersion (ORD) as a powerful technique for studying molecular conformation.

The fact that chemists concentrated their attention on rotation at the sodium D line allowed them to overlook the results of Drude and Cotton in 1895 to 1905 which showed the relationship of ORD to absorption of light. Lowery was the first chemist to realize that there was much more chemical information in the ORD than in simple measurements at the sodium D line. His book remains from its 1935 publication as one of the best surveys of the field (Lowery, 1935).

Truly extensive use of ORD in organic chemistry was made possible by the introduction of the commercial photoelectric polarimeter in the 1950's. The experimental techniques developed in that era are quite extensive and the results have been collected by Djerassi (1960). In more recent years, the introduction of the recording spectropolarimeter and dichrograph have again extended the usefulness of the phenomenon. These modern instruments allow us to measure ORD curves in a few hours rather than days. Also, they give us considerably greater sensitivity and wavelength range than their manual predecessors and have greatly extended the usefulness of optical activity measurements to chemistry and biology.

2. Units

Optical rotation is defined as the angle of rotation of plane polarized light by a sample and is given the symbol α. For optically active molecules in isotropic solution, Biot noticed that the rotation was proportional to the path length of the sample and its concentration. Hence, he defined the specific rotation,

$$[\alpha] = \alpha/l'c' \tag{1}$$

The angles are measured in degrees, l' is the path length in decimeters, and the concentration c' is in g/cm^3. The original definition of Biot, with these rather unwieldy units, has been preserved to the present day. Likewise the molar rotation is

$$[\phi] = [\alpha]M/100 = 100\alpha/cl \tag{2}$$

where M is the molecular weight of the solute, c is the concentration in moles/liter, and l is the path length in centimeters. A factor correcting for the index of refraction of the solvent is sometimes included in published reports of the rotation. The unit of rotation $[m']$ is equal to our $[\phi]$ times $3/(n^2 + 2)$. When comparing results from different laboratories, one should make certain which units are being used.

3. Definition of ORD

The optical rotatory dispersion (ORD) is simply the dependence of $[\phi]$ on wavelength of light λ. There has always been considerable interest in the general functional form that this dependance takes. Biot offered a simple law of inverse squares, $\alpha = k/\lambda^2$, based on his qualitative measurements. He recognized certain substances departing widely from this formula as anomalous. Subsequently, more accurate investigations have led to various proposed formulas to represent the dispersion more quantitatively. Based on a model for an optically active medium consisting of electrons oscillating in helical paths, Drude proposed in 1905 that

$$[\phi(\lambda)] = \sum_m k_m/(\lambda^2 - \lambda_m^2) \tag{3}$$

The sum extends over the oscillators m of various rotatory powers k_m and characteristic wavelengths λ_m. A large number of compounds are found to follow a single-term Drude equation in regions of the spectrum where they do not absorb light.

We may get a picture of what an ORD curve looks like by plotting a single-term Drude equation (Fig. 1). The solid line represents the equation in the wavelength region where it is valid. At $\lambda = \lambda_0$, the rotation reappears with the opposite sign. We will give a more quantitative discussion of ORD curve shape below, but as a plausible construction, let us draw a smooth dotted line connecting the two branches of the curve. This dotted line shows what happens to the ORD in the region

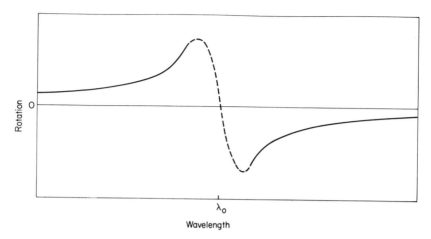

FIG. 1. ORD curve; Drude equation in the region of transparency (———————).
A plausible construction joining the two branches of the Drude curve (-------------).

of λ_0. This is the absorption region for the molecule, where the Drude equation does not apply. This reversal in the sign of the rotation at $\lambda = \lambda_0$ is called a Cotton effect and we will discuss it much more extensively below. We leave the subject of ORD for now, with the reminder that it is just this Cotton effect that gives us most of the information on molecular structure.

B. CIRCULAR POLARIZATION

1. Definition

Let us now take a second way of looking at optical rotation. This alternative point of view will allow us to introduce the concept of circular dichroism (CD) as well as provide a picturesque model for the optically active molecule.

We have been discussing plane polarized light qualitatively and we now wish to represent it mathematically. We introduce a Cartesian coordinate system in which the light beam is directed along the z axis (in the direction of unit vector $\mathbf{k}$). In this system, we may represent the electric vector of a beam traveling in the $\mathbf{k}$ direction and polarized with the electric vector along the x axis (unit vector $\mathbf{i}$) as

$$E_\mathrm{p}(\mathbf{r},t) = E_0\mathbf{i} \cos \omega(t - z/c) \qquad (4)$$

This beam has an angular frequency ω and is polarized in the $\mathbf{i}$ direction.

In order to get our second point of view on optical activity, we decompose this beam into two circularly polarized components, E_l and E_r.

$$E_{l,r} = \tfrac{1}{2}E_0[\mathbf{i} \cos \omega(t - z/c) \pm \mathbf{j} \sin \omega(t - z/c)] \tag{5}$$

We see immediately that the sum of $E_l + E_r = E_p$, the plane polarized beam. Thus the decomposition is valid. For a physical picture of the circularly polarized beams, take E_l as an example. If we pick a position on the z axis and watch the beam go by, we see that the plane of polarization rotates a full circle in every period, $\tau = 2\pi/\omega$. Likewise, if we pick a fixed time and move along the z axis in the direction of the light beam, we see that plane of polarization rotates in a clockwise direction once in every wavelength, $\lambda = 2\pi/\omega$. For E_r, the plane of polarization rotates in a counter-clockwise sense when viewed along the direction of light travel. Thus, the electric vectors, E_l and E_r describe helices in space having opposite handedness. The sum of these two oppositely circularly polarized beams is a plane polarized beam.

2. Circular Birefringence

Continuing to consider our plane polarized beam as a sum of $E_l + E_r$, imagine that the indices of refraction in the sample for E_l and E_r differ, i.e., $n_l \neq n_r$. Thus there will be a phase shift between E_l and E_r, and the plane of polarization will be rotated by the sample. The resulting optical rotation was shown by Fresnel in 1825 to be

$$\alpha = \pi l/\lambda \, (n_l - n_r) \tag{6}$$

where α is in radians. This description of optical rotation, which is completely equivalent to the one based on plane polarization, suggests the name circular birefringence as a synonym for optical rotation; that is, the two circularly polarized beams are differently refracted.

3. Model for Optically Active Molecules

We might now ask, "What is it which causes the left and right circularly polarized beams to be treated differently by the optically active medium?" In answer, let us take as a model for the optically active sample a solution of right-handed helices. The right-handed beam E_r will see something different in these molecules than does the left-handed beam E_l. The former describes a left-handed helix in space and will tend to cut across the electron paths of the right-handed helical molecule, while the latter beam describes a right-handed helix in space and moves along the right-handed helical molecules. Since these two beams are treated differently by the right-handed helical model, it is not surprising that $n_l \neq n_r$ and we observe optical rotation.

Mathematical models related to the one we have just described based on electrons constrained to move in a helical path have been used in practical calculations on certain molecules (Tinoco and Woody, 1964; Maestro *et al.*, 1967). On the other hand, most practical calculations for molecules and for polymers are based on a different molecular model, which we will discuss more extensively below (Kirkwood, 1937; Tinoco, 1962). But as a qualitative description of the optically active molecule, the electron moving on a helix is actually quite general. To any pair of optical enantiomers, we can assign a helical handedness of opposite sign. Unfortunately, it is not possible, in general, to use this helical sense to assign the Cotton effects of the molecule.

4. Circular Dichroism

Now that we have been introduced to circular polarization and circular birefringence, we may ask, "What if the optically active molecule absorbs at the wavelength of the experiment?" Since E_1 and E_r see different indices of refraction, we expect them to see different absorption coefficients also. The difference between the extinction coefficients for left- and right-handed circularly polarized light is called circular dichroism (CD)

$$\Delta\epsilon = \epsilon_1 - \epsilon_r \tag{7}$$

The CD is generally measured by passing alternately right and left circularly polarized light through the sample and measuring the difference in their optical densities (see Chapter 4).

In the literature, we find the CD reported in units of $\Delta\epsilon$, and also in units of molar ellipticity $[\theta]$. In order to understand the relationship between these two measures of CD, let us return to consideration of the rotation of the plane polarized light by an optically active sample which is also absorbing light. Plane polarized light enters the sample; the plane of polarization is rotated and, in addition, its two circularly polarized components are differently absorbed due to circular dichroism. The light which emerges is no longer truly plane polarized due to the unequal amplitudes of the two emerging circular components. Light of this type is intermediate between plane and circular polarization and is called elliptically polarized light. The ellipticity of the light θ is the ratio of the minor to the major axis of the ellipse. It is related to the difference in extinction coefficients by

$$\theta \text{ (rad)} = (2.303cl'/4) \, \Delta\epsilon \tag{8}$$

where the factor 2.303 accounts for the fact that ϵ is measured in a logarithmic base 10 rather than base e. Again, history demands that the units of length be decimeters as in the definition of rotation. Circular dichroism is often reported in units of molar ellipticity in degrees/decimeter,

$$[\theta] = \theta \quad (\text{rad})(180/\pi)(100/cl) \tag{9}$$

where the concentration is in mole/liter and l is now the path length in centimeters. From the above considerations, we see that reports in the literaure of $\Delta\epsilon$ or $[\theta]$ are equivalent, as they are simply proportional,

$$[\theta] = 3300\Delta\epsilon \tag{10}$$

5. ORD in Absorbing Media

At this point we should remark that optical activity is an extremely small effect when compared to bulk index of refraction or bulk absorption, that is, $\Delta\epsilon$ is rarely more than $10^{-4} \times \epsilon$, so that the ellipticity θ is quite small. Thus, the elliptically polarized light emerging from an optically active sample showing both ORD and CD is very nearly plane polarized. Therefore, one can still measure optical rotation even in an absorption region with no special difficulty. By the same token, in measuring the CD, as a difference in absorption for right and left circularly polarized light, one is measuring a difference of one part in 10^{-4}. Unlike most ordinary spectrophotometers, modern CD machines are able to perform this measurement with good accuracy.

C. PROPERTIES OF CD AND ORD CURVES

1. CD Curve Shape

The dependence of $[\theta]$ on the wavelength of light is best understood by comparison with the ordinary bulk absorption curve shape. The simplest curve is that of a single Cotton effect. An example which looks much like an isolated absorption band is seen in Fig. 2a. The CD falls to zero outside the absorption band and may be approximately represented by a Gaussian function. Of course, ϵ_r may be greater than ϵ_l, leading to CD bands of negative sign. In fact, one often finds a negative band adjacent to a positive one, giving rise to a double Cotton effect such as the one pictured in Fig. 2b. More complicated combinations of bands may be found, but they can generally be understood as a combination of the above cases (Tinoco, 1968).

2. Relationship of ORD to CD

The relationship of the anomaly or change in sign in the ORD curve and the appearance of a CD band in optically active compounds was described by Aimé Cotton in 1895. In honor of this discovery, we now

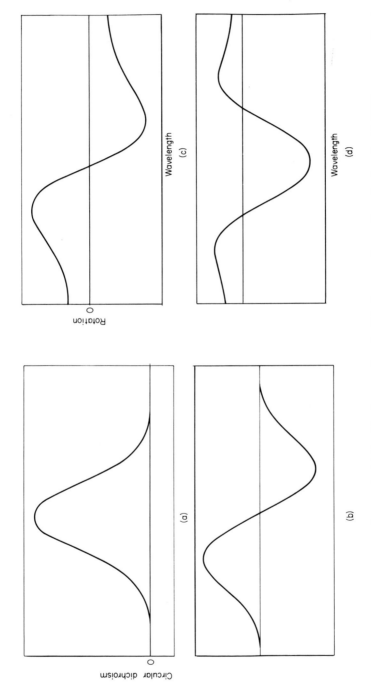

Fig. 2. (a) Single Gaussian CD band. (b) Double Cotton effect in CD, the sum of two Gaussian bands. (c) ORD corresponding to a single Gaussian CD band. (d) ORD of a double Cotton effect.

describe a maximum in the CD at $\lambda = \lambda_0$ (Fig. 2a) or a reversal in the sign of the rotation at $\lambda = \lambda_0$ (Fig. 1) as a Cotton effect at λ_0. The underlying basis for the interrelation of the CD and ORD may be found in the more general relationship between the index of refraction and the absorption. A special case of the general dispersion relation, connecting the real and imaginary parts of the complex susceptibility, is the Kronig–Kramers transform relating the CD and ORD spectra (Moffit and Moscowitz, 1959).

$$[\Phi(\lambda)] = 2/\pi \int_0^\infty \frac{[\theta(\lambda')]\lambda'\, d\lambda'}{\lambda'^2 - \lambda^2} \tag{11}$$

This equation tells us that if we know the CD spectrum $[\theta(\lambda')]$ over the range of all wavelengths, we may determine the optical rotation at any wavelength by carrying out the indicated integration. One may repeat the calculation for a number of wavelengths, yielding $[\Phi(\lambda)]$, the ORD curve. Although, in principle, it appears that one must know the CD $[\theta(\lambda')]$ throughout the entire electromagnetic spectrum from $\lambda = 0$ to $\lambda = \infty$, in practice it is possible to limit the integration to a practical range of wavelengths in the visible or ultraviolet containing a few Cotton effects (Thiery, 1968).

Likewise, when one has an ORD curve and wishes to know the CD, he may utilize the Kronig–Kramers relationship reciprocal to Eq. (11),

$$[\theta(\lambda)] = \frac{-2\lambda}{\pi} \int_0^\infty \frac{[\Phi(\lambda')]\, d\lambda'}{\lambda'^2 - \lambda^2} \tag{12}$$

This relationship is used much the same as Eq. (11) except for one difficulty. The ORD $[\Phi(\lambda')]$ does not fall to zero outside the range of absorption as does $[\theta(\lambda')]$. Thus, limiting the wavelength of integration may, in some cases, overlook contributions to the integral. Fortunately, however, $[\Phi(\lambda')]$ may be well represented by a one-term Drude equation, Eq. (3), outside the wavelength of absorption. This formula may be integrated without difficulty (Emeis et al., 1967).

There are several practical numerical schemes for interpreting ORD curves which utilize various methods for computing the Kronig–Kramers integrals. We will discuss them below in connection with specific biopolymers. In addition, the integrals can be done quite generally by numerical integration. No particular assumptions are necessary in these procedures, so they are applicable to numerous situations. There are a number of programs available for routine use and they require very modest computing facilities (Bush, 1969).

The foregoing discussion is meant to make clear that there is no real difference in the molecular information obtained in measurement of a CD or an ORD curve. Both experiments are measuring the same property, even though the experiments may seem rather different. Each measurement has its place and we will outline in some detail below situations in which one may be preferred over the other. But, in general, there is little profit in presenting both measurements side by side for a single sample without comment. Computation of the Kronig–Kramers transforms is not difficult and one should at least compare the two measurements to test for self-consistency.

3. ORD Curve Shape

A qualitative indication of the appearance of an ORD curve in the Cotton effect region has been given in Fig. 1. A more rigorous discussion has been given by Moscowitz (1960). He begins with a representation of the CD curve as a Gaussian function centered at $\lambda = \lambda_a$.

$$[\theta(\lambda)] = [\theta_a^0] \exp - (\lambda - \lambda_a)^2/\Delta_a^2 \tag{13}$$

He then shows how the Gaussian function may be integrated using Eq. (11) to give the ORD curve. The resulting ORD curve for a single Cotton effect is shown in Fig. 2c and that for a double Cotton effect in Fig. 2d. No simple functional form has been found to represent the ORD curve as well as the Gaussian curve represents the CD or absorption curve. Thus, Moscowitz has used the Kronig–Kramers transform of a Gaussian successfully to represent ORD curves. This technique has also been extended to multiple Cotton effects found in polypeptides as will be discussed below (Carver et al., 1966b).

4. Rotational Strength

We shall conclude this general introductory section with a discussion of the rotational strength. This parameter is a measure of the rotatory power of an absorption band. The primary interest in this parameter is theoretical, since it may be calculated from molecular electronic theories of optical activity as well as from experiment. The rotational strength of an electronic transition from ground state, 0, to some electronic excited state, a, is given from experiment by (Moscowitz, 1960),

$$R_{0a} = \frac{3hc}{8\pi N} \int_a \frac{[\theta(\lambda)] \, d\lambda}{\lambda} \tag{14}$$

N is the number of absorbing molecules per milliliter. The integration is carried out over a given absorption band a and the CD of that band must be isolated from that of other electronic transitions. In the case of overlapping CD bands, this requires that the bands be separated into components. This is often done by decomposing a complex CD spectrum into Gaussian components, each having the form of Eq. (13). The rotational strength of such a Gaussian band is (Moscowitz, 1960)

$$R_{0a} = 1.233 \times 10^{-42} \{[\theta_a{}^0] \, \Delta_a/\lambda_a\} \tag{15}$$

D. ORIGIN OF OPTICAL ACTIVITY

1. Chromophores

We have emphasized in the foregoing development that the optical activity arises from the rotational strength ascribable to a given absorption band and the Cotton effect associated with it. Through analysis of various chromophores, we will get the maximum information about molecules in general, and about biological polymers in specific. The discussion of current techniques used in the investigation of biological polymers, which follows, will be largely in terms of their chromophores. The chromophores occurring in biological polymers generally see an asymmetric environment and thus give rise to Cotton effects. These Cotton effects may be then analyzed to give structural information on their environment.

Since chromophores usually involve π electron systems having planar symmetry, the isolated chromophores themselves are rarely optically active. In most of the biological systems we will discuss below, the chromophores are made optically active by their asymmetric environment. The basis for understanding how an asymmetric environment induces optical activity in a chromophore was given by Kirkwood (1937). This approach is based on perturbation theory and has been extended by Tinoco (1962). This theory allows one to calculate the optical activity of a polymer given two kinds of information. First, one needs to know certain electronic properties of the monomers such as electric dipole moments, transition dipole moments, and charge distributions. These may be determined in some cases from experiments on the monomers or in other cases by molecular calculations on the monomers. These monomer properties will then be common to all polymers containing the same monomer chromophore. Second, one must have the polymer geometry. But this is just the polymer conformation which one generally seeks from physical measurements such as optical activity. So, in practice, one assumes various model geometries for a polymer and then calculates the optical activity in order to compare with experiment. A good assumption about the model geometry should give reasonable results when compared to experiment.

In a number of cases, quantitative calculations of optical activity have been made on the basis of various polymer structure models. These calculations are often helpful in formulating empirical rules and schemes for the study of polymer conformations. The quantitative agreement of the calculations with experiment, however, is rarely impressive and most of the practical techniques are quite empirical in nature. Most of this chapter will be devoted to these empirically based techniques for specific biological polymers.

2. Biological Molecules

Most of the biological uses of optical activity have been found in study of polymeric materials such as polypeptides and polynucleotides and their natural counterparts, DNA, RNA, and proteins. One may question the wisdom of plunging into a study on the optical activity of such enormous macromolecules when, in fact, the optical activity of small molecules, such as glucose, is rather poorly understood. The response is that it is considerably easier to interpret the asymmetric interaction of an array of identical amide chromophores in a polypeptide than it is to understand a small and apparently simple molecule such as glucose.

The key lies in the fact that the chromophores of polypeptides are found in the wavelength region ($\lambda = 190$–230 nm) which is relatively amenable to our present ultraviolet instrumentation. Glucose simply does not have chromophores that are in the accessible range of $\lambda > 190$ nm. One of the factors pivotal to the successful interpretation of optical activity is the existence of chromophores in the wavelength region accessible to our instruments. Polynucleotides, with strong absorption at 260 nm, are good candidates, and the success in using their Cotton effects to study conformation has been considerable. Some nonpolymeric substances, such as mononucleotides, have been studied although progress at interpretation in that area has been less impressive. In the survey of biological molecules which follows, we will not attempt an exhaustive review but rather give examples illustrative of the kind of information that can be found in optical activity measurements. For details on many of these techniques, the reader should consult relevant review papers.

II. Polynucleotides, RNA and DNA

A. The Monomer Properties

1. Monomer Absorption Spectra

We will begin our discussion of polynucleotides with a discussion of the mononucleotides and their chromophoric properties. The chromophores in this case are the heterocyclic purine and pyrimidine base derivatives. Most commonly occurring and most extensively studied are adenine, cytosine, guanine, thymine, and uracil. These are common bases in RNA and DNA but some rarer derivatives are found, especially in transfer RNA. The ultraviolet absorption spectra have been investigated by a number of workers (Clark and Tinoco, 1965; Voet et al., 1963). The absorption bands are summarized in Table I. Although the absorption bands differ among the various bases, one can see that all have strong absorption bands in the region 240–270 nm. These bands are due to π-π^* electronic transitions in the heterocyclic bases. There is no contribution of the ribose phosphates to the absorption in the ultraviolet accessible to ordinary spectropolarimeters (Voet et. al., 1963). The ultraviolet absorption spectra of a corresponding nucleoside, nucleotide, and deoxynucleotide are all much the same, the ribose phosphate perturbing the chromophore only weakly.

Most of the optical activity measurements are made at neutral pH where the nucleotide bases are uncharged. However, at pH's where the bases are ionized, the electronic properties are generally modified and the absorption spectrum shifted or otherwise changed. In interpreting the optical activity of an ionized base, one should refer to the absorption spectrum appropriate to that species. The absorption spectra at various pH's are given by Voet ot al. (1963).

Some workers have proposed that in addition to the strongly allowed π-π^* transitions in nucleotides, there are also weaker n-π^* transitions involving promotion of the nonbonding electrons of nitrogen into the π system of the ring (Miles et al., 1969; Ts'o, 1970). It is difficult to document this proposal from absorption spectra alone due to the strong absorption of the π-π^* transitions. We will have occasion to mention the contribution of such n-π^* transitions to the optical activity of certain polynucleotides. At this time, assignments of n-π^* transitions must be considered tentative.

2. Optical Activity of Mononucleotides

The nucleic acid chromophore, as is the case with most chromophores, is planar as a result of the π electrons. Hence, the base alone will show no optical activity. The addition of the ribose to form the nucleoside perturbs this planarity and the sugar induces some optical activity in the base chromophore. One might expect that the sign and magnitude of the Cotton effects induced in the base chromophores might bear some relation to the location of the ribose in relation to the base.

Such information on the conformation of the mononucleotides could be quite useful in understanding their chemical and biological properties.

TABLE I

ULTRAVIOLET ABSORPTION BANDS[a] FOR BASES AND NUCLEOSIDES IN WATER AT NEUTRAL pH

Compound	Ref.[b]	Long wave region		Middle region		Short wave region			
		λ_{max}	$\varepsilon_{max} \times 10^{-3}$	λ_{max}	$\varepsilon_{max} \times 10^{-3}$	λ_{max}	$\varepsilon_{max} \times 10^{-3}$	λ_{max}	$\varepsilon_{max} \times 10^{-3}$
Adenine	a	260.5	13.4	—	—	207	23.2	<185	—
Adenosine	a	259.5	14.9	—	—	206	21.2	190	19.8
Guanine	a	275	8.1	246	10.7	—	—	196	22.1
Guanosine	a	276	9.0	252.5	13.7	—	—	188	26.8
9-Methyl Hypoxanthine	b	—	—	248	11.0	201	21.0	?	—
Inosine	c	—	—	248	12.2	?	—	—	—
Cytosine	a	267	6.1	230	6.3	—	—	196.5	22.5
Cytidine	a	271	9.1	230	8.2	—	—	198	23.2
Uracil	a	259.5	8.2	—	—	202.5	9.2	—	—
Uridine	a	261.1	10.1	—	—	205	9.8	—	—
Thymine	a	264.5	7.9	—	—	205	9.5	—	—
Thymidine	a	267	9.7	—	—	206.5	9.8	—	—

[a] The bands are inferred from absorption maxima and correlations among analogous compounds, following Ts'o (1970).

[b] The references are a: Voet et al., 1963; b: Clark and Tinoco, 1965; c: Ts'o et al., 1969.

However, experimental studies in this area have been hampered by the very small magnitude of the Cotton effects, especially in the case of purine nucleotides. The perturbation of the ribose on the base is weak, so the departure from planarity is only slightly apparent.

On the other hand, measurements have been interpreted with some success. Yang and Samejima (1968a) have summarized the results of measurements on various nucleosides, nucleotides, and their deoxy analogs. One generally sees a weak negative Cotton effect in the purines, centered near the absorption maximum. There are no pronounced or systematic differences between the ribose and deoxy ribose compounds, indicating a close similarity in the conformation of the ribose sugar relative to the base. Likewise, the effect of the charged phosphate is not evidenced by any profound differences between the nucleosides and their related nucleotides. The pyrimidines show a somewhat stronger positive Cotton effect and again there are no pronounced differences among the nucleosides, nucleotides, and the deoxy analogs.

Attempts to use optical activity to correlate mononucleotide conformation have met with some success. Cyclic sugars, such as ribose, can exist in two anomeric configurations about C-1; ordinary nucleotides, RNA, and DNA are found in a β anomer form. Assignment of α and β anomeric nucleosides by the sign of their Cotton effects was undertaken by Emerson et al. (1067). The rule which they propose is that purine α nucleosides give positive Cotton effects, while the β anomers give negative bands. The opposite situation occurs in pyrimidines, the α nucleosides giving a negative Cotton effect and the β anomers giving positive Cotton effects for the strong absorption band. The reversal in sign between anomers apparently results from the fact that the sugar is placed on opposite sides of the base in the two configurations.

In order to assist in discussion of the conformation of the sugar relative to the base, a convention has been introduced to define the angle of rotation about the glycosyl bond between the base and C-1' of the sugars. (Sundaralingham and Jensen, 1965; Donahue and Trueblood, 1960). Steric considerations restrict this angle to two regions, known as syn and anti, representing the two different sides of the base on which the sugar may be found. It is the anti conformation that is found in Watson–Crick DNA. Attempts to correlate NMR and X-ray crystallography with optical activity have led to the tentative conclusion that pyrimidine nucleosides generally prefer the anti conformation by a substantial energy, and the barrier to rotation about the glycosyl bond must be quite high (Haschemeyer and Rich, 1967). The case with the purine nucleosides seems less conclusive due to the small magnitudes of the Cotton effect. The rotation about the glycosyl linkage seems less restricted than in pyrimidines and thus the perturbations of the sugar on the base are averaged out, leading to the small Cotton effect.

Attempts to correlate the natural nucleosides with the optical activity of cyclic nucleosides in which the sugar was covalently held in some known configuration have led to some generalizations about the sign of the Cotton effect and the nucleoside conformations (Emerson et al., 1966; Ulbricht et al., 1966). Unfortunately, there is some doubt as to the validity of comparison with the cyclonucleosides, due to the perturbation on the chromophore introduced by the covalent linkage. Rules on the sign of the Cotton effect and the syn-anti conformation must be considered as tentative.

A number of other workers have reported data bearing on the question of correlation of Cotton effects with nucleoside conformation (Ikehara et al., 1967; Klee and Mudd, 1967). See the review of Ts'o (1970) for a further discussion. The interpretations of these results have not always been in harmony on the spatial relationship between the sugar and the base. Such information is probably contained in the optical activity curves but conclusive results on interpretation are difficult. Theoretical work in this area, which has been lacking, might help to settle these questions.

In contrast to the rather low optical activity of mononucleotides under most conditions, the Cotton effects of aggregated monomers can become enormous. GMP can form a gel at a concentration of 1% and in a salt concentration of 0.5 M, which is sufficient to screen charge repulsions. Brahms and Sadron (1966) have reported CD for these aggregates having a $\Delta\epsilon_{max}$ of 80. This is larger than the CD of most polynucleotides and reflects the naturally strong tendency of guanine residues to aggregate.

B. Oligonucleotides

1. Oligo Adenylic Acids

We have mentioned in the introduction that we often have much better success in interpreting the optical activity of polymeric macromolecules than with small molecules such as their component monomers. Such is the case with polynucleotides and natural nucleic acids. Our understanding of the interactions giving rise to Cotton effects in polynucleotides, while by no means complete, has yielded a number of interesting results and techniques for studying polynucleotide conformation.

As an example of how these interactions appear in optical activity, we have reproduced the CD of AMP, ApA, and poly A in Fig. 3. In this figure, we report the ellipticity $[\theta]$ per residue. This convention

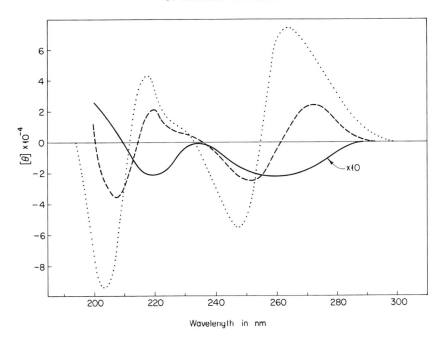

$[\theta] \times 10^{-4}$

Wavelength in nm

FIG. 3. CD of adenylic acids. AMP in water, pH 7, $\times$ 10 (Miles *et al.*, 1969) (———————). ApA in 0.01 M tris, pH 7.4, 23°C (Bush and Scheraga, 1969) (--------------). poly A in 0.01 M tris, pH 7.4, 23°C (Bush and Scheraga, 1060) (. . . . . .   )

is quite common in all optical studies on various biopolymers such as polypeptides and proteins and allows us to compare directly contributions of each chromophore.

In contrast to the small negative Cotton effect seen in AMP, the dinucleoside phosphate Ap(3′–5′)A shows a much larger Cotton effect having a different curve shape. Such a curve is known as a double Cotton effect (cf. Fig. 2b). It is thought to arise from the coupling of the two strong π-π^* transitions in the two adenine chromophores. This CD curve can be interpreted as evidence for base stacking in the dinucleoside with a right-handed skew sense (Warshaw *et al.*, 1965). The effects of n-π^* transitions are not obvious in these spectra. The possibility of their occurrence and interpretation will be discussed in connection with the poly dA spectrum below.

We notice also in Fig. 3 that the general curve shape is similar for ApA and poly A, indicating that the relative conformation of the residues in poly A at neutral pH is similar to that in ApA. For the infinite polymer, we would expect just twice as many interactions among nearest neighbors as for the dinucleoside, giving rise to just twice the rotational strength. This is approximately true and implies a nearest neighbor mechanism for the coupling.

Both the interaction giving rise to the optical activity and the interaction stabilizing the stacked structure are of a nearest neighbor type. From the studies of the ORD of various oligomers of adenylic acid as a function of temperature, Poland et al. (1966) concluded that the melting of poly A is essentially a noncooperative, i.e., that each base stack melts independently of the others.

Stacked polynucleotides exhibit a decrease in absorption intensity over their unstacked counterparts or their monomers. This phenomenon, called hypochromism, may be also used as a measure of polynucleotide stacking. Results on the melting of oligonucleotides similar to those obtained from ORD and CD (Brahms et al., 1966) were obtained by Leng and Felsenfeld (1966) by measuring the absorption intensity as a function of temperature for oligo A. The melting curves found by these two methods are not always in agreement since they give a different measure of the stacking. Optical activity is sensitive to the handedness of the stacking skew sense while hypochromism is not. If polynucleotide melting involves more states than simple stacked and unstacked ones, these two measures of stacking may differ and can be used to derive information on the process. Attempts to extract this information have not met with definitive success (Davis and Tinoco, 1968; Glaubiger et al., 1968).

The picture which emerges from the optical activity studies, considered with numerous other physical measurements for poly A at neutral pH in water, is that of a loosely stacked single strand structure with the bases stacked in a right-handed helical sense. At acid pH however, the situation is quite different. At these pH's, the bases become cooperatively protonated as a double helix is formed. It has been investigated by X-ray scattering from fibers and its structure is known to be parallel stranded with base planes tilted with respect to the helix axis (Rich et al., 1961).

The optical activity of the double stranded acid poly A is quite different from that of the neutral structure. From studies on the CD of various oligomers, Brahms et al. (1966) could show that the double helix formed with oligomers of six residues or more. Unlike that of the single strand, which melts slowly and noncooperatively, the melting curve of the double helix has a sharp melting point. Since in the acid polymer the base chromophore is protonated, a direct comparison of the Cotton effects between neutral and acid poly A is not possible as a result of the differing chromophore in the two cases.

We also point out that recent CD results have shown that this acid double helix of poly A can apparently exist in two forms having the same hypochromicity but different optical activity (Hanlon and Major, 1968). This situation of subtle differences in helix geometry which are seen in optical activity is one to which we will return. Optical activity is often sensitive to rather minor variations in the geometric arrangement of the chromophores in the polymer. Such minor differences may be difficult to detect with any other physical measurement except X-ray crystallography. On the other hand, if such differences are important to biological activity, their discrimination by ORD or CD could be quite useful.

The kind of minor geometric differences which one would expect to detect by optical activity can be quite well characterized. The interactions which give rise to Cotton effects are quite short range in nature, i.e., they extend only a few angstrom units. In other words, the gross conformation of a polymer such as end-to-end distances will not be important to optical activity but will be better reflected in hydrodynamic properties or light scattering. Optical activity is sensitive to local conformation, especially among nearest neighbors or at least residues that are not greatly separated in space. Thus, there is no inconsistency in calling neutral poly A a statistical random coil in a hydrodynamic sense, yet a relatively ordered, stacked polymer in the optical sense.

2. Dinucleoside Phosphates

The CD curves for ApA and poly A at neutral pH are characteristic of a double Cotton effect. Such an effect, with approximately equal negative and positive CD bands, is called "conservative" since the rotational strength is conserved throughout the band. Such behavior is predicted by simple coupling theories of polynucleotide rotation (Bush and Tinoco, 1967; Miles and Urry, 1967). Departure from this simple picture is seen in the CD of the dinucleoside, CpC (Fig. 4). The CD is predominantly positive throughout the ultraviolet. The large positive Cotton effect at 270 nm is followed by a very weak negative band at 230 nm. Not shown in Fig. 4 is a strong positive band at shorter wavelength. The theoretical explanation for this kind of single Cotton effect, or nonconservative band, may be found in the interaction of the cytosine absorption bands with the polarizability of its neighboring groups (Bush and Brahms, 1967).

Such a difference in the CD does not necessarily imply that ApA and CpC have a different geometry. The appearance of nonconservative bands depends on the directions of transition moments and other electronic properties and they may appear, for example, in polynucleotides having base planes either perpendicular to the helix axis or with tilted

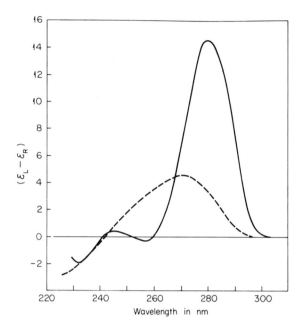

FIG. 4. CD of CpC and CMP in 4.7 M KF at neutral pH, $-17°C$. CpC (⎯⎯⎯⎯⎯)
CMP (⎯⎯⎯⎯⎯) (Bush and Brahms, 1967).

bases (Bush and Brahms, 1967). The case of natural RNA, having a
random sequence of bases, is quite different and in that case it is likely
that the appearance of nonconservative bands may have some significance for base tilting as we will discuss below (Johnson and Tinoco,
1969).

Measurements of the CD of a series of oligo C's and poly C as a
function of temperature gives a similar picture to that found for poly
A (Brahms et al., 1967a). The oligomers and poly C at neutral pH
form a single-stranded stacked structure which melts noncooperatively.
As in the case of oligo A, the oligomers of C longer than six residues
form a double-stranded structure in acidic solutions (Brahms et al.,
1967a).

The ORD of all the ribose dinucleoside phosphates composed of the
natural RNA bases A, C, U, and G have been reported by Warshaw
and Tinoco (1966). These curves are interpreted as indicating that dinucleosides generally exist in right-handed stacked structures with differing degrees of stacking due to differences in the energies of interaction
of the various bases (Bush and Tinoco, 1967). Warshaw and Tinoco
(1966) give the ORD curves for these dimers at acidic and basic as
well as at neutral pH. The addition of charges to the residue causes

conformational changes, as well as changes in the electronic properties of the dinucleotides. The ORD curves for the various dimers at different pH's are quite distinctive, even for dinucleotides of identical composition differing only in base sequence. Thus, the ORD curves at various pH's are extremely useful for analytical purposes including determination of base sequences. Cheng (1968) has reported the ORD of thirteen dinucleosides in the presence of mercurials and claims the curves are quite distinctive and, therefore, useful for analytical purposes.

Brahms and co-workers (Bush and Brahms, 1967; Brahms et al., 1967b, 1969) have reported CD curves for a number of dinucleosides as well as temperature studies on many of them. Quite generally, one observes a decrease in rotational strength with temperature, indicating melting of the stacked structure. If we assume that the dinucleoside can exist in only a stacked and unstacked state, these melting curves can be analyzed to yield an enthalpy of stacking. Using this treatment, Brahms et al. (1967b) have reported ΔH for stacking in the range -6 to -8 kcal. From the temperature dependance of the ORD, Davis and Tinoco (1968) report ΔH for a greater number of dinucleosides ranging from -4.8 to -8 kcal.

In the CD experiments of Brahms et al. (1967a,b), the solutions contain high concentrations of KF ($4.7\ M$). This salt is included to allow measurements over a wider range of temperature without freezing of the buffer solution. In several cases, it has been demonstrated that the high salt concentrations do not change the CD curves or melting properties (Bush and Brahms, 1967; Davis and Tinoco, 1968). On the other hand, there are indications that this may not be general. The Kronig–Kramers transform of the ORD data on GpC of Warshaw and Tinoco (1966) does not agree with the CD data of Brahms et al. (1967b) (see Fig. 5). This discrepancy has been confirmed by CD measurements of GpC under the conditions of both low salt and 4.1 M KF (Bush and Scheraga, 1968) as may be seen in Fig. 5. It is possible that the tendency of guanine residues toward aggregation is evidenced in the higher salt concentrations.

3. Trinucleosides

The most extensive use of the dinucleoside ORD curves has been made in the correlation of the ORD of longer oligomers. On the assumption that the ORD curve of an oligomer would be governed mainly by interactions of nearest neighbors, Cantor and Tinoco (1965) were able to explain the ORD curves of a number of trinucleoside diphosphates. Their empirical equation assumes that a trimer rotation receives contributions from its component monomers plus a contribution from each of its two

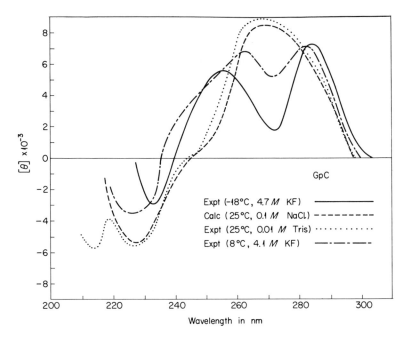

FIG. 5. Effect of salt (KF) on the CD of GpC at neutral pH. Experimental at $-18°$C in 4.7 M KF, 0.01 M tris, neutral pH (Brahms *et al.*, 1967b) (————). Calculated using Eq. (12) from ORD measured at 25°C in 0.1 M salt, pH 6.9 (Warshaw and Tinoco, 1966) (— — — — —). Experimental at 25°C in 0.01 M tris, pH 7.2 (Bush and Scheraga, 1968) (· · · · · · · · · · ·). Experimental at 8°C in 4.1 M KF, 0.01 M tris, pH 7.2 (Bush and Scheraga, 1968) (—·—·—·—·—·—).

nearest neighbor interactions as evidenced in the ORD curves of dinucleoside phosphates. For the trinucleoside IpJpK, the rotation per residue is

$$[\Phi_{IJK}(\lambda)] = \tfrac{2}{3}[\Phi_{IJ}(\lambda)] + 2[\Phi_{JK}(\lambda)] - [\Phi_J(\lambda)] \qquad (16)$$

In this equation, $[\Phi_{IJ}(\lambda)]$ is the molar rotation per residue of the dinucleoside phosphate *IpJ*, and $[\Phi_J(\lambda)]$ is that of the mononucleoside *J*. Within this approximation, they were able to represent the rotation of a trinucleoside diphosphate rather well, especially in the longer wavelength region of the Cotton effect spectrum (see Fig. 6). This means that to interpret the ORD of a single-stranded ribonucleotide, one need know only the 16 ORD curves of the component dinucleoside phosphates and that of the four mononucleosides. These curves have been catalogued in the "Handbook of Biochemistry" (Sober, 1968) by Warshaw.

In general there is a measurable sequence dependence for the trinucleoside curves due to the nonidentity of the ORD of dinucleoside

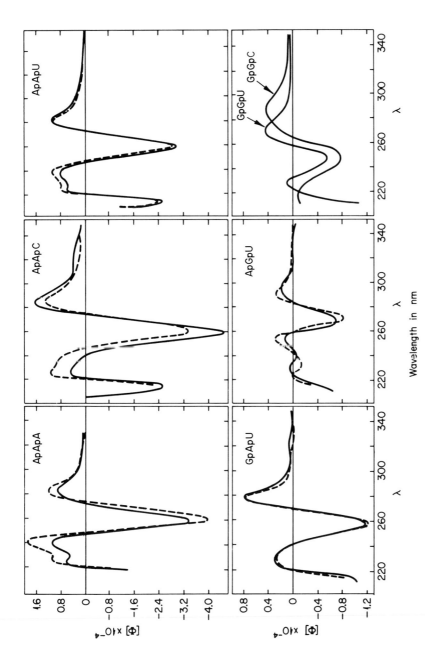

FIG. 6. ORD of trinucleoside diphosphates at pH 7. Experimental (————). Calculated from Eq. (16) (-----------). (Cantor and Tinoco, 1965).

sequence isomers. Thus, ORD may be of considerable use in assigning
sequences of oligomers which appear in sequence studies on RNA
(Mandeles, 1967). To aid in sequence studies, Cantor and Tinoco (1967)
have calculated ORD curves for all 64 trinucleosides of the common
RNA bases. Likewise, if one has a series of tetramers, he could calculate
the ORD curves from the data of Warshaw and Tinoco (1966) in order
to distinguish them. It is not certain just how long an oligomer could
be sequenced simply with knowledge of the base composition and ORD.
Certain sequences are quite easily distinguished (e.g., AGU vs. GAU
in Fig. 6), so the method could probably be useful in longer oligomers
containing them. It seems probable that for oligomers longer than a
few nucleotides or for mixtures, some automated method of recording
and comparing the curves might be needed.

In principle, one could also use the same approach in the CD of
oligomers using the CD curves of Brahms *et al.* (1967b) for dinucleosides
to construct the oligomer CD using a nearest neighbor assumption. Probably the ORD will continue to be used for this purpose since its more
complex curve shape allows one to distinguish differences more readily.

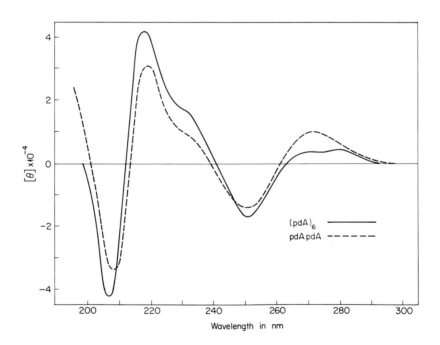

FIG. 7. CD of oligo-deoxyadenylates at pH 7.4, 23°C (Bush and Scheraga, 1969).

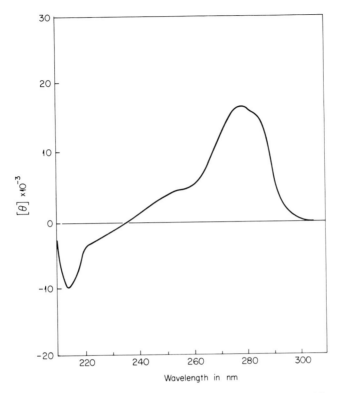

FIG. 8. CD of deoxy CpC at pH 8.5, 1.5°C (Adler *et al.*, 1968).

C. DEOXYOLIGONUCLEOTIDES

The above discussion of the nearest neighbor additivity property re-
lates to oligoribonucleotides only. The situation for oligodeoxynucleotides
has not been so well studied, but certain patterns are clear. First, the
nearest neighbor additivity rules seem less general. For example, the
ORD data of Vournakis *et al.* (1967) on oligodeoxy A and the CD
data of Bush and Scheraga (1969) show that the curves for the longer
oligomers differ in curve shape from that of the deoxy dinucleoside (see
Fig. 7). These differences do not mean that the interactions giving rise
to the optical activity in longer oligodeoxynucleotides are necessarily
long range in nature. They may result from the fact that the conformation of the dimer is different from that of longer oligomers, or it may
be due to electronic perturbations present in the oligomer which are
not found in the dimer (Bush and Scheraga, 1969). We will discuss
the possible causes of these differences in connection with poly dA below.

   In contrast to the difference between oligo dA and dApA is the case of deoxycytidilic acid. Adler *et al.* (1967, 1968) have studied the ORD and CD of deoxy CpC and poly dC. Comparison of Figs. 8 and 9 reveals that, although the peak magnitudes for dCpC and poly dC differ, the curve shapes are similar. Also, there is a similarity between ribo CpC and deoxy CpC (cf. Figs. 4 and 8). This similarity is not common to all analogous ribo and deoxy dinucleosides. Cantor *et al.* (1970) have investigated the CD spectra of all sixteen common deoxy dinucleosides and find that many differ significantly from the ribo analogs. They conclude that the deoxydinucleosides have stacked base conformations which are generally right handed, although the specific geometries apparently differ from the ribo dimers in many cases. The dimer CD spectra differ from the sum of the monomer spectra, indicating that base–base interactions are important. In addition, they find dependence on the sequence as in the ribodinucleosides.

Cantor *et al.* (1970) have also reported the CD of some deoxy trinucleotides to test the possibility of a nearest neighbor additivity formula for the CD of deoxy oligonucleotides of the form of Eq. (16) for ORD. As might be expected from the oligo dA case (Fig. 7), the results are less reliable than in the oligoribonucleotides. The structure

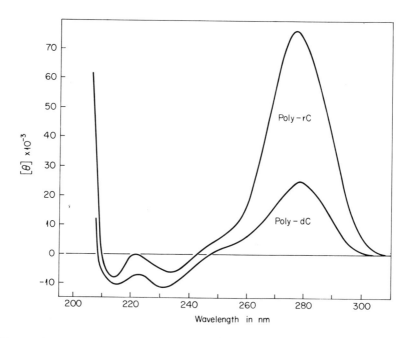

FIG. 9. CD of poly-ribo C and poly-deoxy C at pH 10.2, 2 *M* NaCl (Adler *et al.*, 1968).

and optical activity of the deoxy oligomers is less well understood than that of the ribo analogs but apparently the absence of the 2′ hydroxyl in the former causes certain geometric differences in the deoxy dinucleoside and longer oligonucleotides.

D. SYNTHETIC POLYNUCLEOTIDES

1. Single Strands and Identical Strands

We have already discussed some aspects of synthetic polynucleotide optical activity and structure in connection with oligonucleotides having similar structures. As previously discussed, poly rA, at neutral pH, forms a loosely stacked right-handed helix. Since the stacking is noncooperative, the breaks in stacking are distributed along the polymer and it is not rigid. The acid double helix of poly A is cooperative, showing a sharp melting in the ORD (Holcomb and Tinoco, 1965). Poly C (Fig. 9), although it shows a single rather than a double Cotton effect, is interpreted as having a structure similar to poly A. Poly C also forms a double-stranded helix cooperatively at acid pH which has an optical activity different from that of the neutral form (Brahms et al., 1967a).

Poly U has been investigated both by ORD and by hypochromicity studies (Sarkar and Yang, 1965a; Michelson and Monny, 1966). At room temperature, poly U gives very weak Cotton effects. This is interpreted as indicating that the uracil residues do not stack as do adenine residues in poly A. A similar conclusion about the reluctance of uracil residues to stack has been made by Warshaw and Tinoco (1966). At temperatures as low as 5 to 10°C, poly U undergoes a fairly sharp, reversible transition to an apparently ordered structure. At these temperatures, the ORD increases, giving a fairly large Cotton effect. There is no general way to tell the number of strands in a polynucleotide structure from ORD alone, but the poly U structure is probably multistranded since the melting curve is sharp (Michelson and Monny, 1966).

Poly G has presented certain problems due to the strong tendency toward aggregation seen in guanine residues. The ORD and CD of poly G have been reported by Green and Mahler (1970). The number of strands is uncertain but it is probably 2 or 3 as a result of the strong attraction of the guanine residues. An interesting feature of the poly G spectrum is the appearance of a Cotton effect at about 295 nm (see Fig. 10). This is far from the strong absorption band at 272 nm and may be due to an n-π^* transition as we will discuss below. The strong aggregation phenomena makes preparation of oligonucleotides of G difficult, but there has been some recent success with GpG (Podder and Tinoco, 1969).

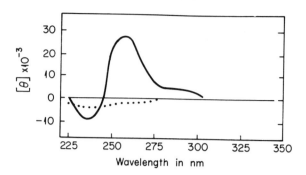

FIG. 10. CD of GMP and Poly G in 0.05 M KF, 0.001 M EDTA, pH 5.5 at 27°C. Poly G (—————), GMP (.........) (Green and Mahler, 1970).

In addition to the synthetic polynucleotides composed of the common nucleotide bases A, U, C, and G, there are a number of interesting studies on polynucleotides made of various derivative bases. These show varying degrees of thermal stability and tendency to form double and multiple strands. For a review, see the work of Yang and Samejima (1968a).

2. Multiple Strands

Although formation of Watson–Crick helices is best understood in DNA, similar base pairing is thought to occur in RNA. The helix geometry and hence the optical activity of RNA is different from that of DNA helices. Synthetic polynucleotides having Watson–Crick complementary strands generally form double strands readily and have been used as models of base pairing in RNA. Problems arise in these studies because some complexes of geometries other than Watson–Crick often appear.

The ORD of mixtures of poly A and poly U has been studied by Sarkar and Yang (1965a). These two polynucleotides associate, depending on conditions, to form either a two-stranded complex or an A plus 2U, three-stranded complex. That the complexes are multistranded is implied by the sharpness of their melting curves in ORD. The ratios of the components are determined from mixing curves. The detailed geometry of these two structures is not well understood, but the A + U two-stranded complex is presumably related to a Watson–Crick type base pairing between the strands. Thus, its ORD is taken as a model for A-U base paired regions in RNA as will be discussed below.

The two-stranded complex poly G plus poly C has also been studied by Sarkar and Yang (1965b). This complex also shows sharp melting and is presumably in the correct Watson–Crick base paired configuration

(Pochon and Michelson, 1965). The optical activity of this two-stranded polymer can serve as a model for G-C base pairs in natural RNA.

In our discussion of RNA, we will show how the appearance of base pairing effects the optical activity of RNA. Unfortunately, for synthetic polynucleotides, there is no simple way to tell the number of strands from ORD alone. For unlike strands, mixing curves in ORD, CD, or hypochromicity may be used to determine the ratio of components in the complex. Sharp melting curves, characteristic of cooperative processes, indicate the likelihood of multiple strands. Single stranded polynucleotides melt over a broad range due to noncooperativity (Poland et al., 1966). Unfortunately, these criteria are somewhat ambiguous and a number of synthetic polynucleotide structures, such as poly U at low temperatures and poly G, remain in doubt.

E. POLYDEOXYNUCLEOTIDES

1. Single Strands

The ORD and CD of poly deoxy C has been compared to that of its ribo analog by Adler et al. (1967, 1968). The curve shapes for the two polymers are similar, indicating the likelihood of a similar geometric arrangement of the residues (see Fig. 9). There is a disparity in the magnitudes of the Cotton effects. The simplest hypothesis is that the geometry is similar but that the degree of stacking is less in the deoxy polymer. The influence of the 2′ hydroxyl group in the structure of the ribo polymers has been the subject of much controversy, and it is quite possible that differences other than simply the degree of stacking exist between the poly rC and poly dC. We point out a fact to keep in mind in these as well as other comparisons on analogs such as the ribo and deoxy series. The difference between the 2′OH and 2′H should not alter the chromophoric properties of the nucleotide bases. Thus, any differences in optical activity between ribo and deoxy series must be due to differences in geometry of the arrangement of the bases.

2. Differences from Polyribonucleotides

As we have previously mentioned, the ORD and CD of poly deoxy A and the longer oligomers differ from that of the dinucleoside and from poly rA (Ts'o et al., 1966; Vournakis et al., 1967; Bush and Scheraga, 1969). This is in contrast to the situation of poly dC whose optical activity resembles that of the dimers and ribo polymer. In Fig. 11 we compare the CD of poly rA with that of poly dA. Especially in the longer wavelength region, there are distinct differences in these CD curves that are not easy to explain. Bush and Scheraga (1969)

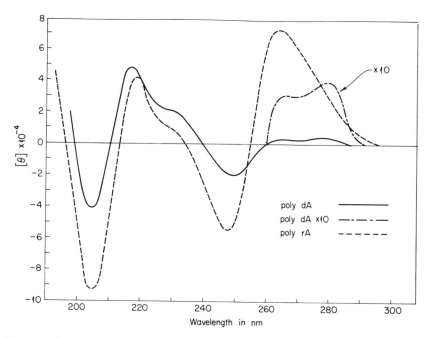

Fig. 11. CD of poly-deoxy A and poly-ribo A in 0.01 M tris, pH 7.4, 23°C (Bush and Scheraga, 1969).

have proposed that the longer wavelength CD band at 280 nm is due not to a π-π^* transition but to a weaker nitrogen nonbonding n-π^* transition at that wavelength. They further propose that the variations in the magnitude of this CD band among various adenine polymers are due to the solvent sensitivity of the n-π^* transition and that the different CD's represent differing solvation between the dinucleoside and the internal residues of the polymer.

3. Contribution of n-π^* Transitions to Optical Activity

The hypothesis that n-π^* transitions contribute to the optical activity of polynucleotides remains unverified. But if it could be shown to be correct and subject to interpretation, there is a distinct possibility that different kinds of structural information could be obtained on polynucleotides from the optical activity due to n-π^* transitions. These transitions are more localized in the heterocyclic ring structure than are the delocalized π-π^* transitions and hence ought to differ in their dependence on geometry. Preliminary calculations in this laboratory have shown that CD bands of a n-π^* transition have appreciable rotational strength and are more sensitive to local geometry than those of π-π^* transitions (Bush, 1970).

If n-π^* transitions appear in the CD of adenine polymers, it is likely that they will appear in other polynucleotides as well. In the CD poly G (Fig. 10), it is seen that a Cotton effect is also seen in the long wavelength region about 295 nm. At this wavelength, it is unlikely to arise from the π-π^* transitions, but rather from a long wavelength n-π^* transition (Green and Mahler, 1970). More will be said about these long wavelength transitions in connection with natural RNA.

F. NATURAL RNA

1. General

Much of the relevance of optical activity for biochemistry comes from its ability to give information about natural biological structures. With the advent of modern spectropolarimeters, the measurement is fairly simple. Moreover, the method has the advantage over some other physical methods that it requires only a small sample. One usually uses a sample of optical density near one in a 1-cm cell which contains as little as 1 ml of solution. Also, the sample is not degraded by the experiment. An additional feature is that there may be a number of impurities in the sample yet interpretation is still possible. The only requirement is that the impurity not absorb too much light in the wavelength range of the experiment, and that if the impurity is optically active, that there be some way of correcting for it.

The difficulty in using optical activity measurements to study conformations of RNA, DNA, or any other natural polymer is one of interpretation. The experiments often tell us that two structures differ, but it is sometimes difficult to conclude exactly what the difference is. We will outline below some of the successful techniques that have been developed for natural polynucleotides. We will also indicate some of the interesting problems which are as yet unresolved but whose solution could be profitable to biochemistry.

2. Nearest Neighbor Additivity

We have discussed above how the stacking interaction in oligonucleotides may be observed by optical activity measurements. Since the optical activity of trinucleotides can be interpreted as coming from its monomers plus nearest neighbor interactions as represented by the component dinucleosides, one might expect that any single-stranded RNA could be understood just from its nearest neighbor interactions. Such was shown to be the case by Cantor et al. (1966) for TMV RNA. The

frequency of nearest neighbors in this RNA is unknown, but there are indications from hydrolysis data that it is random. Using a random
sequence assumption, Cantor *et al.* (1966) computed the ORD of TMV
RNA from that of dinucleoside phosphates using the formula

$$[\Phi_{\mathrm{RNA}}(\lambda)] = \sum_i X_i \left\{ 2 \sum_j X_j[\Phi_{IJ}(\lambda)] - [\Phi_I(\lambda)] \right\} \qquad (17)$$

X_i and X_j are the fraction of base I and J in the RNA. The rotations
are the same as in Eq. (16). They showed that the ORD curve calculated
on these assumptions agreed reasonably well with that measured for
salt free TMV RNA as shown in Fig. 12. However, the more common
conformation of TMV RNA, as well as most other RNA's, is found
in solutions of 0.1 M salts. In such a solution, the phosphate charge
repulsion is screened, and the RNA forms a hairpin loop configuration
with the chain folding back on itself and base pairing. In Fig. 12, this
base pairing is seen to cause a shift in the Cotton effect to the blue
which reflects the contribution of base pairs to the ORD. Base pairing
is not found in dinucleosides and shorter oligomers (Jaskunis *et al.*,
1968), so one must go to longer polynucleotides to find a model for
this contribution. Cantor *et al.* (1966) could explain the blue shift resulting from the formation of base pairs by the addition of equal amounts
of A-U and G-C base pairs to the ORD as represented by the ORD
of the models of poly A + poly U and poly G + poly C (Sarkar and
Yang, 1965a,b).

These experiments imply a technique for using ORD to determine
the extent of base pairing in RNA. If one can calculate the ORD from
the nearest neighbor frequency and the dinucleoside ORD curves, then

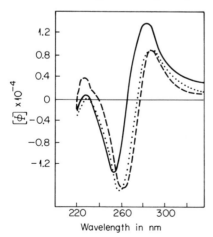

FIG. 12. ORD of TMV RNA at neutral pH. Experimental in 0.15 M salt
(———); experimental in the absence
of salt (...........). Calculated from Eq.
(17) (_ _ _ _ _ _). (Cantor *et al.*, 1966.)

it must be free of base pairing. Base pairs are reflected in blue shifts of the Cotton effects, with G-C pairs causing a larger shift than A-U pairs.

This approach has been used in transfer RNA of known sequence (Cantor et al., 1966; Vournakis and Scheraga, 1966). In these RNA's not all the bases are A, U, G, and C, some being rare derivatives for which the ORD of the dinucleosides is unknown. However, using reasonable assumptions, the ORD could be calculated from model structures such as the clover leaf model (Vournakis and Scheraga, 1966).

The 5 S ribosomal RNA is a short RNA of 122 residues whose sequence has been determined in mammalian KB cells (Forget and Weissman, 1967) and also for E. coli (Brownlee et al., 1967). Unlike transfer RNA, it is composed only of the four common RNA bases so the validity of the nearest neighbor calculation of the ORD based on the 16 dinucleosides is strictly valid. Cantor (1968) has compared his ORD measurement on E. coli 5 S RNA to calculations based on structures with varying amounts of base pairing and concluded that a 5 S RNA model containing approximately 43 base pairs is consistent with his ORD results. Such an estimate should be considered approximate, but it does give a good indication of base pairing in this unknown structure.

3. Nucleoprotein Complexes

The ORD assay for base pairing may also be used in ribonucleoprotein complexes. Bush and Scheraga (1967) investigated the ORD of ribosomes, their subunits, and ribosomal RNA under salt free as well as 0.1 M salt conditions. The ORD of the ribosomal protein was not large in the 260 nm region of the RNA Cotton effect (see Fig. 13). From comparison of the RNA Cotton effect in whole ribosomes with the Cotton effects of free RNA with and without salt, they were able to conclude that the RNA in the ribosomes is extensively base paired, as is found in RNA in 0.1 M salt solutions. Their ORD curve for salt free RNA could be calculated from nearest neighbor frequencies and dinucleoside ORD curves by the method of Cantor et al. (1966).

The same general approach was used to interpret the ORD of tobacco mosaic virus (TMV) (Bush and Scheraga, 1967). This structure is known from X-ray scattering to hold the RNA in a large helix of 23 Å pitch. In this conformation, base pairing would not be expected. The virus is 95% protein so that the protein contribution dominates even the RNA Cotton effect region at 260 nm. However, RNA free ghosts are easily prepared by aggregation of the TMV protein under appropriate conditions. Subtraction of the protein ghost ORD curve from that of the native TMV led to an ORD curve of the RNA as it is found in

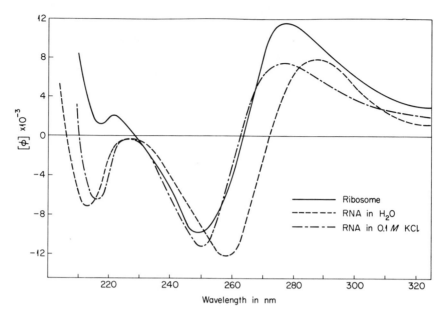

FIG. 13. Molar rotation per residue of RNA. 70 S *E. coli* ribosomes in 0.01 *M* magnesium acetate, 0.01 *M* tris, pH 7.4, 8°C (————————), *E. coli* rRNA in 10^{-4} *M* EDTA, pH 6.8, 24°C (------------), and of *E. coli* rRNA in 0.1 *M* KCl, 0.01 *M* tris, pH 7.4, 6°C (—·—·—·—·—·—.). (Bush and Scheraga, 1967.)

the virus. This curve is then compared to the salt free RNA ORD curve and that calculated by Cantor *et al.* (1966) from dinucleoside phosphates. These curves are all found to be in reasonable agreement, indicating the absence of base pairing in TMV RNA in the virus.

We conclude that the ORD of RNA-protein complexes such as viruses can be used to measure base pairing. In some cases it is necessary to have a ghost protein free of RNA but in the form of the virus. This is then used to estimate the contribution of the protein to the ORD of the virus.

4. Long Wavelength Cotton Effects

There is a small Cotton effect appearing in RNA which we have not previously discussed since it does not appear in synthetic polynucleotides. It was first reported by Sarker *et al.* (1967) and appears in the CD of RNA at about 295 nm (see Fig. 14). This small effect is generally overshadowed by the strong positive band at 260 nm in RNA. However, since the absorption of RNA is low in the region of this 295 nm Cotton effect, one can increase the path length or concentration to achieve a good signal-to-noise ratio. A similar effect was seen in DNA at 305

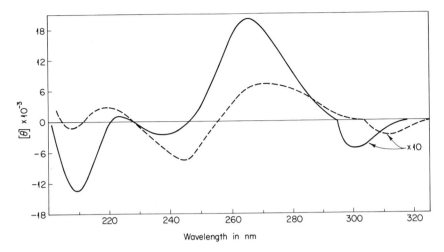

FIG. 14. CD of rRNA from *E. coli* (————) and salmon sperm DNA (-------------),
both in 0.005 M tris, 5×10^{-4} M Mg^{++} and 0.1 M KCl. The small negative Cotton
effects at 290 to 325 nm are $\times 10$ (Sarkar and Yang, 1967).

nm by the same authors (Sarkar *et al.*, 1967), but they were unable
to find it in various single- and double-stranded synthetic polynucleotides. The effect has subsequently been reported in a number of other
natural RNA's but its origin remains unknown. It is possible that it
may be due to n-π^* transitions in one or more of the bases and that
it is analogous to the positive bands in poly dA (Fig. 11) and poly rG
(Fig. 10).

G. DNA

1. Single and Double Cotton Effects

The CD curves of DNA differ in a general way from those of typical
RNA curves as shown in Fig. 14. The latter shows a single Cotton effect
while the former shows a double Cotton effect of decreased magnitude,
These generalizations are valid for both single- and double-stranded
RNA's and DNA's (Brahms and Mommaerts, 1964). Thus, optical activity is able to distinguish these two structures, which are quite distinct
in their biological roles. These differences must represent a regular geometric difference since there is essentially no difference in the chromophoric properties of the component nucleotides.

It appears that the essential difference appearing in the transition
from a double Cotton effect such as that in normal DNA to a strong
single Cotton effect in RNA involves tilting of the base planes with

respect to the helix axis. In DNA form B, the base planes are perpendicular to the helix axis. This arrangement causes the interactions giving rise to the strong single Cotton effect to cancel, leaving only the small double Cotton effect seen in DNA (Johnson and Tinoco, 1969). In RNA, the tilting of the base planes with respect to the helix axis causes a large single Cotton effect to appear. This hypothesis is consistent with experiment (Yang and Samejima, 1968b). Recent calculations show that this generalization should be true for any polynucleotide of random sequence of bases, such as RNA and DNA (Johnson and Tinoco, 1969). The applicability of this principle extends to both single- and double-stranded polymers but not to synthetic polynucleotides having a regular sequence (Johnson and Tinoco, 1969). Additional experimental verification will be needed but it appears likely that this property may be a useful one in establishing the existence of tilting of base planes in various structures.

2. Nearest Neighbor Additivity

Attempts have been made to compute the optical activity of single-stranded DNA from that of the component dinucleosides (Cantor et al., 1970). They used the same approach for single-stranded DNA as used in the ORD of single-stranded RNA described above [see Eq. 17]. Since the nearest neighbor additivity principle does not seem to work too well in oligodeoxynucleotides, it is surprising that the method gives a reasonable result for single stranded DNA in salt free solutions. The nearest neighbor additivity approach has not been adequately tested in DNA, but it will probably find less general employment than in RNA.

3. Solvent Induced Conformational Changes

The effects of various solvents in causing conformational changes in DNA has been the subject of a number of recent studies. Ethylene glycol has been shown to induce a single negative Cotton effect in DNA. The double-stranded structure is retained (Green and Mahler, 1968, 1970). Also ethanol has been shown to induce a single Cotton effect in DNA, having a positive sign similar to the Cotton effect in RNA (Brahms and Mommaerts, 1964). In addition, Tunis and Hearst (1968) have reported that the ORD of DNA is modified in various solutions of high salt strength. Comparison of their ORD curves for DNA in 6.8 M LiCl with the ORD of RNA-DNA hybrids leads them to conclude that the DNA in high salt is not in the A form. No salt effects are seen in RNA optical activity curves up to the concentrations which produce precipitations (approx. 3 M) (Tunis and Hearst, 1968).

4. DNA Virus

ORD curves for DNA bacteriophage show some interesting deviations from those of DNA in solution (Maestre and Tinoco, 1967). The tight packing required to store the large viral DNA inside the virus protein coat apparently causes changes in the geometry of the DNA helix. Maestre and Tinoco (1965) were able to correlate the changes in ORD among a series of phages with the DNA per unit volume of the phage. They interpreted this as indicating differing amounts of distortion required to pack the DNA inside the coat. The modifications in the DNA Cotton effects seen in high salt concentrations by Tunis and Hearst (1968) are similar to those seen in bacteriophage by Maestre and Tinoco (1967). If the changes in DNA ORD caused by high salt concentrations are caused by changes in the number of residues per turn as suggested by Tunis and Hearst (1968), then this could imply a twist of the DNA in phage.

5. Dye Binding to DNA

Whenever a planar symmetric chromophore is perturbed by an asymmetric environment, it may become optically active. Thus, dye molecules, when bound to DNA, may show Cotton effects at the wavelength of the dye absorption. Interpretation of these Cotton effects is complicated by the fact that binding to DNA occurs by more than one mechanism and the dye molecules may polymerize. Thus, the results are quite sensitive to pH and ionic strength. The binding of acridine orange to DNA has been extensively studied. The dye absorption undergoes a shift on binding (Blake and Peacocke, 1966; Yamaoka and Resnik, 1966). The acridine orange binding to native DNA differs according to the dye-polymer ratio apparently due to competing mechanisms of binding. There have also been studies of the effects of metal ion binding to DNA. The binding is complicated and apparently induces conformational changes in the DNA as well as direct perturbations of the electronic properties of the nucleotide bases. For a survey of the dye and metal ion binding and ORD of DNA, see the review by Yang and Samejima (1968a).

III. Polypeptides and Proteins

A. CHROMOPHORES

Optical activity has been used for some time in the study of polypeptide and protein conformation. Measurement of ORD in the visible and near ultraviolet gives some information on polypeptide conformation, especially on conformational transitions. Measurements in the amide chromophore region (190–230 nm) are much more informative and are now quite easily performed with modern commercial instruments. In recent years, there has been considerable progress in the interpretation of polypeptide and protein Cotton effects.

In the case of polypeptides, the monomer does not possess the chromophore of the polymer. The amide appears only with the formation of the peptide bond. The electronic structure of the chromophore has been extensively studied. The longest wavelength band appears at approximately 225 nm and is described as an n-π^* transition involving promotion of the oxygen nonbonding electrons into the amide π system. The transition is electrically forbidden and its small absorption is nearly obscured in the spectrum by the much stronger π-π^* absorption band at 190 nm. Both transitions contribute to the large Cotton effects characteristic of α helical polypeptides.

There is also at least one other transition observed between the n-π^* and π-π^* in the gas phase spectra of amides. This transition has been reported in the absorption and CD of optically active amides in the gas phase (Basch *et al.*, 1968). It has been assigned to n-σ^* in which the upper state is of the Rydberg type and spatially extended. Therefore this transition is not observed in the spectra of condensed phases (Basch *et al.*, 1968). Although this transition is sometimes mentioned in connection with polypeptide optical activity, we do not feel it makes any contribution in liquids or solids, and thus is unimportant in polypeptides and proteins.

B. α Helix

1. π-π^* Contribution

It was Moffit (1956) who first recognized that the strong π-π^* transitions of the amides would be expected to be strongly coupled in helical polypeptides. This interaction leads to a splitting of the π-π^* transition into various bands due to the interactions among the identical amides, by a mechanism analogous to exciton splitting in molecular crystals. The absorption spectrum of the α helix is split into components, with the longer wavelength band polarized parallel to the helix axis. The short wavelength component is doubly degenerate and polarized perpendicular to the helix axis. These polarizations have been experimentally verified by Gratzer *et al.* (1961).

Moffit's (1956) analysis, while correct for the absorption, was later shown to be in error for the optical activity (Moffit *et al.*, 1957). For the optical activity, the bands have subsequently been subjected to improved analysis. The π-π^* band at 190 nm is split into CD bands, a

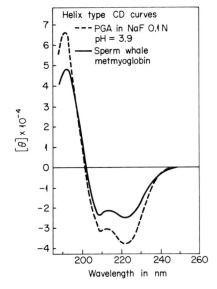

FIG. 15. CD spectrum of poly-L-glutamic acid in α helical form (..........). Sperm whale myoglobin (——————) (Quadrifoglio and Urry, 1968a).

negative band appearing in the α helix at 208 nm, a positive band at 191 nm, and another negative band at 185 nm (Woody and Tinoco, 1967; Woody, 1968). These bands are illustrated in Fig. 15 of the CD of the α helix. Since the interaction causing the splitting depends on the polymer geometry, these bands may appear at slightly different wavelengths in different polypeptide structures as will be seen below.

2. n-π* Contribution

In addition to the CD bands due to the π-π* bands, we see a negative Cotton effect at 224 nm in the α helix due to the n-π* transition. Schellman and Oriel (1962) first explained how this band, which is scarcely visible in the absorption, contributes strongly to the optical activity. Its strength results from the fact that, although it is electrically forbidden, it is magnetically allowed. The mechanism leading to its rotational strength involves the static field of the helical polymer and is thus quite different from the mechanism leading to the rotation of the strong π-π* band (Schellman and Oriel, 1962; Woody and Tinoco, 1967).

3. Experimental Curves

As seen in Fig. 15, the pattern of CD for the right-handed α helix is two adjacent negative bands at 224 and 208 nm, followed by a positive band at 191 nm. Since the two bands at 224 and 208 nm are of quite different origin, it is not unreasonable to observe that their relative magnitudes are sensitive to small changes in the α helix geometry. Although the details of the effects of geometry changes are not well understood, differences among various polypeptides and proteins are observed. These differences could yield useful information if interpreted. Some calculations have been carried out recently on the n-π^* and π-π^* rotations of dipeptides in various geometries (Bayley *et al.*, 1969), and also the n-π^* rotation for regular helices differing slightly from the α helix (Vournakis *et al.*, 1968). As yet, no clear interpretation of differences in the CD through depths at 208 and 224 nm has emerged.

The α helix is a rigid structure involving hydrogen bonding between a carbonyl oxygen and the NH of another group separated by three residues. It is this rigid structure that gives each amide residue an identical, asymmetrical environment and gives rise to the large Cotton effects characteristic of α helices. Both the CD and the ORD of the α helix are characteristic, and the spectra of polypeptides and proteins are often analyzed to determine α helix content as will be discussed below.

C. RANDOM COIL

When a synthetic polypeptide undergoes a transition to the random coil form, the CD curve is greatly altered, as shown in Fig. 16. This conformation is characterized by a weak positive band at 216 nm and a negative Cotton effect at 198 nm. There seems to be some variability in this conformation and an extremely weak negative band is observed

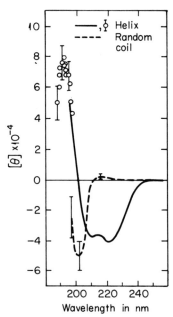

FIG. 16. CD spectrum of poly-L-glutamic acid in 0.1 M NaF. Helix, pH 4.3 (————); random coil, pH 7.6 (-------------) (Holzwarth and Doty, 1965).

at 234 nm under certain conditions (Carver *et al.*, 1966b). It has been proposed that this latter band is a characteristic of an extended chain rather than a truly random coil form (Tiffany and Krimm, 1969). In a truly random conformation, one would expect the asymmetric perturbations to approximately cancel, leading to very weak Cotton effects. While this is true of longer wavelength bands, the 198 nm CD band is considerably stronger than would be expected. The reason for this discrepancy is not well understood. Although the chain appears statistically random over large distances as measured by hydrodynamics, it may be that there is a preferred disposition of the amides to their near neighbors. As we have mentioned above, in connection with neutral polyadenylic acid (see Section II.B.1), optical interactions are short range, while the random coil in polypeptides is random in a long range sense. A statistical coil in the hydrodynamic sense may have local order, giving rise to optical activity. One may conclude that whatever the random coil conformation is, it does give nonvanishing contributions to the optical activity, especially in the π-π^* region.

D. OTHER REGULAR STRUCTURES

1. β Pleated Sheets

Under appropriate conditions, poly-L-lysine is found in a form in which the chain folds back on itself to form an antiparallel, β pleated sheet conformation. Sarkar and Doty (1966) have characterized this structure by infrared absorption and report ORD and CD. Similar data on poly-L-lysine ORD are reported by Davidson and Fasman (1967). Figure

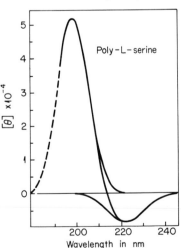

FIG. 17. CD spectrum of poly-L-serine in 80% trifloroethanol, 20% water, 25°C, with resolution into two Gaussian bands (Quadrifoglio and Urry, 1968b).

17 shows the CD of poly-L-serine in a β conformation as given by Quadrifoglio and Urry (1968b). These authors point out that the optical activity of β structures is variable depending on the specific polypeptide and experimental conditions. This property makes more difficult the assignment of the contribution of a partial β structure to a polypeptide or a protein. The pattern of Cotton effects for this β structure shows a negative CD band at 222 nm and a larger positive band at 197 nm. For a further discussion of these assignments, see Deutsche et al. (1969).

2. Polyproline

Poly-L-proline forms two characteristic helices. Polyproline I is a right-handed helical form with 3.33 residues per turn. Polyproline II requires that the peptides be in the trans rather than the cis conformation, as in polyproline I. It is right handed and has a helical repeat of 3 residues per turn. From the ORD analysis by Carver et al. (1966b), the position of the CD bands can be determined. Polyproline I has a weak negative band at 236 nm. This band is somewhat stronger than the band at this wavelength for random coil polyglutamic acid. There is a positive band at 214 nm and a negative band at 198 nm. Polyproline II, with its different geometry, shows a small positive band at 225 nm and a negative band at 207 nm. For further discussion of these bands see Deutsche et al. (1969).

E. INFLUENCE OF SIDE CHAINS IN POLYPEPTIDES

The optical activity due to amide backbone transitions is common to all polypeptides and proteins. However, amino acids having chromophoric side chains contribute Cotton effects that are of interest in the study of the structure of both polypeptides and proteins. In Table II

TABLE II

ULTRAVIOLET AMINO ACID ABSORPTION BANDS[a]

	λ_{max}	$\epsilon_{max} \times 10^{-3}$	λ_{max}	$\epsilon_{max} \times 10^{-3}$	λ_{max}	$\epsilon_{max} \times 10^{-3}$
Tryptophan, pH 5	280	5.2	219	35	196	21
Tyrosine, pH 5–6	275	1.2	224	7.8	192	47
Tyrosine, pH 13	293	2.5	239	11.6	?	
Phenylalanine, pH 5	257	0.19	206	8.0	187	58
Methionine, pH 5	—	—	207	1.9	—	—
Histidine HCl, pH 5	—	—	212	6	—	—

[a] From Wetlaufer, 1962.

are summarized some of the absorption properties of side chains likely to be important in polypeptides.

The aromatic amino acid, tyrosine, has absorption bands centered at 275, 224, and 196 nm. The band at 196 nm has an enormous extinction coefficient and will generally make large contributions to Cotton effects at this wavelength. Since the effects at 224 and 196 nm lie in the range of the amide Cotton effects, they are capable of introducing confusion as may be seen in the case of poly-L-tyrosine. Bands are seen at 270, 248, and 224 nm. Although the 224 nm band has been assigned as amide n-π* and taken as an indication of a right-handed α helix (Beychok and Fasman, 1964), this assignment may be criticized on the grounds that the 224 nm band in the side chain would be expected to interfere. Although the handedness of the poly-L-tyrosine helix could be determined by the CD measurements, a convincing interpretation based on sound theoretical grounds is lacking.

Other aromatic chromophores whose influence in optical activity could be considerable are histidine, typtophan, and phenylalanine (Urry, 1968). Phenylalanine is a weak absorber but may make some interesting contributions to the polypeptide Cotton effects. Its CD spectrum in the 260 nm region shows considerable vibronic structure (Horowitz et al., 1969; Blaha and Fric, 1968). The histidine chromophore apparently makes contributions to the CD of poly-L-histidine at 212 and 202 nm but the conformation of the polymer is not known (Urry, 1968). Using low temperature measurements of CD, Strickland et al. (1969) were able to characterize the vibronic structure of tryptophan derivatives. The spectra of these models were then used to interpret the CD spectrum of the tryptophan residues in a protein. There is considerable work yet to be done in characterizing side chain Cotton effects in polypeptides.

The Cotton effects due to the presence of cystine disulfide bridges in proteins and polypeptides appear to be quite interesting. Coleman and Blout (1968) have reported optically active bands in model compounds. The absorption is generally weak compared to the optical activity and the bands are often difficult to detect in the absorption. Coleman and Blout (1968) found Cotton effects for L-cystine at low pH at 252, 218, and 187 nm. Although the wavelength of these disulfide Cotton effects is variable depending on conditions of solvent, the Cotton effects may be important in the 240–260 nm region, especially in proteins containing disulfides bridges but few aromatic groups.

Our knowledge is far from complete in characterizing the effects of side chain chromophores in optical activity. For the present, there are several facts that should be kept in mind when dealing with optical activity of polypeptides and proteins having chromophoric side chains.

For side chain chromophores with long wavelength bands, one can get an idea of the rigidity of the side chains by looking at side chain bands which do not overlap the amide bands. Problems of interpretation will occur if side chain bands occur in the same wavelength region as the amide CD bands. In addition, coupling between the amide and the side chain chromophores, which is often the source of the rotational strength of the side chain absorption bands, will also have an effect on the rotational strength of the amide bands. One must be extremely cautious in interpreting optical activity curves of polypeptides containing absorbing side chains.

F. CYCLIC PEPTIDES

Although studies on constrained peptide structures, such as the natural cyclic peptide antibiotics and synthetic cyclic peptides, have not been extensive, these compounds promise to show an enormous variety of Cotton effects. In spite of the fact that they contain only a few residues and their conformations are apparently quite unrelated to those of other synthetic polypeptides, even the simplest cyclic peptides may show Cotton effects ranging from α helix to random coil types. Since NMR often gives quite explicit information on the conformation of cyclic peptides, they may be able to serve as model structures for various regions of proteins, augmenting our catalog of ORD curves which may be ascribed to specific conformations. In addition to their usefulness as models for peptide backbone conformation, cyclic peptides seem to be especially promising for the study of side chain Cotton effects in proteins.

In cyclic peptides, the usual interpretation of optical activity often gives misleading results. The CD of the cyclic decapeptide, gramicidin-S has been reported by Quadrifoglio and Urry (1967). It is reminiscent of the α helix CD showing both a n-π^* band at 217 nm and π-π^* bands at 208 and 188 nm. This highly constrained cyclic peptide cannot have any true α helix. However, the conformation is rigid and the effects of chromophore coupling and static fields are such as to give rotational strengths similar to those of the α helix. Pysh (1970) has calculated the ORD of gramicidin-S for three proposed model geometries. Comparison with experimental data leads him to favor the model of Stern *et al.* (1968) which is based on NMR studies and which does not resemble an α helix.

This fact illustrates that rather different geometric arrangements may lead to similar CD curves. Such a conformation could possibly appear in a protein, leading to the erroneous conclusion that it was an α helical run of about six residues. The cyclic peptide antibiotic tyrocidine also gives an ORD curve having α helical features (Craig, 1968). But other cyclic peptide ORD curves reported by Craig (1968) are quite different. Likewise, Coleman and Blout (1967) give ORD curves for several cyclic peptides which show a variety of Cotton effects.

There have been a few studies on the optical activity of synthetic cyclic peptides. Balasubramanian and Wetlaufer (1967) report the ORD of the cyclic tetrapeptide of L-alanine. Its optical activity is quite weak and is not obviously related to other polypeptide curves. Blaha and Fric (1968) have reported the ORD and CD of some cyclic hexapeptides containing both D and L residues arranged so the molecule retains its optical activity. These peptides show appreciable Cotton effects and also exhibit the side chain Cotton effects of phenylalanine.

It is probable that the study of synthetic cyclic peptides by NMR and by ORD will be quite profitable, since structural evidence from the two methods can be correlated. Kopple *et al.* (1969) have reported structures for some cyclic hexapeptides based on NMR data. The structure involves a ring held rigid by two trans-annular hydrogen bonds. In spite of this rigidity, the CD of cyclo-Gly$_5$-L-Leu shows Cotton effects more like those of a random coil (see Fig. 18). The addition of other side chains to these cyclic hexapeptides does not change the hydrogen bonding scheme as determined by NMR, but the Cotton effects are en-

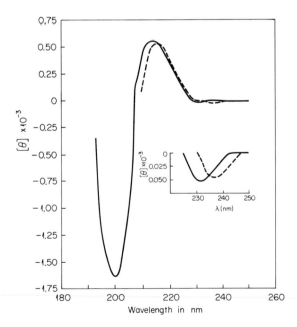

FIG. 18. CD spectrum of cyclo-gly$_5$-L-leucine in water (——————) and in methanol (--------------) (Ziegler and Bush, 1971).

hanced by aromatic side chain interactions and also by reduced conformational mobility of the amide backbone (Ziegler and Bush, 1971).

G. Techniques for the Study of Polypeptide Structure

Since the optical activity of the α helical polypeptides is so distinctive, several techniques have been developed for using ORD and CD data to estimate the α helical content of polypeptides.

1. ORD outside the Cotton Effect Region

(a) *Moffit and Yang Method.* The earliest and best known method is that due to Moffit and Yang (1956). As originally formulated, their equation was believed to rest on certain theoretical foundations that proved faulty. The method is, in fact, based on rather general considerations and has been extensively verified experimentally. It may be considered empirical. The equation of Moffit and Yang for the ORD of a polypeptide outside the absorption region is

$$[m'] = a_0/(\lambda^2 - \lambda_0^2) + b_0/(\lambda^2 - \lambda_0^2)^2 \tag{18}$$

The equation was found to fit the ORD in the range of 300 to 600 nm for both α helix and random coil under a wide range of conditions.

The Moffit–Yang equation has three parameters to be fit to the ORD: λ_0, a_0, and b_0. In practice, λ_0 is adjusted to give linear plots of

$$[m'](\lambda^2 - \lambda_0^2) \text{ versus } 1/(\lambda^2 - \lambda_0^2) \tag{19}$$

from which one may determine a_0 and b_0. A choice of $\lambda_0 = 212$ nm has been found to apply to most polypeptides. This λ_0 represents a type of average wavelength at which Cotton effects characteristic of polypeptides are found. Moffit and Yang (1956), as well as subsequent investigators, find $b_0 = -630$ for right-handed α helices and $b_0 = 0$ for random coil. a_0 is found to depend on solvent conditions and varies widely. b_0 has been used as a measure of helix content in unknown polypeptide structures. If an unknown structure is found to have b_0 near -630, then a right-handed α helix is suspected. $+630$ for b_0 indicates a left-handed α helix, while intermediate values can be interpreted either as evidence for mixtures of helical sense or of random coil. These latter choices may be tested by studies of the helix coil transition or by hydrodynamic measurements. For more detail on the use of this method, see Yang (1967).

(b) *Modified Two-Term Drude Method.* An alternative to the Moffit–Yang method for the analysis of the ORD outside the Cotton effect region has been suggested by Shechter and Blout (1964). It is based on the Drude equation with Cotton effects placed at the known absorption wavelengths. They propose a four-term Drude equation, with terms at 193 nm and 225 nm representing the α helical form and terms at 198 and 225 nm representing the random coil form. They have cast this four-term equation into the form

$$[m'] = A_{193}(193)^2/(\lambda^2 - 193^2) + A_{225}(225)^2/(\lambda^2 - 225^2) \qquad (20)$$

From plots of A_{193} vs. A_{225} for polypeptides of various helix contents, they can establish characteristic values for α helix and random coil for these two constants in different solvents.

The relative merits of the modified two-term Drude equation approach of Shechter and Blout (1964) and the method of Moffit and Yang have been the subject of some controversy, but for the most straightforward application, their results are similar. If one sets $\lambda_0 = 212$ nm, the Moffit–Yang and Shechter–Blout methods become identical, since both are essentially two parameter formulas. Thus the A_{193} and A_{225} parameters have been shown to be linearly related to b_0 and a_0 by Carver et al. (1966a).

(c) *General Comments.* There are certain polypeptides in which any method based on ORD in the 300–600 nm region is most likely to lead to erroneous conclusions. In cases where side chain Cotton effects are significant, they may dominate the ORD in the region of 300 to 600 nm and lead to errors. Surely the polytyrosine ORD curve could not be expected to give a b_0 indicative of the amide chromophore, since the strong Cotton effects at 248 nm will cause b_0 to be positive, regardless of what the amide chromophore contribution is.

The b_0 method, or any other based on data outside the Cotton effect, is most useful in cases where, because of solvent absorption or some other limitation, measurements cannot be made in the region of the amide chromophore Cotton effects (190–230 nm). When possible, one should use methods based on CD or ORD in the Cotton effects to determine helix content. There are several methods which we will discuss.

2. ORD in the Cotton Effect Region

(a) *Direct Method.* The availability of ORD measurements in the Cotton effect region now affords us much more direct methods of estimating α helix content which should be used when possible. The simplest method of interpreting the ORD in the Cotton effect region has been described by Greenfield et al. (1967). They present model ORD curves based on poly-L-lysine in the α helix and random coil forms. In addition to synthesizing ORD curves containing varying amounts of these two structures, they also include contributions of poly-L-lysine in the β form. These synthesized curves may be compared with observed ORD curves for polypeptides of unknown structure and a good match sought. This method gives qualitatively reasonable results but does not always give a perfect match with observed ORD, especially for proteins. Such results can be interpreted as an indication of structures other than ideal α, β, or random coil.

(b) *Computer Curve Fitting Method.* Carver *et al.* (1966b) have proposed a different method for interpretation of the ORD in the Cotton effect region. In their method, the ORD is fitted by a nonlinear least squares procedure to the Kronig–Kramers transforms of Gaussian CD bands [see Eqs. (11, 13)]. These bands may then be identified with the Cotton effects in the experimental CD spectrum. The fitting is generally done on a computer and the quality of the fit can be characterized by the residual mean square error. For α helical models, the ORD can be well fitted by three Cotton effects; at 224, 208, and 192 nm. These effects are also seen in the CD of α helical polypeptides (cf. Fig. 15).

For the random coil, they indicate that three Cotton effects give a good fit with $\lambda = 198$, 217, and 235 nm. The long wavelength band at 235 nm is extremely weak, but it has been independently detected in CD curves, so it is apparently not an artifact of their method. Their method has also been applied to the polyproline conformations and agrees well with the CD. For any polypeptide having an unknown mixture of conformations, the relative contributions from the contributing structures may be estimated from the amplitudes of the Gaussians needed to obtain a good fit to the experimental ORD.

This method is equivalent to a measurement of the ORD, followed by a transformation to the CD by a numerical integration of the Kronig–Kramers transform [Eq. (12)]. The resulting CD curve may then be decomposed into Gaussian bands centered on the absorption wavelengths of known bands characteristic of helix, random coil, β structure, or any other accepted model. The decomposition into Gaussian bands may be accomplished either by manual trial and error or by the use of an analog computer such as the DuPont Curve Resolver.

3. Circular Dichroism

It will be recognized that the above procedure could be shortened somewhat by measurement of the CD directly. One avoids the Kronig–Kramers calculation and moves directly to a decomposition of the CD into bands characteristic of models such as α, β, and random coil. Such model CD curves have been given in Figs. 15–17. This approach is the simplest one available and is to be recommended unless one has a good reason why it will not work in a specific case.

H. Proteins

1. Backbone Conformation

The methods outlined above for studying the structure of polypeptides may all be applied to the study of protein structure. The methods may be used to determine the relative contributions to the structure of a given protein of α helix, β structure, and random coil or disordered structure. The most common method is the b_0 method of Moffit and Yang (1956). This is not necessarily the best method but is the oldest and apparently most familiar. The availability of commercial instruments capable of giving good optical activity measurements in the Cotton effect region makes it unnecessary to use measurements outside the absorbing region to infer Cotton effects. One can measure directly the Cotton effects and compare them with the magnitude of Cotton effects thought to characterize model conformations in polypeptides (Greenfield *et al.*, 1967).

There are several advantages to this latter approach. First, one needs less protein sample. The Cotton effect rotations are larger than those in the 300–600 nm region. One should use a protein concentration such that the optical density is not much over one at the wavelength of the measurement. Such an optical density will lead to the optimum signal-to-noise ratio.

Second, since one is looking directly at the Cotton effects, he can determine by inspection if there are any special problems in his sample. For example, side chain Cotton effects, if they are prominent, will be detected at wavelengths of side chain absorption outside the amide chromophore region. Similarly, one can see directly the Cotton effects of the amide region, determining their wavelengths for comparison with the wavelengths of the Cotton effects of model polypeptides. This latter advantage is most obviously gained in the CD measurements, where assignment of Cotton effects to absorption bands is most straightforward, but it can also be gained from ORD curves through the medium of the Kronig–Kramers transformation.

2. Examples of Protein Studies

We will discuss a few proteins as illustrations. For a more complete review of the literature on optical activity of proteins, see Yang (1967). Also, a very extensive review of the ORD of proteins is found in the monograph of Jirgensons (1969). He has assembled ORD data on a wide spectrum of proteins, classifying them in a useful way.

The case of myoglobin is quite simple since the structure is known from X-ray to be largely α helix (cf. Fig. 15). Holzwarth and Doty (1965) compare the CD of sperm whale myoglobin in the amide region to that of α helical polypeptides. Their results agree with a 70% α helical content, with the remainder, random coil. While the α helix content is consistent with the known structure, the meaning of 30% random coil for a rigid structure remains unclear.

In analyzing the CD of ribonuclease, Schellman and Lowe (1968) point out the significance of the fact that the n-π^* CD band in this protein is not at the characteristic wavelength of the α helical polypeptides. The magnitude of the effect agrees with a helical content of 15% which is consistent with X-ray structure, but the wavelength is shifted from 224 nm to 217 nm in ribonuclease. Since the wavelength of the n-π^* is known to be sensitive to environment, they attribute this shift to the effects of conformation and solvation on wavelength. The effects of the environment in shifts of the n-π^* in the region of 235 to 218 nm are quite interesting but have not been fully evaluated.

The discussion by Timasheff et al. (1967) on β lactoglobulin is indicative of the extent of information available from optical activity measurements on proteins. By correcting the ORD outside the amide region for side chain Cotton effects, they were able to get the ORD of the amide alone. They were then able to calculate b_0 free from the side chain Cotton effects. This b_0 was then used to estimate contributing α, β, and random structures. Moreover, they could construct the CD curve for the protein using these estimates and the CD curves of the appropriate model polypeptides. They concluded that β lactoglobulin should have 10% α helix, 40% β structure, and 50% "disordered" structure.

Another example of careful analysis of optical activity data on a protein is that Kasarda et al. (1968) on the wheat protein, α gliadin. From studies on CD and ORD over a range of conditions affecting aggregation and conformation, they were able to make conclusions about both the amide backbone and the side chain regions of the molecule undergoing conformational changes.

3. General Comments

There are certain general difficulties which arise in any scheme which attempts to analyze protein CD or ORD curves into contributions from α helical regions, random coil, and β structure. The first is that there is no obvious reason to expect to find any random coil or disordered regions in a rigid globular protein of fixed conformation. There is no a priori reason to feel that random coil polyglutamic acid resembles anything in a protein. Likewise, even the inclusion of β structures in the models contributing to proteins cannot tell the full story of the range of possible conformations of a polypeptide chain in a protein. Recent studies by Schellman and Nielsen (1967) show that distorted helical structures may make quite a large contribution to protein optical activity. Moreover, the short helical runs found in most proteins probably are not equivalent to the long runs found in synthetic polypeptide α helices. This has been shown in the calculation of Woody and Tinoco (1967) who demonstrated that only for runs of longer than 20 residues do the contributions approximate that of long runs of helix.

The optical activity of cyclic peptides shows that loops in a peptide or protein can produce a wide variety of Cotton effects. Thus, it is not too surprising that optical activity curves of certain proteins yield results which do not agree with known structures in polypeptides. The synthesis of CD and ORD curves from α helix and random coil, even with the addition of the β structure, must be taken as approximate at best. These structures do not describe all possible protein regions, and even if appropriate models for all protein regions were available, ORD simply does not contain enough information to determine their relative proportions. Optical activity curves must be analyzed in connection with other experiments in order to give reliable estimates of protein structure. No completely reliable method of analysis exists. Whether one uses b_0, ORD or CD, he must keep an open mind with regard to side chain effects, other special chromophores, and to constrained or distorted chain conformations. It is the challenge offered by such effects as these that makes the examination of a new protein interesting.

In light of these words of caution, it is surprising that the estimates of α helix content from optical activity measurements agree with those determined from X-ray structures. Only in those cases where side chain absorption is especially important, as in carbonic anhydrase, or in cytochrome C where the heme group absorption is important, do the estimates of helix content differ greatly from that found by X-ray structures (Deutsche et al., 1969).

4. Nonamide Cotton Effects in Proteins

The side chain Cotton effect has been discussed above mainly as an interfering effect in determining the amide backbone Cotton effects. But in fact their Cotton effects can be used to give information on local environmental changes in proteins. Side chain Cotton effects will be subject to changes in the nearby groups which perturb them.

Before the side chain Cotton effects of a protein can be evaluated, the side chain must be identified from its chromophoric properties and from knowledge of the amino acid content of the protein (see Table

II). Aromatic side chains, such as tyrosine and tryptophan, are best characterized by CD in the near ultraviolet (Beychok, 1967). Side chains absorbing at wavelengths longer than 230 nm are easily characterized in the CD since Cotton effects can be directly assigned to absorption bands. Use of CD in such cases is preferred to ORD since the strong amide activity present in ORD makes it difficult to detect the less abundant side chain chromophores.

The most commonly exploited long wavelength effects are those due to tyrosine and tryptophan. However, especially in proteins having few aromatic residues, disulfides may become important. Disulfide bridges have weak absorption but apparently significant optical activity in proteins. Pflumm and Beychok (1969) have given an interesting discussion of the changes in disulfide Cotton effects in the reoxidation of the disulfide bridges in ribonuclease. The review of Beychok (1967) gives details on a number of side chain CD curves in proteins.

Other special protein chromophores may make significant contributions to the optical activity. Prosthetic group chromophores often absorb in the visible and may be studied by CD, quite free from any effects of amide backbone bands. The most familiar examples are the heme containing proteins. For a full discussion, see the review of Ulmer and Vallee (1965) on the ORD of Cotton effects of prosthetic groups in enzymes and also the review of the CD by Beychok (1967).

IV. Polysaccharides

The two classes of biopolymers which have received the most study by optical activity are polypeptides and polynucleotides. However, there are distinct possibilities that other biopolymers might also be amenable to analysis by optical activity, if well characterized model systems can be established. The most promising case seems to be that of polysaccharides.

A. CHROMOPHORES

1. N-Acetylamino Polysaccharides

A number of structural polysaccharides have the 2-amino substituent. In many important cases, this amino group is N-acetylated to give an amide (Brimacombe and Webber, 1964). Due to the asymmetric environment of the sugar, this amide is perturbed and gives rise to Cotton effects analogous to those of the polypeptide. Since the optical activity arising from the amide chromophore in polypeptides has been extensively studied, it should be possible to apply some of that knowledge to interpretation of the Cotton effects of the N-acetylamino polysaccharides. Theoretical studies along these lines are under way in this laboratory.

In general, one would expect both the n-π^* and π-π^* transitions of the amide to give Cotton effects. Also, since the amides themselves are spatially separated from one another, one would expect very little exciton splitting of the π-π^* transition. Thus, we would not expect to see Cotton effects due to π-π^* transitions at wavelengths of 195 to 205 nm as are seen in polypeptides. Since the amide is attached to C-2, we would expect the Cotton effects of the amide of one sugar to be sensitive to the substituents at C-3 and C-1. Thus, the Cotton effects of the amide may give information on the relative conformation of the amide in one sugar and that of its C-1 linked neighbor in disaccharides and polysaccharides.

2. Carboxylic Acids

Also appearing as a chromophore in structural polysaccharides are carboxyl and carboxylate residues. Unfortunately, the electronic structure and absorption bands of these chromophores have received less attention than that of the amide. Based on studies of uronic acids, Listowsky et al. (1969) have assigned bands in the 210–235 nm region to n-π^* transitions of the carboxyl group.

B. EXPERIMENTAL

1. N-Acetylamino Saccharides

That optical activity may be used empirically to differentiate various intersaccharide linkages and substituent positions has been shown by Kabat et al. (1969). For 2-acetylamino sugars, the anomeric configuration at C-1 has been shown to have a strong influence on the optical activity in the 220 nm region (Kabat et al., 1969).

These authors showed that methyl-α-N-acetyl-D-glucosamine has a CD quite different from that of the β anomer. The Cotton effect they report in the 210–215 nm region of the CD, which they assign to the amide n-π^*, is seen to shift in wavelength in quite an interesting way among anomers of N-acetylglucosamine and -galactosamine. The cause of these shifts remains obscure, but could be related to different solvent surroundings, giving rise to solvent shifts of the n-π^* transition. Moreover, Kabat et al. (1969) propose rules for recognizing various C-3 and C-4 derivatives from characteristics of the optical activity.

Not only are there differences among various anomers, C-3 and C-4 substituents as shown by Kabat et al. (1969), but there are also certain specific effects due to the polymeric nature polysaccharides. Studies on various natural polysaccharides containing various residues and linkages have led Stone (1969) to the conclusion that "the simple derivatives may not be sufficient models to explain the ORD and CD of polysaccharides." Although it is sometimes difficult to interpret the optical activity of natural polysaccharides whose chemical structure is not fully elucidated, it does seem that there may be a conformation dependance for these Cotton effects. From optical studies, Stone (1969) proposed a conformation for chordrioitin sulfates stabilized by hydrogen bonds between adjacent residues.

2. Other Sugars

In contrast to the 2-acetylamino sugars, the location of the chromophore in uronic acids is distant from the C-1. Thus, the anomeric configuration has been shown to have only a small effect on the Cotton effects of the carboxylate chromophore (Listowsky et al., 1969). These authors propose that substituents at C-4 can be distinguished, and that the CD bands they observe in the 210 and 235 nm region are due to carboxyl n-π^* transitions. They attribute the appearance of these two bands at distant wavelengths to differing solvation of the chromophores, as a result of two different conformations in equilibrium in solution. There is definitely a need for further exploration of the uronic acid Cotton effects, but the optical activity promises to yield extensive information.

Listowsky and Englard (1968) have reported CD curves for some sugars such as glucose and galactose having no π electron chromophore. They observed Cotton effects at the extreme short wavelength limit of current instrumentation at 190 nm. In these bands, no maximum was observed. The bands are apparently due to the σ-σ^* transitions on the ether oxygen. They showed that the sign of the Cotton effect could be correlated with the axial substituents at the C-4 position.

3. Dye Studies

As in the case of other optically active biopolymers, dyes bound to the asymmetric polysaccharide may exhibit induced Cotton effects at the dye absorption frequency. Interpretation of such results is often difficult due to the problems associated with determining the geometric nature of the dye binding, in addition to polymer grometry. Stone and Moss (1967) propose a helical model structure for heparin based on hypochromism and ORD of methylene blue complexes. Those studies indicate interactions among the dye molecules as a consequence of binding to the heparin.

V. Concluding Remarks

A. RELATIVE MERITS OF ORD AND CD

In the introduction, we emphasized that both ORD and CD are measuring the same essential molecular properties, but there are certain practical differences which may make one measurement desirable over the other in certain cases.

For example, in cases where one wishes to make correlations between the electronic transitions of the chromophores and optical activity, CD is the method of choice. Its curve shape is less complex than that of ORD and resembles an absorption band in shape. Also, a simple Gaussian curve [Eq. (13)] may be used to represent a band in decomposition of a complex curve. Since the CD drops to zero outside the band, contributions to the CD are observed only within the absorption band. There is less overlapping and assignment is much simpler than in ORD. If one wishes to make electronic assignments and only the ORD is available, one can employ the Kronig-Kramers transformation [Eq. (12)].

CD is also to be preferred over ORD in observing a small Cotton effect which is very close in wavelength to a large one. Sarkar et al. (1967) have observed a very small negative CD band in RNA at 295 nm (see Fig. 14). This band is largely obscured by the much stronger band at 260 nm and was unnoticed in ORD where it appears as only a slight modification of the curve shape in the 295 nm region. It may be discerned in Kronig–Kramers transformations of the ORD, but to accurately document this effect, one must measure very accurately the ORD or CD in the 280–300 nm region. Such accurate measurements are possible using longer path length cells or higher concentrations of RNA, since the absorption strength is low in this wavelength region. A similar situation arises in the 280 nm region of poly rA (see Fig. 11) where the ORD as originally published was extremely difficult to interpret, due to small Cotton effects at long wavelength (Vournakis et al., 1967).

The same considerations apply to studies on side chain chromophores in proteins. In a typical protein, containing a few aromatic chromophores, the amides greatly outnumber the side chain chromophores, and thus one must use larger samples to get good signal-to-noise ratios in the longer wavelength side chain absorption bands. In ORD, the amide background rotation is large and interferes with determination of side chain Cotton effects in the 260–280 nm region. CD does not suffer from the strong amide background of the ORD.

The ORD will find its best uses in empirical correlations because its more complex curve shape contains numerous features. In addition, Cotton effects beyond the range of the measurement contribute to the ORD and may thus be included in the empirical correlation. One might compare the CD of two samples and find them to be quite similar, while their ORD would easily distinguish them due to differences in Cotton effects beyond the range of measurement.

Also ORD can be used in cases where the CD bands are not accessible due to instrumental limitations or due to solvent absorption. For example, one can use the Moffit–Yang treatment for determination of polypeptide α helix content even in the presence of a solvent having strong absorption in the amide chromophore region.

B. Faraday Effect

In 1845, Faraday discovered that in the presence of a magnetic field parallel to the light path, optically inactive material showed rotation. Shortly thereafter, a number of investigators attempted to relate the effect to molecular structure. They drew their inspiration in this effort from the spectacular successes achieved in relating natural optical rotation to molecular dissymmetry. Their lack of success emphasizes the fact that, despite the apparent similarities, the two effects are quite different.

Natural optical activity has been productive of a number of techniques for investigating chemical structure and polymeric conformation as witnessed by the foregoing pages. That such productivity is less likely from magnetically induced optical activity may be seen from the following qualitative argument. If one imagines the dinucleoside phosphate ApA as two planar symmetric chromophores held by the ribose phosphate in a skewed conformation, the asymmetric perturbation of each chromophore on its neighbor is held in place as the molecule tumbles in solution. Hence the sign and magnitude of the natural Cotton effects are quite sensitive to the geometric relationship of the two chromophores. If, however, the perturbation giving rise to the activity is a magnet fixed in the instrument, this sensitivity to molecular geometry is not present (Moscowitz, 1967). Thus, the Faraday effect will not be useful in studies of polymer conformation as has been the case for natural rotation. The effects of polymer conformation will be no more pronounced in Faraday effect than they will be in ordinary absorption (Harris, 1967; Tinoco and Bush, 1964).

We do not wish to leave the impression that the Faraday effect is without interest in biological problems. On the contrary, recent successes on the experimental aspects of Faraday effect show how the technique should be used. It has been found that measurements are most profitably made, not on the rotation, but on the magnetic circular dichroism which is related to Faraday rotation in the same way natural CD is related to natural optical rotations [see Eqs. (11, 12)] (Briat and Djerassi, 1968). MCD measurements have been found to be useful in the assignment of electronic spectra. In the difficult problem of the assignment of the absorption bands in the nucleic acid bases, MCD has yielded useful evidence (Voelter et al., 1968).

We conclude that the most interesting applications of Faraday effect in biological research will be in assignment of electronic transitions or in special cases where changes in polymer conformation cause changes in wavelengths of bands which cannot be detected by absorption. An example might be that of an n-π^* transition whose wavelength was sensitive to polymer conformation, but whose absorption could not be detected because of the inherently weak oscillator strength of n-π^* transitions.

C. INFRARED AND FAR ULTRAVIOLET

In our discussion, we have been considering optical activity caused by chromophores with characteristic wavelengths in the visible and in the ultraviolet at wavelengths less that 190 nm. There are a number of interesting questions which could be investigated by extension of the short wavelength limit into the far ultraviolet (λ less that 190 nm). In the vacuum ultraviolet, as this region is known, many new chromophores become accessible and it is possible that interesting information on structure could be found. However, the very presence of numerous chromophores in the vacuum ultraviolet presents certain problems. Oxygen and nitrogen absorb in this region so the apparatus must be evacuated. Solvents become opaque so one is restricted largely to measurements of molecules in the gas phase or thin films deposited on quartz plates. Although CD and ORD experiments in the vacuum ultraviolet are possible, the experimental difficulties are formidable enough to make it unlikely that a large volume of results of biological interest will appear in the immediate future.

At the opposite end of the spectrum, in the infrared, the prospects are likewise obscure but somewhat more promising. One might imagine that vibrational transitions in asymmetric molecules might exhibit Cotton effects in the infrared region of the spectrum. Recent calculations bearing on this question show that for a helical array of vibrators, a small rotational strength would be found in the fundamental absorption region (Deutsche and Moscowitz, 1968). Although the rotational strength of the helical polymer is small, the absorption intensity in the infrared is also much smaller than that in the ultraviolet. The ratio of rotational strength to absorption intensity is similar to that seen in the ultraviolet.

Thus, if one could design an infrared CD instrument capable of measuring millidegrees of ellipticity, Cotton effects ought to be observed. Polarameters have been designed and built for measurement of ORD in the infrared, but they achieved sensitivity of tens of millidegrees, and no Cotton effects could be documented (Wyss and Gunthard, 1966).

D. Optical Activity of Oriented Polymers

An area which has not been fully investigated and which could be quite interesting to biological research is that of the optical activity of oriented systems. Tinoco (1962) has emphasized that the optical activity of oriented polymers depends on the direction of the light incidence, and that one gains much more information on polymer structure from the optical activity of the oriented system. Woody and Tinoco (1967) have calculated the rotational strengths for oriented polypeptide helices and pointed out that such measurements may be used to distinguish α and 3_{10} helices. Unfortunately, experiments on oriented systems have not been forthcoming. Only a few measurements on the ORD of oriented polypeptide helices outside the Cotton effect region have been reported (Yamaoka, 1964).

However, if experiments in the Cotton effect region could be carried out on model systems, they might prove extremely useful in biological systems. It is possible that optical activity measurements on oriented protein molecules could give certain kinds of very explicit information, especially on the orientation of side chains or other special chromophores. Likewise, in polynucleotides, the question of tilting of the base planes with respect to the helix axis could be profitably investigated. In measurements of the optical activity of an oriented polynucleotide helix, a geometry such as that of DNA form B with the bases perpendicular to the helix axis would have no rotational strength at all in the π-π^* transition region for light incident perpendicular to the helix axis (Tinoco et al., 1963).

Acknowledgments

I would like to acknowledge Drs. Charles Cantor, Paul Ts'o, and Myron Warshaw for kindly making available manuscripts in advance of publication.

References

Adler, A., Grossman, L., and Fasman, G. D. (1967). *Proc. Nat. Acad. Sci. U.S.* **57**, 423.

Adler, A., Grossman, L., and Fasman, G. D. (1968). *Biochemistry* **7**, 3836.

Balasubramanian, D., and Wetlaufer, D. B. (1967). *In* "Conformation of Biopolymers" (G. N. Ramachandran, ed.), p. 147. Academic Press, New York.

Basch, H., Robin, M. B., and Keubler, N. A. (1968). *J. Chem. Phys.* **49**, 5007.

Bayley, P. M., Neilsen, E. B., and Schellman, J. A. (1969). *J. Phys. Chem.* **73,** 228.

Beychok, S. (1967). *In* "Poly-α-Amino Acids" (G. D. Fasman, ed.), p. 293. Dekker, New York.

Beychok, S., and Fasman, G. D. (1964). *Biochemistry* **3,** 1675.

Blaha, K., and Fric, I. (1968). *In* "Peptides-1968" (E. Bricas, ed.), p. 40. North-Holland Publ., Amsterdam.

Blake, A., and Peacocke, A. (1966). *Biopolymers* **4,** 1091.

Brahms, J., and Mommaerts, W. F. H. M. (1964). *J. Mol. Biol.* **10,** 73.

Brahms, J., and Sadron, C. (1966). *Nature (London)* **212,** 1309.

Brahms, J., Michelson, A. M., and VanHolde, K. E. (1966). *J. Mol. Biol.* **15,** 467.

Brahms, J., Maurizot, J. C., and Michelson, A. M. (1967a). *J. Mol. Biol.* **25,** 465.

Brahms, J., Maurizot, J. C., and Michelson, A. M. (1967b). *J. Mol. Biol.* **25,** 481.

Brahms, J., Maurizot, J. C., and Pilet, J. (1969). *Biochim. Biophys. Acta* **186,** 110.

Briat, B., and Djerassi, C. (1968). *Nature (London)* **217,** 918.

Brimacombe, J. S., and Webber, J. M. (1964). "Mucopolysaccharides." Elsevier, Amsterdam.

Brownlee, G. G., Sanger, F., and Barrel, B. G. (1967). *Nature (London)* **215,** 735.

Bush, C. A. (1969). Program ORPD-02. Perkin-Elmer Program Exchange Library. Perkin-Elmer Corp., Norwalk, Conn. 06852.

Bush, C. A. (1970). *J. Chem. Phys.* **53,** 3522.

Bush, C. A., and Brahms, J. (1967). *J. Chem. Phys.* **46,** 79.

Bush, C. A., and Scheraga, H. A. (1967). *Biochemistry* **6,** 3036.

Bush, C. A., and Scheraga, H. A. (1968). Unpublished results.

Bush, C. A., and Scheraga, H. A. (1969). *Biopolymers* **7,** 395.

Bush, C. A., and Tinoco, I. (1967). *J. Mol. Biol.* **25,** 601.

Cantor, C. R. (1968). *Proc. Nat. Acad. Sci. U.S.* **59,** 478.

Cantor, C. R., and Tinoco, I. (1965). *J. Mol. Biol.* **13,** 65.

Cantor, C. R., and Tinoco, I. (1967). *Biopolymers* **5,** 821.

Cantor, C. R., Jaskunis, S. R., and Tinoco, I. (1966). *J. Mol. Biol.* **20,** 39.

Cantor, C. R., Warshaw, M. M., and Shapiro, H. (1970). *Biopolymers* **9,** 1079.

Carver, J. P., Shechter, E., and Blout, E. R. (1966a). *J. Amer. Chem. Soc.* **88,** 2562.

Carver, J. P., Shechter, E., and Blout, E. R. (1966b). *J. Amer. Chem. Soc.* **88,** 2550.

Cheng, P. Y. (1968). *Biochem. Biophys. Res. Commun.* **33,** 746.

Clark, L. B., and Tinoco, I. (1965). *J. Amer. Chem. Soc.* **87,** 11.

Coleman, D. L., and Blout, E. R. (1967). *In* "Conformation of Biopolymers" (G. N. Ramachandran, ed.), p. 123. Academic Press, New York.

Coleman, D. L., and Blout, E. R. (1968). *J. Amer. Chem. Soc.* **90,** 2405.

Craig, L. C. (1968). *Proc. Nat. Acad. Sci. U.S.* **61,** 152.

Davidson, B., and Fasman, G. D. (1967). *Biochemistry* **6,** 1616.

Davis, R. C., and Tinoco, I. (1968). *Biopolymers* **6,** 223.

Deutsche, C. W., and Moscowitz, A. (1968). *J. Chem. Phys.* **49,** 3257.

Deutsche, C. W., Lightner, D. A., Woody, R. W., and Moscowitz, A. (1969). *Annu. Rev. Phys. Chem.* **20,** 407.

Djerassi, C. (1960). "Optical Rotatory Dispersion." McGraw-Hill, New York.

Donohue, J., and Trueblood, K. N. (1960). *J. Mol. Biol.* **2**, 363.

Emeis, C. A., Oosterhof, L. J., and deVries, G. (1967). *Proc. Roy. Soc. Ser. A* **297**, 54.

Emerson, T. R., Swan, R. J., and Ulbricht, T. L. V. (1966). *Biochem. Biophys. Res. Commun.* **22**, 505.

Emerson, T. R., Swan, R. J., and Ulbricht, T. L. V. (1967). *Biochemistry* **6**, 843.

Forget, B. G., and Weissman, S. M. (1967). *Science* **158**, 1695.

Glaubiger, D., Lloyd, D. A., and Tinoco, I. (1968). *Biopolymers* **6**, 409.

Gratzer, W. B., Holzwarth, G., and Doty, P. (1961). *Proc. Nat. Acad. Sci. U.S.* **47**, 1785.

Green, G., and Mahler, H. R. (1968). *Biopolymers* **6**, 1509.

Green, G., and Mahler, H. R. (1970). *Biochemistry* **9**, 368.

Greenfield, N., Davidson, B., and Fasman, G. D. (1967). *Biochemistry* **6**, 1630.

Hanlon, S., and Major, E. O. (1968). *Biochemistry* **7**, 4350.

Harris, R. A. (1967). *J. Chem. Phys.* **47**, 4481.

Haschemeyer, A. E. V., and Rich, A. (1967). *J. Mol. Biol.* **27**, 369.

Holcomb, D. N., and Tinoco, I. (1965). *Biopolymers* **3**, 121.

Holzwarth, G., and Doty, P. (1965). *J. Amer. Chem. Soc.* **87**, 218.

Horowitz, J., Strickland, E. H., and Billups, C. (1969). *J. Amer. Chem. Soc.* **91**, 184.

Ikehara, M., Kaneko, M., Muneyama, K., and Tanaka, H. (1967). *Tetrahedron Lett.* p. 3977.

Jaskunis, S. R., Cantor, C. R., and Tinoco, I. (1968). *Biochemistry* **7**, 3164.

Jirgensons, B. (1969). "Optical Rotatory Dispersion of Proteins and Other Macromolecules." Springer, New York.

Johnson, W. C., and Tinoco, I. (1969). *Biopolymers* **7**, 727.

Kabat, E. A., Lloyd, K. O., and Beychok, S. (1969). *Biochemistry* **8**, 747.

Kasarda, D. D., Bernardin, J. E., and Gaffield, W. (1968). *Biochemistry* **7**, 3950.

Kirkwood, J. G. (1937). *J. Chem. Phys.* **5**, 479.

Klee, W. A., and Mudd, S. H. (1967). *Biochemistry* **6**, 988.

Kopple, K. D., Ohnishi, M., and Go, A. (1969). *Biochemistry* **8**, 4087.

Leng, M., and Felsenfeld, G. (1966). *J. Mol. Biol.* **15**, 455.

Listowsky, I., and Englard, S. (1968). *Biochem. Biophys. Res. Commun.* **30**, 329.

Listowsky, I., Englard, S., and Avigad, G. (1969). *Biochemistry* **8**, 1781.

Lowery, T. M. (1935). "Optical Rotatory Power," Longmans, Green, New York (reprinted, 1964, Dover, New York).

Maestre, M. F., and Tinoco, I. (1965). *J. Mol. Biol.* **12**, 287.

Maestre, M. F., and Tinoco, I. (1967). *J. Mol. Biol.* **23**, 323.

Maestro, M., Moccia, R., and Taddei, G. (1967). *Theor. Chim. Acta* **6**, 80.

Mandeles, S. (1967). *J. Biol. Chem.* **242**, 3102.

Michelson, A. M., and Monny, C. (1966). *Proc. Nat. Acad. Sci. U.S.* **56**, 1528.

Miles, D. W., and Urry, D. W. (1967). *J. Phys. Chem.* **71**, 4448.

Miles, D. W., Robbins, M. J., Robbins, R. K., and Eyring, H. (1969). *Proc. Nat. Acad. Sci. U.S.* **62**, 22.

Moffit, W. (1956). *J. Chem. Phys.* **25**, 467.

Moffit, W., and Moscowitz, A. (1959). *J. Chem. Phys.* **30**, 648.

Moffit, W., and Yang, J. T. (1956). *Proc. Nat. Acad. Sci. U.S.* **42**, 596.

Moffit, W., Fitts, D., and Kirkwood, J. G. (1957). *Proc. Nat. Acad. Sci. U.S.* **43**, 723.

Moscowitz, A. (1960). *In* "Optical Rotatory Dispersion" (C. Djerassi, ed.), p. 150, McGraw-Hill, New York.

Moscowitz, A. (1967). *Proc. Roy. Soc. Ser. A* **297**, 16.

Pflumm, M. N., and Beycok, S. (1969). *J. Biol. Chem.* **244**, 3982.

Pochon, F., and Michelson, A. M. (1965). *Proc. Nat. Acad. Sci. U.S.* **53**, 1425.

Podder, S. K., and Tinoco, I. (1969). *Biochem. Biophys. Res. Commun.* **34**, 569.

Poland, D., Vournakis, J. N., and Scheraga, H. A. (1966). *Biopolymers* **4**, 223.

Pysh, E. (1970). *Science* **167**, 290.

Quadrifoglio F., and Urry, D. W. (1967). *Biochem. Biophys. Res. Commun.* **29**, 785.

Quadrifoglio, F., and Urry, D. W. (1968a). *J. Amer. Chem. Soc.* **90**, 2755.

Quadrifoglio, F., and Urry, D. W. (1968b). *J. Amer. Chem. Soc.* **90**, 2760.

Rich, A., Davies, D. R., Crick, F. H. C., and Watson, J. D. (1961). *J. Mol. Biol.* **3**, 71.

Sarkar, P. K., and Doty, P. (1966). *Proc. Nat. Acad. Sci. U.S.* **55**, 981.

Sarkar, P. K., and Yang, J. T. (1965a). *J. Biol. Chem.* **240**, 2088.

Sarkar, P. K., and Yang, J. T. (1965b). *Biochemistry* **4**, 1238.

Sarkar, P. K., and Yang, J. T. (1967). *In* "Conformation of Biopolymers" (G. N. Ramachandran, ed.), p. 197. Academic Press, New York.

Sarkar, P. K., Wells, B., and Yang, J. T. (1967). *J. Mol. Biol.* **25**, 563.

Schellman, J. A., and Lowe, M. (1968). *J. Amer. Chem. Soc.* **90**, 1070.

Schellman, J. A., and Nielsen, E. B. (1967). *In* "Conformation of Biopolymers" (G. N. Ramachandran, ed.), p. 109. Academic Press, New York.

Schellman, J. A., and Oriel, P. (1962). *J. Chem. Phys.* **37**, 2114.

Shechter, E., and Blout, E. R. (1964). *Proc. Nat. Acad. Sci. U.S.* **51**, 695, 794.

Sober, H. A. (1968). "Handbook of Biochemistry." Chem. Rubber Co., Cleveland, Ohio.

Stern, A., Gibbons, W. A., and Craig, L. C. (1968). *Proc. Nat. Acad. Sci. U.S.* **61**, 734.

Stone, A. L. (1969). *Biopolymers* **7**, 173.

Stone, A. L., and Moss, H. (1967). *Biochim. Biophys. Acta* **136**, 56.

Strickland, E. H., Horowitz, J., and Billups, C. (1969). *Biochemistry* **8**, 3205.

Sundaralingham, M., and Jensen, L. H. (1965). *J. Mol. Biol.* **13**, 914.

Thiery, J. (1968). *J. Chim. Phys. Physicochim. Biol.* **65**, 98.

Tiffany, M. L., and Krimm, S. H. (1969). *Biopolymers* **8**, 347.

Timasheff, S. N., Susi, H., Townsend, R., Stevens, L., Gorbunoff, M. J., and Kumosinski, T. F. (1967). *In* "Conformation of Biopolymers" (G. N. Ramachandran, ed.), p. 173. Academic Press, New York.

Tinoco, I. (1962). *Advan. Chem. Phys.* **4**, 113.

Tinoco, I. (1968). *J. Chim. Phys. Physiochim. Biol.* **65**, 91.

Tinoco, I., and Bush, C. A. (1964). *Biopolym. Symp.* **1**, 235.

Tinoco, I., and Woody, R. W. (1964). *J. Chem. Phys.* **40**, 160.

Tinoco, I., Woody, R. W., and Bradley, D. F. (1963). *J. Chem. Phys.* **38**, 1317.

Ts'o, P. O. P. (1970). "The Monomeric Units of Nucleic Acids—Bases, Nucleosides and Nucleotides." Dekker. New York to be published.

Ts'o, P. O. P., Rapaport, S. A., and Bollum, F. J. (1966). *Biochemistry* **5**, 4153.

Ts'o, P. O. P., Kondo, N. S., Robbins, R. K., and Broom, A. D. (1969). *J. Amer. Chem. Soc.* **91**, 5625.

Tunis, M. J., and Hearst, J. E. (1968). *Biopolymers* **6**, 1218.

Ulbricht, T. L. V., Emerson, T. R., and Swan, R. J. (1966). *Tetrahedron Lett.* p. 1561.

Ulmer, D. D., and Vallee, D. L. (1965). *Advan. Enzymol. Relat. Subj. Biochem.* **27,** 37.

Urry, D. W. (1968). *Annu. Rev. Phys. Chem.* **19,** 517.

Voelter, W., Records, R., Bunnenberg, E., and Djerassi, C. (1968). *J. Amer. Chem. Soc.* **90,** 6163.

Voet, D., Gratzer, W. B., Cox, R. A., Doty, P. (1963). *Biopolymers* **1,** 193.

Vournakis, J. N., and Scheraga, H. A. (1966). *Biochemistry* **5,** 2997.

Vournakis, J. N., Poland, D., and Scheraga, H. A. (1967). *Biopolymers* **5,** 403.

Vournakis, J. N., Yan, J., and Scheraga, H. A. (1968). *Biopolymers* **6,** 1531.

Warshaw, M. M., and Tinoco, I. (1966). *J. Mol. Biol.* **20,** 29.

Warshaw, M. M., Bush, C. A., and Tinoco, I. (1965). *Biochem. Biophys. Res. Commun.* **18,** 633.

Wetlaufer, D. B. (1962). *Advan. Protein Chem.* **17,** 304.

Woody, R. W. (1968). *J. Chem. Phys.* **49,** 4797.

Woody, R. W., and Tinoco, I. (1967). *J. Chem. Phys.* **46,** 4927.

Wyss, H. R., and Gunthard, H. H. (1966). *J. Opt. Soc. Amer.* **56,** 888.

Yamaoka, K. (1964). Ph.D. Thesis, Univ. of California, Berkeley, California.

Yamaoka, K., and Resnik, R. A. (1966). *J. Phys. Chem.* **70,** 4051.

Yang, J. T. (1967). *In* "Poly-α-Amino Acids" (G. D. Fasman, ed.), p. 239. Dekker, New York.

Yang, J. T., and Samejima, T. (1968a). *Progr. Nucl. Acid Res. Mol. Biol.* **9,** 223.

Yang, J. T., and Samejima, T. (1968b). *Biochem. Biophys. Res. Commun.* **33,** 739.

Ziegler, S. M., and Bush, C. A. (1971). *Biochemistry* (in press).

Author Index

Numbers in italics refer to the pages on which the complete references are listed.

A

Abbate, M. J., 319, *340*
Abbe, E., 14, *70*
Abermann, R., 125, *152, 156*
Adler, A., 371, 372, 375, *404*
Adler, I., 181, *264, 270*
Agar, A. W., 105, 109, 117, *152, 153*
Akselsson, R., 258, *262*
Allen, G., 333, *340*
Alvarez, C. J., *274*
Ambler, A. P., 281, 284, *343*
Andersen, C. A., 163, 164, 195, 202, 247, 250, 259, *261, 263, 266, 267, 268, 270, 272, 274*
Anderson, C. H., *273*
Anderson, T. F., 130, *152*
Angell, C. L., 338, *342*
Appleton, T. C., 245, 248, *261*
Avigad, G., 399, 400, *406*

B

Bachmann, L., 125, *152, 156*
Baden, V., *268*
Badonnell, M. C., *269*
Bahr, G. F., 245, *261*
Bailey, E., 320, *342*
Bailey, G. F., 333, 335, *340*
Baker, R. F., 108, *154*
Balasubramanian, D., 391, *404*
Balasubramanian, K. A., 338, *340*
Balimont, P., *273*
Banfield, W. G., *268*
Bang, S., *273*
Barber, V. C., 140, *152*
Barker, S. A., 324, *340*
Barnes, R. B., 307, *342*
Barrel, B. G., 379, *405*

Basch, H., 384, *404*
Baud, C. A., *269, 272, 273*
Baud, J. P., *273*
Bauman, R. P., 319, *340*
Bax, D., *273*
Bayley, P. M., 386, *405*
Beaman, D. R., *269*
Beaven, G. H., 320, *342*
Beckwith, R., 326, 328, *342*
Beer, M., 128, *152*, 321, *340*
Behringer, J., 339, *340*
Bellamy, L. J, 281, 283, 285, 320, *340*
Bendit, E. G, 322, *340*
Benedict, A. A., 325, *340*
Berkley, C., *268*
Bernardin, J. E., 396, *406*
Bernstein, H. J., 333, *340*
Berry, J. P., *269*
Besic, F. C., *273*
Bethe, H. A., 190, *261*
Beychok, S., 398, 399, *405, 406*
Beycok, S., 398, *407*
Bigelow, W. C., *268*
Billups, C., 389, *406, 407*
Birks, L. S., 207, *261, 268, 272, 273*
Bishop, H., 194, *261*
Blaha, K., 389, 391, *405*
Blaise, G., 260, *262*
Blake, A., 383, *405*
Blout, E. R., 307, 319, 320, *340, 344*, 356, 387, 388, 389, 391, 392, 393, 394, *405, 407*
Bollum, F. J., 375, *407*
Boothroyd, B., *272*
Borcherds, P. H., 96, *153*
Born, M., 23, 27, 66, *70*
Boult, E. H., 126, 139, *152*
Bourne, E. J., 324, *340*
Boyde, A., 140, *152*, 250, 251, *261, 272, 273, 274*

409

Subject Index

A

Abbe theory, 59

Aberrations
  of magnetic and electrostatic lens,
    93–95
    astigmatism, 94–95
    chromatic, 95
    spherical, 93–94
  of optical systems, 9–13, 34–35
    astigmatism, 11–12
    chromatic, 12–13
    coma, 11
    spherical, 10–11
    transverse color, 12

N-Acetylamino polysaccharides, optical
  activity studies on, 398–400

Achromatic objectives for light microscopes, 33–38

Achromatization of optical systems, 5

Adenine, optical activity studies on, 359

Air, refractive index of, 4

Airy patterns (Airy discs), in light microscopy, 20–24, 68

Alanine
  cyclic peptide of, optical activity, 391
  Raman studies on, 337

Aluminum oxide, in support films for
  electron microscopy, 119

Amici type objective, for light microscope, 36

Amino acids
  IR spectroscopy of, 319–322
  Raman studies on, 337, 338
  UV absorption bands of, 388

Amphipleura pellucida, use in testing
  microscope objectives, 30

Amplitude, of a spherical wave, 17

Analyzing electron microscope, description and uses of, 151

Aplanatism, in optical systems, 13–14,
  36

Apochromats, description of, 38

Apochromatic objectives, for light microscope, 38–39

Arteries, microprobe analysis of mineralization of, 272

Artifacts, in electron microscopy, 128–133

Astigmatism
  correction in electron microscope,
    115–116
  in eye, 32
  from magnetic and electrostatic lens,
    94–95
  in optical systems, 11–12

ATP, vibrational spectrum from, 280

Attenuated total reflectance spectroscopy, 311

B

Bacteria, electron microscopy of, 122,
  130, 132

Bacteriophage, DNA of, optical activity
  of, 383

Bases, optical activity studies on,
  358–362

Beckman IR spectrometers, 298–299

Bedacryl, as film substrate for electron
  microscopy, 122

Beryllium, in support films for electron
  microscopy, 119

Bias voltage, for electron microscope,
  97

Binocular vision, of eye, 32–33

"Blind" photography, in electron microscopy, 106

Blood, internal reflection spectroscopy,
  325

Bohr frequency rule, 278

Bone
  internal reflection spectroscopy of, 325
  microprobe analysis of mineralization
    of, 272–274

Buffers, for use in IR spectroscopy, 310

C

Camera, pinhole type, principles of, 2–3
Carbohydrates, IR spectroscopy of, 324
Carbon dioxide, IR absorption of, 288
Carbon films, for electron microscopy, 119, 121–122
Carotenoids, IR spectroscopy of, 324
Catalase, for electron microscope calibration, 107
Cells, for IR spectroscopy, 303, 309, 315, 324–325
Chloroform, as solvent for IR spectroscopy, 302–305
Chromatic aberration
  of magnetic and electrostatic lenses, 95
  in optical systems, 12
Chromophores, optical activity from, 357–358
Chromosomes, high voltage electron microscopy of, 150
α-Chymotrypsin, Raman studies on, 339
Circular birefringence, description of, 351
Circular dichroism, 347–408
  curve shape, 353
  merits compared to ORD, 401–402
  of polypeptides and proteins, 383–398
  of polysaccharides, 398–400
  principles of, 352–353
  relationship to ORD, 353–356
  in UV and IR regions, 403–404
Circular symmetry, see Spherical aberration
Collodion films, for electron microscopy, 119–120
Coma, as aberration in optical systems, 11, 13
Compensating eyepiece, for light microscope, 43–44
Complex amplitude of light, theory of, 17–20
Compound microscope, optics of, 14–15
Condensers
  for electron microscopes, 98–99
  for light microscopes, 33, 45–47
Converging lens, 6
Cotton effect, 354
  definition of, 350

Coverglass, correction for, in light microscopy, 42
Crown glass, refractive index of, 4
Cyclic peptides, optical activity studies on, 390–392
Cysteine, Raman studies on, 338
Cytosine, optical activity studies on, 359

D

Deoxyoligonucleotides, optical activity studies on, 371–373
Deuterium oxide
  IR spectroscopy of, 310
  Raman spectra of, 334
Dichroic ratio, definition of, 315
Diffracting crystals, for X-ray spectroscopy in microprobe analysis, 172–176
Diffraction gratings, for X-ray spectroscopy in microprobe analysis, 178–179
Diffraction patterns, by electron microscope, 104–105
Dinucleoside phosphates, optical activity studies on, 365–367
Disc of least confusion, in optical systems, 11
Distortion, in optical systems, 10
DNA
  IR spectroscopy of, 322
  optical activity studies on, 381–383, 404
    dye binding, 383
  Raman spectroscopy of, 336, 339
Drude method, modified two-term method for study of polypeptide structure, 392–393
Dyson's long working distance objective, 42

E

EDTA, use in IR spectroscopy, 310
Egg, scanning electron microscopy of, 139
Electron beams, effects on biological material, 131–132
Electron diffraction patterns, by electron microscope, 104–105